Visual Cognition

An Invitation to Cognitive Science
Daniel N. Osherson, Series Editor

ısual Cognition

ın Invitation to Cognitive Science
Second Edition

Volume 2

Edited by Stephen M. Kosslyn and
Daniel N. Osherson

A Bradford Book
The MIT Press
Cambridge, Massachusetts
London, England

This book was set in Palatino by Asco Trade Typesetting Ltd., Hong Kong and printed in the United States of America.

Library of Congress Cataloging-in-Publication Data

An invitation to cognitive science.—2nd ed.
      p.   cm.
   "A Bradford book."
   Includes bibliographical references and indexes.
   Contents: v. 1. Language / edited by Lila R. Gleitman and Mark
Liberman—v. 2. Visual cognition / edited by Stephen M. Kosslyn and
Daniel N. Osherson—v. 3. Thinking / edited by Edward E. Smith
and Daniel N. Osherson.
   ISBN 0-262-65045-2 (set).—ISBN 15042-5 (v. 2 : hardcover).
—ISBN 65042-8 (v. 2 : pbk.).
   1. Cognition. 2. Cognitive science. I. Gleitman, Lila R.
BF311.I68   1995
153—dc20   95-10924
                                                    CIP

Contents

List of Contributors

Irving Biederman
Department of Psychology
University of Southern California

Fred Dretske
Department of Philosophy
Stanford University

Martha J. Farah
Department of Psychology
University of Pennsylvania

Melvyn Alan Goodale
Department of Psychology
University of Western Ontario

Grant Gutheil
Department of Psychology
Cornell University

Zijiang J. He
Department of Psychology
University of Louisville

Stephen M. Kosslyn
Department of Psychology
Harvard University

Eileen Kowler
Department of Psychology
Rutgers University

Ken Nakayama
Department of Psychology
Harvard University

Harold Pashler
Department of Psychology
University of California, San Diego

Shinsuke Shimojo
Department of Psychology
University of Tokyo

Elizabeth S. Spelke
Department of Psychology
Cornell University

Gretchen Van de Walle
Department of Psychology
Cornell University

Foreword

The book you are holding is the second of a four-volume introduction to contemporary cognitive science. The work includes more than forty chapters, written by linguists, psychologists, philosophers, computer scientists, biologists, and engineers. The topics range from muscle movement to human rationality, from acoustic phonetics to mental imagery, from the cerebral locus of language to the categories that people use to organize experience. Topics as diverse as these require distinctive kinds of theories, tested against distinctive kinds of data, and this diversity is reflected in the style and content of the chapters.

The authors of these volumes are united by their fascination with the mechanisms and structure of biological intelligence, especially human intelligence. Indeed, the principal goal of this introductory work is to reveal the vitality of cognitive science, to share the excitement of its pursuit, and to help you reflect upon its interest and importance. You may therefore think of these volumes as an invitation—namely, the authors' invitation to join the ongoing adventure of research into human cognition.

The topics we explore fall into four parts, each corresponding to a separate volume. The parts are language, visual cognition, thinking, and conceptual foundations. Each volume is self-contained, so they can be read in any order. On the other hand, it is easiest to read the chapters of a given volume in the order indicated. Each chapter concludes with suggestions for further reading and a set of problems to test your understanding.

It remains only to wish you a pleasant and invigorating journey through cognitive science. May it lead to a life-long interest!

Istituto San Raffaele
July 1995

DANIEL N. OSHERSON
SERIES EDITOR

Visual Cognition: Introduction
Stephen M. Kosslyn

It is difficult to overstate the importance of vision in our lives. We use vision not only to help us identify food when we are hungry and a bed when we are sleepy, but also to read, to observe a friend's facial expressions, to appreciate the beauty of a potential mate or a work of art, to navigate, and to determine whether we can cross the street without being run over by an approaching truck. Vision permeates even our language, which is replete with visual metaphors (as you can see—or at least glimpse—with a moment's reflection). Visual perception guides us in our daily commerce with the environment and provides much of our knowledge about the world.

Thus it is important to understand vision if we are to understand the nature of the mind. Such an understanding will have many other benefits as well. If, for instance, we knew how human vision works, we would be in a position to endow machines with similar capabilities, an accomplishment that would benefit many individuals—from the blind to the mentally retarded—who could enjoy reading machines and "seeing-eye" automobiles. We could also design visual displays that play to our strengths and avoid our weaknesses. As anyone who has read national news magazines knows, there is plenty of room for improvement in even simple charts and graphs—let alone the displays used in nuclear power plants and similar installations. The chapters in *Visual Cognition* outline the high points of this area of cognitive science, focusing on how information is stored and processed during vision and asking questions such as these:

- How do we see surfaces and objects, given that the back of the eye is a two-dimensional sheet?
- What does it mean to "pay attention" to something?
- What kinds of events take place when we recognize faces that do not take place otherwise?
- How do we register shape in a way that allows us to recognize objects under varying circumstances, including when parts are missing or added?

- What is the relation between the systems that allow us to identify objects and those that allow us to pick up a pin on a table top?
- Why do our eyes move as they do when we examine an interesting scene?
- What does it mean to say that one is "seeing with the mind's eye"?
- Which aspects of visual perception are learned, and which are innate?
- What does it mean to "see" something?

This book focuses on what is often called *high-level vision*, the aspects of vision that involve stored information. As appealing as it might sometimes seem, we do not live in a fantasy world; the actual world has a way of imposing itself on us. This intrusive aspect is a consequence of *low-level vision*. Low-level visual processes are driven by the stimulus input; they do not rely on knowledge or belief, as high-level processes do. Low-level processes use information about changes in light intensity to detect edges, about disparities in the images falling on the retinae of the two eyes to compute depth, and so forth. Such processes are important, in large part, because they register the presence of specific shapes at specific parts of the visual field. This is the grist upon which all later processing operates.

In Chapter 1, Nakayama, He, and Shimojo deal with the issue of the representation of surfaces. They argue that this level of visual processing is a critical intermediate stage, poised between image-based, low-level processing and high-level visual processing. Representations of surfaces are needed to manipulate objects (as discussed by Goodale) and to recognize them (as discussed by Biederman). Nakayama, He, and Shimojo also argue for the important role of learning and suggest ways in which the visual system draws inferences about surfaces from images. Chapter 2 addresses central issues about attention, which is the selective aspect of information processing. In it, Pashler illustrates ways in which attention determines what information is passed on for more thorough processing. Attention, which is critical for most visual tasks, may also play a critical role in consciousness. Chapter 3 examines the structure of high-level vision, making the case that there are distinct systems used for different kinds of processing. In particular, Farah presents evidence that faces and shapes of objects are handled by different systems, which probably are implemented in different parts of the brain. In Chapter 4 Biederman explores how we recognize such objects as telephones and airplanes. He focuses on the nature of the mental representations used to store shape and argues that our visual systems decompose objects and scenes into specific types of parts, which are then compared to representations stored in memory.

Chapter 5 shifts gears slightly, focusing not on recognition but on action. We use vision not only to identify objects, but also to guide us when we navigate and reach for objects. In this chapter Goodale shows how systems used in object recognition are complemented by others that govern movements. Chapter 6 focuses on a specific system governing movements, the one controlling eye movements. Kowler shows how much we can learn by observing eye movements, which are an index of attention as well as of goal-driven processing. We see that eye movements tell us much about visual cognition in general. Chapter 7 builds on the previous chapters to consider another function of visual representations—their role in forming and using mental images.

In a sense, the final two chapters rely on all that has come before. In Chapter 8, Spelke, Gutheil, and Van de Walle address the ontology of visual perception, and summarize new information about ancient controversies. They focus on the role of general principles versus kind-specific knowledge in our ability to perceive objects and show how information about the child's abilities not only illuminates this issue but also bears on a wide range of issues about the nature of visual perception. Finally, in Chapter 9, Dretske takes a hard look at the philosophical foundations of the assumptions made in the preceding chapters.

Although the tour through visual cognition offered in this volume is by no means exhaustive, it does convey a clear sense of the progress made in recent research into the more "mental" aspects of visual processing. Our purpose is to present the key discoveries and insights in various areas of research on visual cognition and to show how varied types of research fit together. Even more, we intend to demonstrate how the essential interdisciplinary nature of cognitive science is allowing modern investigators to make headway on problems that have plagued thinkers for thousands of years.

Visual Cognition

Chapter 1
Visual Surface Representation: A Critical Link between Lower-level and Higher-level Vision

Ken Nakayama, Zijiang J. He, and Shinsuke Shimojo

One of the most striking things about our visual experience is how dramatically it differs from our retinal image. Retinal images are formed on the back of our eyeballs, upside down; they are very unstable, abruptly shifting two to four times a second according to the movements of the eyes. Moreover, retinal images are sampled very selectively; the optic-nerve fibers that send information to the brain sample more densely from the central area than from peripheral portions of our retinae. Yet, the visual scene appears to us as upright, stable, and homogeneous. Our perception is closely tied to surfaces and objects in the real world; it does not seem tightly related to our retinal images.

The goal in this chapter is to illuminate some of the most elementary aspects of perception as a way of arguing that an indispensable part of perception is the encoding of surfaces. We believe that a surface representation forms a critical intermediate stage of vision poised between the earliest pickup of image information and later stages, such as object recognition. In addition, it is probably the first stage of neural information processing, the results of which are available to us as conscious perceivers.

Why do we think surfaces are so important? The visual part of our brain is not an abstract or neutral information transmission system but one that must capture significant and recurring aspects of our visual environment. Early stages of our visual brain must begin to encode what is most general about our visual environment, providing information about diverse scenes, many of which will differ greatly one from another in terms of specific objects and their layout.

The surface-representation level may provide this necessary intermediate stage for the development of more complex visual processing—for

Thanks are due to the Life Sciences Directorate of AFOSR for support of the research reported here, to Dr. Charles Stromeyer for a painstaking review and criticism of the manuscript, and to Dr. Nava Rubin, Satoru Suzuki, and Emre Yilmaz for their constructive comments.

1

locomotion across a world of surfaces and for manipulation and recognition of objects that are defined by surfaces.

One of the most important characteristics of a world defined by surfaces is that it is three dimensional; ordinarily it has a ground plane below and is accompanied by other assorted surfaces, many of which occlude each other. This means that we cannot expect to see just one surface at a time along any given direction of gaze. Often we see multiple surfaces in local regions of visual space, with closer objects at least partially covering those behind. Thus many surface regions have no counterpart in the retinal image. Yet, remarkably, we do not feel much loss of information when part of a surface is rendered invisible by occlusion; we do not see invisible surface regions as nonexistent. This suggests that we are making "unconscious inferences" (Helmholtz 1910) about literally invisible entities. In the two-dimensional drawing shown in Figure 1.1, we encounter a small set of closed forms that are almost impossible for us to perceive as simply two dimensional. Even without recognizing the lines or patches as parts of familiar objects, we automatically see the configuration as part of a scene in depth and infer that patch x is in front of patches y and z. More important, we infer that patches y and z make up the same surface and that this surface continues behind surface x.

Where in the brain are such inferences made? If we use the word inference, of course, we invite all kinds of possibilities. Is it the kind of inference that we associate with ordinary thinking? Or is it something more visual, linked more specifically to the visual system? We are persuaded by the latter view and shall argue that such inferences are tightly and exclusively tied to visual processing. Our view is that such inferences are embedded in the visual system and can occur at surprisingly early stages, almost independent of our knowledge about familiar objects.

Before continuing our description of surfaces and surface representation, however, we pause to outline briefly what is generally understood about lower-level and higher-level vision as a general context for our results.

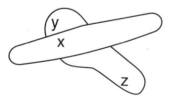

Figure 1.1
Three two-dimensional image areas, labeled x, y, z. These do not combine to form an obviously recognizable object. Nevertheless, region x is perceived to be in front, covering regions y and z, which are perceived to be part of the same surface completing behind x.

First, consider the anatomy of the brain, especially the brains of primates. One of the most startling conclusions to emerge over the past forty years is that approximately 50 percent of the cerebral cortex of the macaque monkey is devoted to vision; the estimated percentage in humans is only slightly smaller (Zeki 1978; Allman and Kaas 1974; Van Essen et al. 1990). At first glance this might seem disproportionate, given the apparent ease and simplicity of seeing, in comparison to, say, thinking, language, or problem solving. Seeing seems so automatic that it might lead us to assume that it requires much less processing. Yet, again, the past forty years of brain research have begun to indicate otherwise, that vision is an extremely complex process, so complex that it is now conceivable that it occupies a sizable fraction of our brains.

Let us look at some specific details. The visual system of the macaque monkey, an animal whose low-level visual capacities are remarkably similar to our own (DeValois et al. 1974), is increasingly understood as an elaborate hierarchical system subserving diverse ultimate functions. The animal's retina contains over one hundred million photoreceptors that send over a million axons to the cerebral cortex via the thalamus in fairly precise register; thus, different parts of the visual field have their exact counterparts in the striate cortex, the first visual cortical receiving area. Surprisingly, more than twenty such separate maps of the retina are projected onto the cortex (Maunsell and Newsome 1987).

What might all these additional visual areas be used for? Little is known. Yet, there is now some evidence that these higher-level visual areas can be divided into at least two streams that serve different higher-order visual functions. In a provocative theoretical speculation—based primarily on anatomy and the results of lesion studies in monkeys and clinical cases in humans—Ungerleider and Mishkin (1982) have suggested that these many cortical areas can be roughly categorized into several substreams that point to important sets of disparate functions for vision. A ventral stream is important for object recognition; damage here leads to an inability to recognize objects in monkeys and to severe losses of object recognition in human patients. A dorsal stream is more specialized for determining the position of points in space or the spatial relations between them. Others (e.g., Goodale et al., this volume) have suggested that this dorsal system might best be described as relating to spatially guided motor behavior, for example, reaching and grasping. These two streams are depicted in Figure 1.2. For the moment, we consider them in order to characterize the major higher-order functions of vision and their anatomical substrates. (See also chapters 3, 4, 5, and 7, this volume.)

These higher-level functions must have input from lower-order visual processes, which, in turn, must receive inputs from the retina and striate cortex. What kinds of information are necessary to serve as useful input for such diverse higher-order functions?

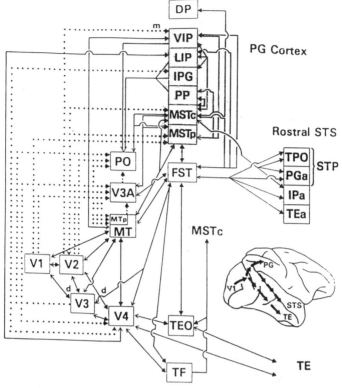

Figure 1.2
Schematic diagram of the connectivity of some of the known cortical areas, grouped into a dorsal and ventral stream. (Reprinted by permission from L. G. Ungerleider and M. Mishkin, Two cortical visual systems. In D. J. Ingle, M. A. Goodale, and R. J. W. Mansfield, eds., *Analysis of visual behavior*, 1982. Copyright 1982 by MIT Press.)

As a start, we might think of the signals arising from well-known classes of visual neurons in the early visual pathway. Work in this area was pioneered by electrophysiological recordings, starting about forty years ago (Barlow 1953; Lettvin et al. 1959; Hubel and Wiesel 1959). By painstakingly recording from one nerve cell at a time, neurophysiologists have pieced together an unusually detailed account of how neurons respond to visual input at various stages of the visual pathway—from the photoreceptors, to the visual cortex, and beyond. Not surprisingly, photoreceptors are sensitive to just a small localized part of the visual field, responding only when light is presented in this very small region. Additional procession is evident, however, when we record the responses of the ganglion cells, the cells that make up the fibers of the optic nerve and convey information from the eye to the brain. Instead of simply responding to luminance in a given region, these cells respond to luminance differences. In general, they respond best when light is flashed in a small circular region of the visual field and not in its immediate local surround.

As one records from the primary or striate visual cortex, additional selectivity becomes evident. Cells here respond to a more specific local geometrical-image property, that of orientation. For example, one class of visual cortical cells responds best to an oriented blob, at, say, 45 degrees, another to an oriented edge of the same orientation, while others respond to other edges, bars, or blobs at other orientations. Thus, if we think of the visual cortex as a whole, there appears to be a complete set of analyzers for each retinal location, each one sensitive to different orientations (and sizes of image regions). The region of the visual field that can influence the firing rate of a cell is called its receptive field. It is clear from analysis of cells and their receptive fields that different aspects of the visual image are coded in different sets of visual neurons.

Although most cells of the striate visual cortex respond more vigorously to one eye or the other, some are binocular. These cells have separate yet highly similar receptive fields mediated through each eye and have the same orientation preference and position in the visual field. Yet careful measurement reveals that for some binocular cells the relative position of the receptive fields in each eye is slightly offset (Barlow et al. 1967; Poggio and Fischer 1979). This is an important discovery because we have known for many years (Wheatstone 1838) that the small difference between image points in two fused photographs or line drawings is the basis of stereoscopic vision. This means that if an animal fixates on a given point in space, different cells will respond differentially to the relative depth of a given visual stimulus, suggesting that a population of disparity-sensitive binocular cells can provide the visual system with a method of encoding stereoscopic depth in a scene.

From what we have said so far, it is evident that the properties of single cells as embedded in the visual system are remarkable; they are selective to complex visual patterns and, even more specifically, to the depth of visual stimuli. In fact, much of the modern work on visual perception assumes that we can understand perception in terms of the properties of these cells; Barlow (1972) has espoused an explicit neuron doctrine for perception. The example of neurons with differing binocular separations seems to go a long way toward explaining how we see stereoscopic depth in natural scenes.

Motion perception seems to be another area in which single-cell recording would be explanatory. In all species studied, cells have been found that are highly sensitive to the direction of image motion. Such cells respond to movement in one direction but not to movement in the opposite direction (Barlow and Levick 1965; Nakayama 1985).

One might conclude from these very impressive findings that perception is simply the working out of the firing patterns of single cells. To understand how we see things, all we need do is continue to explore the response properties of visual neurons. We might think that this level of processing machinery could deliver an adequate representation to the higher functions of object recognition and visuo-motor control. Yet, although we do not deny that some aspects of perception are illuminated by understanding the properties of these cells, they do not adequately explain the specific aspects of perception we shall describe in this chapter. *Our view is that higher functions require, as an input, a data format that explicitly represents the scene as a set of surfaces.*

We have, therefore, divided the remaining portion of this chapter into three sections: Part 1.1 surveys the phenomenology of surface perception; Part 1.2 examines experimental studies showing the importance of surfaces; and Part 1.3 presents our theoretical understanding of the mechanisms of surface perception. In Part 1.1, we consider certain perceptual demonstrations, some of them familiar to the reader, which show how the viewing of very simple patterns is surprisingly revealing of the underlying properties of surface perception. We show that surface perception requires an inferential process residing largely "within" the visual system. These inferences do not require higher-level cognitive processing based on the knowledge of familiar objections. In Part 1.2, our goal is twofold. First, we report on experiments that confirm that phenomenological descriptions, thus adding weight to our previous analysis. Second, we show that the role of surface representation is crucial in a wide variety of visual functions, even those that have been traditionally thought to be directly mediated by the properties of early cortical neurons. The visual functions we studied include visual search, visual object recognition, visual motion perception, and visual texture perception. These studies indicate that seemingly primitive visual functions require, as a prerequisite, the analysis of visual surfaces. We also demonstrate that space perception and visual attention cannot be understood independent of an explicit consideration of a surface representation. In Part 1.3, we suggest a possible site in the brain where surface representation might begin and conclude in a more theoretical vein, suggesting a framework for understanding the perceptual learning of surfaces from images.

1.1   Phenomenological Studies

Experimental phenomenology is a valuable tool for studying perception. It requires the discerning characterization of a person's visual experience in response to well-defined stimuli. Although it is somewhat unusual in a scientific field to dwell on the details of private conscious experiences, it is an essential step to understanding perception. It also gives the study of perception its particular immediacy. Contemporary scientific research in most fields requires complex measuring devices, extensive data collection, and statistical analysis—all of which distance us from the primary data. The study of perception affords the student and researcher alike the opportunity to experience at first hand some of the basic facts of vision. If well conceived, perceptual demonstrations provide viewers with unusually direct access to the nature of their own perceptual machinery.

Some of these demonstrations may be familiar to the reader. They have been marvelled at and endlessly reproduced, gracing textbooks and popular works alike. Yet, despite their wide exposure, some of these demonstrations are often misunderstood, even by experts. Furthermore, they have not been used as part of an overall argument for the existence of a separate stage of visual surface representation.

We start with one of the most famous demonstrations, the Rubin (1921) face-vase phenomenon (see Figure 1.3). Sometimes we see a pair of faces, and sometimes we see a single vase. Additional reflection on what we are seeing leads to several important conclusions. First, the perception is bistable, meaning that we see either the vase as a figure or faces as figures. Second, when one portion of the picture becomes the figure, the other portion degenerates. Yet it doesn't just become less visible; it becomes the background, continuing behind. Third, with each perceptual reversal, there is also a reordering of depth. Whichever portion is seen as the figure always appears to be closer.

Before attempting to explain this demonstration in terms of surface perception, we need to deal with an obvious objection. Maybe the face-vase reversal has nothing to do with surface representation but is mediated at a higher cognitive level, say at the level of object representation. Its bistability may rely on the fact that we all know what faces and vases look like and that we alternate between the two because one can only look at one recognizable object at a time.

This concern is addressed in Figure 1.4, which was also introduced by Rubin. Even though none of the patches on the left or the right are familiar or easily identifiable, the same reversal occurs; and the basic phenomenological effects described for the face-vase figure can be confirmed. This suggests that figure-ground reversal does not depend on such higher levels of processing as object recognition. However, this demonstration, as well as the original Rubin face-vase demonstration, is very different from other well-known classes of ambiguous figures, such as the famous Jastrow Rabbit-Duck illusion in Figure 1.5. Of course, they share the same bistable reversing quality, which suggests a similarity. Yet, there is a fundamental difference. In Figures 1.3 and 1.4, what switches is the patch that is seen as either foreground or background; as described earlier, this also involves a

Figure 1.3
Face-Vase reversing figure. (Adapted from E. Rubin, *Visuall wahrgenommene Figuren* [Copenhagen, 1921].)

Figure 1.4
Reversing figure without familiar objects. (Adapted from E. Rubin, *Visuall wahrgenommene Figuren* [Copenhagen, 1921].)

Figure 1.5
Rabbit-Duck reversing figure. In contrast to the previous two demonstrations, foreground and background do not reverse when the perception reverses, suggesting that this is a different class of figural reversal, one mediated by object-level processes.

reversal of depth perception. With the Rabbit-Duck, however, no such reversal of foreground-background occurs. The figure is always seen in the foreground. What varies is the object perceived. Unlike the two other cases, the Rabbit-Duck involves a reversal at the level of object recognition, requiring object knowledge.

Based on this discussion, we suggest that the figure-ground reversal reflects a more basic, autonomously driven mechanism that is relatively free from top-down, object-level knowledge. In other words, we see evidence of a level of perceptual analysis that is interposed between cells with particular receptive fields, say in the striate visual cortex (as studied by neurophysiologists), and such later stages of visual representation as object recognition.

Figure 1.6 schematizes our placement of the level of visual surface representation as an independent, explicit stage of visual analysis in relation to the overall scheme outlined in Figure 1.2. It is a general purpose, intermediate representation in that it codes enduring aspects of our physical world yet is not concerned with detailed specifics. This surface level

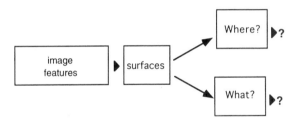

Figure 1.6
Presumed placement of surface representation in relation to lower-level and higher-level visual functions.

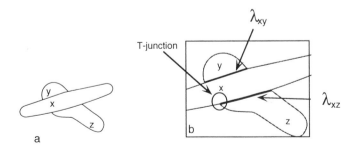

Figure 1.7
Inset of Figure 1.1, detailing the common border λ_{xy} between region **x** and **y** and the common border λ_{xz} between region **x** and **z**. In the parsing of surfaces, the visual system needs to determine which surface regions "own" these common or shared borders. Note existence of T-junction (in circle), which helps to establish depth ordering.

determines whether surfaces are seen as connected or disconnected; folded, straight, or curved; whether they pass in front of or behind; whether they are transparent or opaque. Again, we see this level as distinct from object-level processing, which requires knowledge of specific object or object classes.

1.1.1   Amodal Completion of Occluded Surfaces

Our research has shown that adopting a few simple rules makes surface representation much more comprehensible. For clarity, we initially outline these rules semi-dogmatically, illustrating them with the example presented in Figure 1.7.

Rule 1.   *When image regions corresponding to different surfaces meet, only one region can "own" the border between them.* Thus in Figure 1.7, it is important for the visual system to assign ownership to contours λ_{xy} and λ_{xz}. For example, it needs to decide which image region, **x** or **y**, owns the contour λ_{xy}.

Rule 2.   *Under conditions of surface opacity, a border is owned by the region that is coded as being in front.* In Figure 1.7, this means that region x "owns" the border λ_{xy}.

Rule 3.   *A region that does not own a border is effectively unbounded. Unbounded regions can connect to other unbounded regions to form larger surfaces completing behind.* We call such completion *amodal* completion after Michotte (1964) and Kanizsa (1979).

To see how these rules might play out in actual practice, consider the border between region **x** and region **y** as well as the border between region **x** and **z**. In Figure 1.7, Rule 2 states that the border is owned by the region that is coded as in front. How does the visual system know a region is in front? In this case, the information is supplied by what are known as T-junctions, one of which is circled in Figure 1.7b. This is a junction where three lines meet. Two of the lines are collinear, forming the top of a T; the other line forms the stem of the T. In many natural scenes, such T-junctions are good (but not entirely infallible) clues to depth and occlusion. The top of the T is usually the occluding contour, occluding the stem of the T presumed to continue behind.

Now consider the image patches **y** and **z**. Note that the borders shared with patch **x**, λ_{xy} and λ_{xz}, belong to patch **x**. This means that at this border, regions **y** and **z** are essentially unbounded. Then, according to Rule 3, region **y** and **z** can become connected behind the occluder.

To illustrate these points in a different way, we generate a stimulus (Figure 1.8) in which border ownership changes with the introduction of an occluding figure. Thus, when the individual fragments of the letter *Bs*

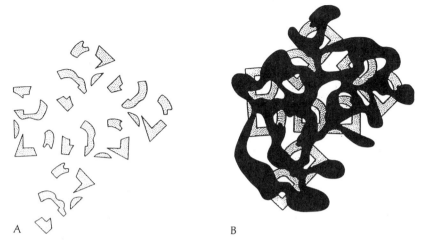

A   B

Figure 1.8
Effect of occluder facilitating the recognition of an object behind. (A) Fragments of the letter B. (B) Same fragments plus the occluder, which makes recognition of the letter much easier. Also, note the existence of T-junctions indicating depth and occlusion. (Reproduced with permission from A. L. Bregman, Asking the "what for" question in auditory perception. In M. Kubovy and J. R. Pomerantz, eds., *Perceptual Organization*, 1981. Copyright 1981 by L. Erlbaum Associates.

are presented, we cannot see the Bs. Only when the occluder is present can we discern the letters. Again, the presence of T-junctions in Figure 1.8a and not in Figure 1.8b justifies the rules we have outlined.

At this point the reader may feel uncomfortable. Sure, the basic ideas are reasonable, but isn't there a kind of logical circularity, particularly because we said that T-junctions indicate occlusion and that they provide information for the stem of the T-junction to continue behind? We have T-junctions in both Figure 1.7a and 1.8b. Isn't there another way of defining depth without T-junctions?

In the remainder of the chapter, we will rely strongly on a fairly obvious and effective method of introducing depth—binocular disparity. For those unfamiliar with stereograms, we include an appendix to the chapter describing various ways of gaining proficiency in the perception of three-dimensional scenes from fused image pairs without using glasses or optical aids. We use stereograms, not because we are interested in binocular disparity or stereopsis itself, but because of the unusual advantages inherent in this method of creating depth. What is particularly useful about binocular disparity is that dramatic changes in depth can be created by tiny shifts in image position. Furthermore, by switching left and right images, we can reverse depth without changing the total amount of information

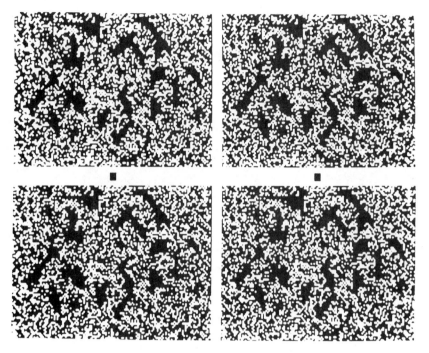

Figure 1.9
Fragments of the letter *B* revealed by stereoscopic depth without T-junctions. First, notice that without stereoscopic fusion, it is essentially impossible to perceive the fragments as comprising parts of *B*s. Crossed fusion of the top two half-images or parallel fusion of the bottom two half-images will show a snake-like figure in front connecting the fragments without the benefit of T-junctions. (Reprinted by permission from K. Nakayama, S. Shimojo, and G. H. Silverman, Stereoscopic depth: Its relation to image segmentation, grouping, and the recognition of occluded objects, 1989, *Perception* 18, 55–68.)

given to the two eyes. So, if our depth hypothesis (i.e., Rule 2) is correct, we should be able to make radical changes in the perceived layout of surfaces with otherwise imperceptible changes in the monocular image.

We should, therefore, be able to show the efficacy of rules 1, 2, and 3 without the benefit of T-junctions. Figure 1.9 is a stereogram showing fragments of *B*s lying in a background plane with a snake-like occluder in front of it. Note that the *B*s are essentially invisible if the pattern is not fused; they are not seen if there is no depth ordering. With stereoscopic fusion, however, something dramatic happens. The *B*s in the background are now clearly visible as individual fragments join to complete the letters behind other surfaces defined stereoscopically. This demonstrates that monocularly defined T-junctions alone do not control the selective completion of surfaces behind occluders. In the next demonstration, in Figure

Figure 1.10
Perception of the letter *C* as influenced by depth. When the figure is normally viewed as a stereogram, we see a *C*, amodally completing behind a small gray rectangular occluder in front. When viewed in the reverse configuration such that the occluder is seen as behind, we perceive two disconnected U-shaped fragments and no longer perceive the fragments as part of a *C*. (Reprinted by permission from K. Nakayama and S. Shimojo, and G. H. Silverman, Stereoscopic depth: Its relation to image segmentation, grouping, and the recognition of occluded objects, 1989, *Perception* 18, 55–68.

1.10, we make the point even more forcefully, by showing that stereoscopic depth can easily overrule existing T-junctions. Without stereoscopic fusion, we see a complete large letter *C* behind a gray rectangular occluder. This is not surprising in light of the arguments presented so far. Because the gray patch (via T-junctions) is perceived to be in front, ownership of the common border (according to Rule 2) is ceded to the rectangle and the remaining image fragments are unbounded, thus completing amodally behind (according to Rule 3).

When stereoscopically fused, no perceived change is expected, because the depth defined by binocular disparity and by the T-junction are in agreement. Both are compatible with interpreting the gray patch as in front, allowing the *C* to remain as highly visible, completing behind the occluder. The reader can verify this by either cross fusing the two left images or parallel fusing the two right images (as described in the Appendix). Perception is very different, however, when the images of the two eyes are reversed such that the gray patch is seen as behind. When this happens, the pieces of the *C* break up into isolated fragments, forming two *Us*—one upright, one inverted, separate and ungrouped. The *C* is no longer visible.

At this point, it should be clear that our perception of recognizable objects can be dramatically influenced by visual surface representation. In addition, we see that perceived depth is extremely important in the perception of objects, although not in the sense usually assumed. Rather than being used to represent the internal three-dimensional structure of the

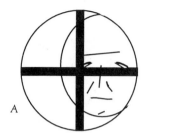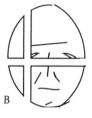

Figure 1.11
(A) Schematic face seen through a window. (B) Face fragments alone (see text).

objects themselves, depth has a more important role: it determines what pieces of an image actually comprise the object to the recognized. Depth is needed to parse objects into wholes or parts, to determine what in an image actually constitutes the parts and boundaries of a single object (Nakayama, Shimojo, and Silverman 1989). In other words, depth dictates perceptual grouping and perceptual segmentation.

Next, we need to deal more specifically with the issue of object recognition. Many contemporary theorists suggest that object recognition requires the matching of stored templates to portions of an image (Biederman 1987; Marr 1982; Nakayama 1990). Examples of ordinary occlusion suggest that there is a profound problem in determining what part of an image will be used for the template matching, a process presumed to occur in object recognition. We cannot simply match templates with the raw, or even filtered, image; because some very spurious matches would be made, preventing the operation of any reasonable recognition mechanism. This problem can be perhaps illustrated by the cartoon shown in Figure 1.11. In A we see a face through a circular, paned window, whereas in B, we see only the visible face fragments. Face recognition has often been seen as a holistic process (see Chapter 3, this volume). An important consideration for this recognition is presumed to be the overall outline of the face and the exact spatial relations between its various parts—not just recognition of the parts themselves. For example, in Figure 1.11b the face is spuriously elongated because the boundary of the window is interpreted as the boundary of the face. How is it then, that we can recognize a face even when it is broken up into pieces and when the outline of the pieces no longer conform to the outline of the face?

This type of problem reinforces our conviction that before the process of object recognition can begin, an object must be separated from the rest of the image and made available to the mechanisms of pattern recognition. This realization further justifies the flow chart outlined in Figure 1.6 and

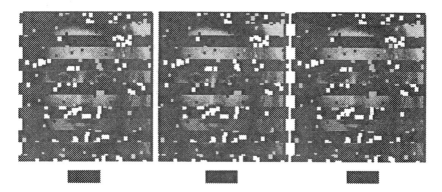

Figure 1.12
Stereogram of a face either in front of or behind occluding strips. Note that the face is more easily perceived when it is behind. (Reprinted by permission from K. Nakayama, S. Shimojo, and G. H. Silverman, Stereoscopic depth: Its relation to image segmentation, grouping, and the recognition of occluded objects, 1989, *Perception* 18, 55–68.)

suggests that we cannot think of object recognition as proceeding from image properties such as those captured by early cortical receptive fields; there needs to be an explicit parsing of the image into surfaces. Without such parsing of surfaces, object recognition cannot occur.

Figure 1.12 is a striking illustration of the importance of depth and surface parsing for object recognition. It shows interrupted strips containing partial images of a face that, viewed stereoscopically, can be seen as either in front of or behind the other interlaced strips. The information available about the hidden face is identical for both depth conditions. Yet, there is an obvious difference in our ability to perceive the face. It is very difficult when the face fragments are in front; but when they are in back, perception is hardly disrupted. It is almost as if all the face is there behind the occluding strips (Nakayama et al. 1989). Again, we see this large difference in the face's visibility as the specific playing out of Rules 1, 2 and 3. When the face fragments are in front, each face strip owns the border between the face and non-face; linkage between the strips does not occur. With the face fragments in back, the common border is owned by the occluding strips in front and the face fragments in back are unbounded, leading to surface completion.

Completion of objects behind nearer objects is ubiquitous in our daily lives. Our demonstrations show that the completion of image fragments behind occluders is not arbitrary but acts according to very specific and highly adaptive rules. It depends on depth and, as a consequence, border ownership, which in turn dictates which image fragments are grouped or segregated.

1.1.2   Completion of Surfaces in Front (Modal Completion of Subjective Surfaces)

Although the need to complete surfaces behind occluders is very frequent in everyday life, occasionally we also need to infer the existence of contours and surfaces in front of other surfaces. This occurs when the luminance difference between a foreground and a background surface is not evident, due to poor illumination or to the chance identity of foreground and background luminance. This situation raises the issue of subjective contours and subjective surfaces.

Thanks to the well-crafted demonstrations of Kanizsa (1979), we are well aware that our brain can create a contour where none exists in the image (see Figure 1.13). Kanizsa describes such contours as examples of *modal*, or *visible* completions. He notes that modal contours and surfaces must complete in front of other surfaces, in contrast to amodal, or invisible completion, which indicates a completion behind other surfaces. Although modal completion has received far greater attention than amodal completion, they have much in common (Kellman and Shipley 1991). Most important, they both qualify as inferences, testimony that our visual system can determine the presence of an edge or surface from incomplete information.

We can ask the same questions about modal completion as about amodal completion. Where do such inferences occur? Are these perceived contours inferences of the sort we make in our daily life or are they inferences made within the confines of the visual system? In the past, these contours were also dubbed *cognitive contours*, implying that thinking or problem solving is involved (Gregory 1972; Rock 1984). Nowadays, the term cognitive contour is little used, and the reasons are important. From what we have said so far, we might argue that the contours seen in Figure 1.13 could have been constructed by some type of top-down inference; that is, we could say that we could reason that a triangle could have covered the adjacent region, thus justifying the term cognitive contour. The same might hold for the sinusoidal contour seen in Figure 1.13b. The

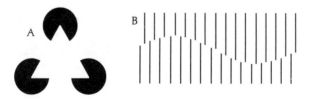

Figure 1.13
Subjective or illusory contours. (A) Kanizsa triangle. (B) Subjective sinusoidal contour formed by offset abutting lines.

almost palpable sense that we *see* the contour certainly argues against this notion of a higher-level inference. But better evidence is needed. In this regard, two additional classes of subjective contours, those driven by binocular disparity and those occasioned by DaVinci stereopsis, are relevant.

If we look at the fish-like silhouettes in Figure 1.14 (adapted from Kanizsa 1979) before fusing them stereoscopically, we can sometimes imagine subjective contours or cognitive contours, with the heads or tails of one "fish" occluding the other "fish." Although there is a tendency to see the broader "heads" covering the narrower "tails," it can reverse. So, our cognitive knowledge or imagination can influence the perception of such contours, particularly when the scene is very impoverished. When fused as a stereogram, however, the specific layout of the subjective contours are immediately apparent. Cross fusing the two half-images on the left and center, we see the tails in front and, automatically, their boundaries as subjective occluding contours. In the opposite stereo case, the heads are seen in front, and we immediately see visual subjective contours bounding them. Furthermore, the perception of these contours is stable and unchanging; one is hard-pressed to argue for some form of deliberate top-down inference in this case. The subjective contours appear to be formed by an efficient, adaptive, and autonomous process driven, in this case, by binocular disparity signals, which overcome higher-order knowledge or expectations about objects.

Even more telling is the case of DaVinci stereopsis. In an earlier study, we created subjective contours in a situation where higher-order inference cannot occur; that is, where no information is available at a conscious level (Nakayama and Shimojo 1990). To understand DaVinci stereopsis, it is necessary to appreciate that some regions of most real-world scenes are

Figure 1.14
Stereo version of Kanizsa "fish." Although this is a flat two-dimensional figure, observers generally see depth even when pairs of images are not fused as a stereogram. Unfused, two different surface arrangements are apparent. Usually one sees the broader "heads" in front with visible subjective contours completing in front of the narrower "tails." At other times, one sees the narrower "tails" in front, bounded by their corresponding subjective contours. When fused stereoscopically, binocular disparity determines the depth placement of the heads and tails accompanied by subjective contours.

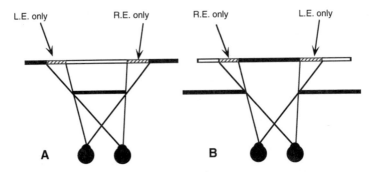

Figure 1.15
Top view of two scenes outlining the geometrical foundations of DaVinci stereopsis. (A) An opaque square in front of a frontoparallel surface. (B) A more distant surface seen through an aperture or window. In each case the differential binocular optical consequences of occlusion are characteristic and invariant. Regions seen by the left eye only are on the farther surface just to the left of a closer occluding surface. Regions seen by the right eye only are on the farther surface just to the right of a closer occluding surface.

visible to one eye or the other but not to both. This can be understood by referring to Figure 1.15. This top-view diagram shows two situations: Figure 1.15a shows a square occluding a wall behind; Figure 1.15b shows a distant wall viewed through a window. Because the closer surface is opaque, there are regions (depicted as the hatched area) that are visible only to the right eye or the left eye.

Such half-occlusions, or unpaired points, arise almost constantly in our everyday life because we are inevitably exposed to the edges of objects at different depths. These unpaired regions lie on more distant surfaces adjacent to the image boundaries of nearer surfaces. What is important for our discussion is the highly constrained nature of this binocular unpairedness. First note the obvious fact that our eyes are horizontally aligned and thus have different viewpoints along a single horizontal dimension. This means, in general, that such half-occlusions occur only when there is a vertical component to an occluding contour; they do not occur for purely horizontal edges. Even more important is the fact that there is an obligatory, nonarbitrary relation between a given unpaired point and the placement of the occluding contour that causes it to be unpaired. Unpaired right-eye-only points are seen only next to occluding contours to their immediate left. Unpaired left-eye-only points can be seen next to occluding contours to their immediate right (see Nakayama and Shimojo 1990; Shimojo and Nakayama 1990).

One might ask the following question. Given that such pairing is ubiquitous in everyday life, what would happen if we were able to insert a few unpaired points in an otherwise identical pair of images? Would this call forth the perception of an illusory subjective contour, and would such contours assume the exact placement dictated by the geometrical considerations just outlined? With these general considerations in mind, we created a stereogram that, although it contains no binocular disparity, is able to create the impression of a sharply defined occluding surface in depth (see Figure 1.16). To understand what is occurring in this stereogram, refer to Figure 1.17, which depicts the surfaces perceived in terms of the exact placement of the right-eye-only, left-eye-only, and binocular points.

Case A, shown in Figure 1.17a is a control condition. To view this case, one simply needs to fuse any of the identical images in the top row of Figure 1.16. Since the images are the same, all points are seen binocularly and there is no binocular disparity. Not surprisingly, one sees only a single flat surface in the picture plane, with no depth. Case B, the main DaVinci demonstration, is exactly the same as case A, except that four half-points have been removed from the binocular image. The physical-stimulus situation is explained in Figure 1.17b, showing the dots remaining. Note that there are two left-eye-only points (depicted by the open symbols) and two right-eye-only points (depicted by the gray symbols) in addition to the rest of the points, which are binocular. This pattern of binocular stimulation simulates a condition in which an invisible opaque surface is placed in front of a surface containing dots, a condition similar to the top view depicted in Figure 1.15a.

To view the DaVinci case the reader must fuse images on the bottom row of Figure 1.16, either cross fusing the left and center images or parallel fusing the center and right image. If fusion is successful, the perceptual consequences should be apparent and dramatic. One sees a phantom black square in the stereogram, the borders of which are exactly depicted in Figure 1.17b. The square, which appears even blacker than the background, lies in front of the rest of the pattern and is bounded by very sharp vertical edges. It is necessary to scrutinize the stereogram carefully to see that the relationship between the perceived square in front is exactly as depicted in Figure 1.17b. Thus, the left edge of the phantom occluding square is perceived to the immediate right of the left-eye-only unpaired points, and the right edge is perceived to the immediate left of right-eye-only unpaired points. When viewing the stereogram the exact placement of the phantom square and the unpaired points can be checked by alternately closing each eye.

Case C, the reversed-eye pattern, takes a little more exposure and practice but is well worth the effort. It simulates what is seen in Figure 1.15b. The proper stereoscopic stimulation can be accomplished by fusing the alternative pair of images in the bottom row of Figure 1.16. A methodological hint: Gazing directly at the location of the presumed window in Figure 1.16 may break fusion because the system may make an inappro-

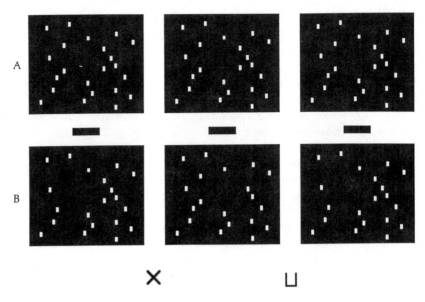

Figure 1.16
DaVinci stereopsis stereogram. (A) The control case: each half-image for binocular fusion is identical. As such the fused image should appear as flat. (B) The identical stimulus as in (A), except that four dots (two from each eye) have been removed. The observer sees a subjective square in front bounded by the unpaired points (as illustrated in Figures 1.15a and 1.17b). When fused in the reversed-eye configuration, observers see an aperture through which is seen a distant surface (as illustrated in Figures 1.15b and 1.17c).

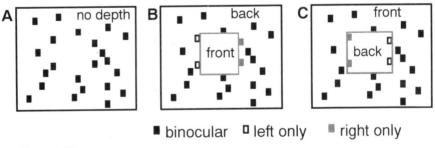

Figure 1.17
Explanation of the DaVinci stereogram, illustrating the relations between perceived surfaces in depth and the exact placement of paired and unpaired points shown in Figure 1.16. (A) There are no unpaired dots, and no depth is seen. In (B) two left-eye-only dots on the left side and two right-eye-only dots on the right side lead to perception of a central square in front. In (C) the configurations of the left-eye, right-eye stimuli are reversed, leading to the perception of a window through which one sees the unpaired dots in back. In all cases, left and right side of the occluding surface is bounded by right-eye-only and left-eye-only points, respectively.

priate convergence eye movement, attempting to fuse the two pairs of unpaired points. This tendency can be overcome by fixating on the horizontal bars between the upper and lower rows and attending to the area of interest.

The perceived configuration in relation to the physical stimulus is outlined in Figure 1.17c. In this situation, instead of a square, one sees a subjective window, revealing the unpaired points, which are now seen as far back, and define the edges of the window. Note that the window is significantly wider than the occluding square seen when the eyes are reversed.

We cannot overemphasize that there is a very specific and unvarying rule about where the subjective contour will lie. According to the diagrams in Figures 1.15 and 1.17, subjective contours should arise to the immediate left of right-eye-only points and to the immediate right of left-eye-only points. Careful examination of the lower stereograms in Figure 1.16 shows that this simple relationship holds for all four situations. This fact is easiest to appreciate when the subjective square is in front but is also apparent to those who can see a subjective window with the unpaired points in back. In each case, the position of the subjective contours in relation to the unpaired point is predictable and determinate and arises from the optical and geometrical constraints imposed by viewing a scene from different vantage points.

The main point is that these findings indicate clearly that vivid subjective contours can be created by information unavailable to conscious experience. We have no awareness of which eye is receiving the unpaired right-eye-only or left-eye-only stimulation, and we are unaware of the geometrical relations depicted in Figure 1.15; yet we see the results of our perceptual machinery—subjective occluding contours at very specific and predicted loci in the display. We believe this demonstration, in particular, lays to rest any view that subjective contours are the result of higher-order, nonvisual inferences. In Part 1.3, we make the point that such visual inferences occur very early in the visual pathway, perhaps as early as the striate cortex (area V1, as shown in Figure 1.2).

1.2   Experimental Studies

Phenomenology, the method used in the studies described so far, is often viewed suspiciously by those unfamiliar with its contributions. In part, this is due to the demand for an objective, not subjective, methodology in psychology and cognitive science. In part, it is due to worries about observer and/or experimenter bias and a lack of quantitative or statistical measurement. That said, however, one must also add that despite all these seemingly valid misgivings, phenomenology survives and even flourishes among a small group of practitioners. Moreover, its results and conclusions often enjoy wide circulation in the scientific and lay community at large.

Why is this so? We discern several possible reasons. First, the results are actually much more objective than one might suppose. With well-crafted demonstrations, perceptual agreement between observers is actually far greater than that obtained in many psychological experiments, which often require statistical analyses of results from large numbers of subjects. Second, of course, is the immediacy and verifiability of the demonstration. All practitioners and interested parties can see the phenomenon for themselves and need not be concerned that the scientific reports are, as they sometimes are, mistaken. Third, the advent of good and cheap media technology, precise printing, and computer graphics technology enables many excellent demonstrations to be widely disseminated. Overall, phenomenology furnishes us with a surprisingly large, rich, personal (yet shared) data base from which to draw systematic theoretical conclusions.

There is, nevertheless, a great need for more objective verification of the sorts of phenomena we describe, not only to validate the method but also to reach out to other areas of knowledge, particularly the brain sciences. Because, for example, we cannot similarly characterize the visual perception of other species, we need to develop more objective experiments that do not rely on the subtle details of perception obtained from verbal reports. We cannot limit ourselves to phenomenology to obtain, for example, a satisfactory neurophysiological explanation of surface perception.

How then, do we convert a phenomenological observation of visual surfaces into one that can be substantiated by objective experiment, one that might also be conducted, if desired, on a laboratory animal? In the studies we describe here we use an indirect route. Instead of asking for a description of the experience of a surface as practiced in the section above, we ask an observer to perform a task we presume depends on surface encoding. In this way, we can evaluate the consequences of a surface representation without relying on the observer's subjective phenomenological judgment.

1.2.1   Two Views of Intermediate Visual Processing

Our goal, however, is more than simply the verification of phenomenological observations. The nature of our results allows us to challenge some widely held beliefs about intermediate visual processing and to replace them with an alternative conception. We argue that many seemingly early visual tasks are actually performed on a surface representation rather than

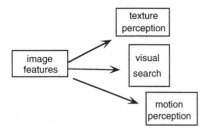

Figure 1.18
Current views regarding the dependence of rapid visual processing (texture perception, visual search, motion perception) on feature processing as mediated by receptive fields of cortical neurons in early cortical areas.

on the image. This position requires us to review briefly the widely held views we oppose. Some of the latter are described pictorially in Figure 1.18.

According to this way of thinking, there are a number of intermediate visual processes that do not require object knowledge, but can perform various rapid visual functions, including texture segregation, visual search, and motion perception. It is generally assumed that these functions operate at the level of simple features or filters in a retinotopic space. What characterizes this approach is the belief that the kinds of operations conventionally thought to be involved in the wiring of receptive fields are also likely to be explanatory in dealing with these perceptual functions.

To review this point, we need to step back and describe how vision scientists conceive of features and filtering and how these processes might be understood in terms of receptive fields. The basic form of the explanation proposed is extremely simple. Receptive fields of retinal ganglion cells, for example, can be understood if we simply assume that they are fed by two classes of convergent yet antagonistic inputs that are spatially delineated: an excitatory center and an inhibitory surround. Light falling on a center region alone will excite the cell. Light falling on the center and surround region, however, will not. We can conceive of these cells as sensitive to local differences in luminance or, more technically, contrast. A similar straightforward convergence is assumed to explain the properties of cortical cells. Converging and excitatory inputs from only correctly located ganglion cells would provide a cortical cell with orientation selectivity (Hubel and Wiesel 1959). Other simple schemes can account for motion sensitivity (Barlow and Levick 1965), as well as more complex receptive field specification such as end stopping (Hubel and Wiesel 1965). As mentioned earlier, convergent input to a cell from similar receptive fields at slightly different offsets in the two eyes for different cells would provide a system whereby stereoscopic depth could be coded by comparing the inputs to different sets of cells. From these general findings Barlow (1972) made a strong conjectural case showing how a system might plausibly code important properties of a visual scene.

1.2.2   Visual Search and Visual Texture Segregation

The psychological/perceptual functions of texture segregation and visual search have been similarly conceived, although at a somewhat higher level of complexity. Thus, for texture segregation, it is assumed that by an analogous summation, and then differencing, of the outputs of cells with oriented receptive fields, a later stage should be able to signal texture boundaries. A similar conception suggests how an odd target in a popout task is identified. A strong indicator that the basis for easy texture segregation and popout must be fairly primitive—and can perhaps be accounted for by these simple mechanisms—is seen by examining the relative ease with which a segregated figure emerges in Figure 1.19a where the texture is defined by differently oriented elements, in contrast to the greater difficulty of seeing the emergence of texture in Figure 1.19b, where the texture elements are defined by the letters *T* and *L*. Although each *L* is easily distinguishable from each *T* in Figure 1.19b, it is apparent that the difference is not sufficient for rapid texture discrimination.

This distinction between simple and more complex features is also apparent in experiments on visual search (see Figure 1.20). Here the observer is asked to find the odd target among distractors, with the number of distractors varied. With only a few distractors, search reaction times for the two target/distractor types are comparable. When many more distracting elements are added, however, performance is degraded only for the *L* among *T*s case, in which reaction times increase markedly. With the

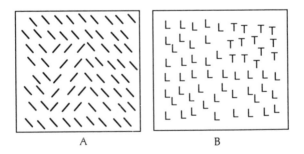

A   B

Figure 1.19
Texture segregation displays. (A) Texture difference determined by oriented lines. (B) Texture difference determined by different letters. Note that the emergence of a differently textured area is more prominent in A than in B.

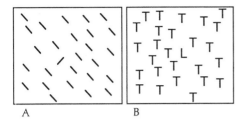

A   B

Figure 1.20
Visual Search. (A) For an odd orientation. (B) For an odd letter.

simple oriented lines no such increase in search time is apparent. Thus, performance seems to be independent of the number of distractors.

These findings on texture segregation and visual search lead to similar conclusions. The basis of rapid visual processing is presumed to lie in spatial primitives at a very rudimentary level of pattern recognition (Julesz 1986; Treisman 1982; Beck et al. 1983). Because of the orientation tuning of receptive fields and the oriented nature of stimuli that are easy to segregate, it might seem natural to see these receptive fields as prime candidates for mediating the very primitive type of pattern recognition required. A number of models of texture segregation and visual search make relatively appropriate predictions of a wide range of phenomenon (e.g., Malik and Perona 1990; Fogel and Sagi 1989). These models, for example, assume hypothetical units that pool the activity of classes of receptive-field types, then take differences in the outputs from these units, effectively obtaining differential excitation if a boundary exists between, say, regions of one orientation and another (as in Figure 1.19). Thus, rather than showing a sensitivity to simple luminance differences (as described for ganglion cells in the retina), these hypothetical units would be sensitive to differences in the density of particular texture elements, thus enabling a system to be selectively responsive to texture boundaries. Models of visual search suggest that a related mechanism can account for the emergence of the odd target among a field of distractors (Julesz 1986; Koch and Ullman 1985).

Even though such models explain much of the data described so far, they cannot explain the classes of phenomena we will describe below. Our motive in doing so is to suggest that surface representation is a necessary intermediate form of perceptual representation, one that forms an appropriate foundation for other visual functions—object recognition, object manipulation, and locomotion. We propose, as an alternative hypothesis to primitive receptive-field outputs, that perceptual function must funnel through a surface representation and that the most rapid visual functions we can measure must also pass through this required stage (as illustrated in

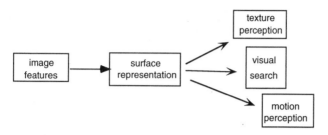

Figure 1.21
Our proposed view, showing that surface representation must precede such perceptual functions as texture perception, visual search, and visual motion (in contrast with the view outlined in Figure 1.18).

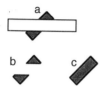

Figure 1.22
Occlusion and grouping: (A) scene in which gray region is seen as rectangle behind (as in C) and not as two disconnected fragments (as in B).

Figure 1.21). Thus the primitives that have been assumed to govern texture segregation—that is, receptive-field outputs—are not the ones responsible for the perceptual phenomenon under study.

We hypothesize that, because we cannot easily perceive the results of operations prior to the stage of surface representation, the latter is the first stage to which we have immediate access as perceivers (He and Nakayama 1992, 1994c). Such a hypothesis provides some strong predictions. It means that because of amodal completion (in accordance with Rules 1, 2, 3), the gray regions depicted in Figure 1.22a, instead of being encoded as two small separate polygonal figures (Figure 1.22b), are likely to be seen as an oriented rectangle in back (Figure 1.22c).

Such reasoning leads to the following question: What level of visual processing governs performance in rapid visual tasks presumed to be important for everyday vision? Is it the shapes of the image pieces themselves or the surface shape as defined by amodal or modal surface completion? Our surface hypothesis, of course, predicts that completed surfaces will be found most important. Earlier views of these processes, on the other hand, predict that the fragmented shapes of the image will dominate. To evaluate the merits of these competing hypotheses, our strategy was to

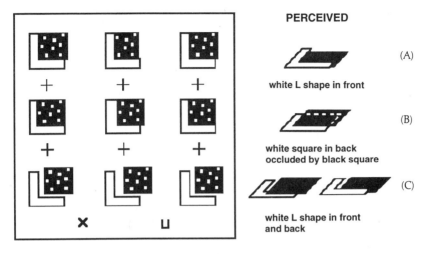

PERCEIVED

(A) white L shape in front

(B) white square in back occluded by black square

(C) white L shape in front and back

Figure 1.23
Elements used in a visual search display in which an observer is to find normal Ls among reversed Ls, or vice versa (reversed Ls among normal Ls). The individual L-shaped elements are presented adjacent to black textured squares and coded stereoscopically either in front of or in back of these squares. Each of the different conditions is depicted in one of three rows, labeled A, B, and C. A stereoscopic view of an individual element is presented in the box on the left, accompanied by a pictorial description of the perceived surface layout of each element on the right (under column labeled Perceived). (A) The L is in front of the squares and appears as an L. (B) The L appears in back and appears as part of a larger figure completing amodally behind. (C) Control condition, in which the L-shape is separated from the black square. Here the L is seen as an L, no matter what the relative depth of the black square or the L. (Modified by permission from Z. J. He and K. Nakayama, 1992.)

conduct experiments in which the stimulation of early cortical receptive fields is more or less unchanged but, by subtle shifts in binocular disparity, we altered depth relations in the display. This change in depth relation can lead, in turn, to the dramatic shifts in surface representation we described in the previous section. Our experiments show how this leads to a large difference in visual performance.

1.2.3   Surface Shape in Visual Search

We start by describing experiments on visual search (He and Nakayama 1992). In Figure 1.23 we show a stereogram of Ls adjacent to a black textured square, which is present in all the displays. When the Ls are in front, it should be clear, they look like Ls (Figure 1.23a). When they are in back, however, they look very different (Figure 1.23b). Stereoscopic depth (in accordance with Rule 2) ensures that the border is owned by the square in front. As a consequence (and in accordance with Rule 3), the L becomes part of an amodally completing surface, continuing behind the black squares. As such, it becomes less L-like and looks almost like a square in back.

If we set up a visual search experiment in which the observer was to find an L among reversed Ls (or vice versa), we would expect that binocular disparity would have little effect on the result if simple image features are important in determining the outcome. If, on the other hand, completed surface shape is important, we would expect the visual search to become more difficult when the Ls are behind. In this situation, both regular Ls and reversed Ls would become part of larger, more indistinguishable surfaces completing behind the rectangular occluder, each appearing as "almost a square." As a consequence, the inverted Ls would become much less distinguishable from the regular Ls in the visual search task.

This prediction is borne out by studies on search reaction times in which we varied the number of distractors for the Ls-in-back versus the Ls-in-front cases. In Figure 1.24, we show that when the Ls are in front, search times are more or less constant for increasing numbers of distractors. For Ls in back, however, it is very different. Search times increase dramatically with greater numbers of distractors.

One might argue, however, that it is easier to see targets when they are in front because there is a perceptual salience for closer targets. Control experiments in which small gaps are placed between the Ls and the squares (as in Figure 1.23c) indicate that the fact of the Ls being in front cannot alone account for the results shown in Figure 1.24 and that the involvement of surface completion is crucial (He and Nakayama 1992).

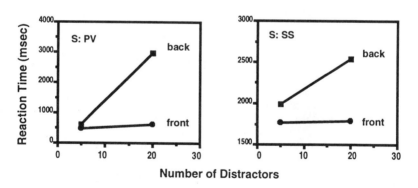

Figure 1.24
Reaction time to see the odd target (and L among reversed Ls, or vice versa), showing dependence of distractor number for Ls-in-front versus Ls-in-back cases. Note that reaction times increase only for the Ls-in-back case. (Reproduced by permission from Z. J. He and K. Nakayama, Surfaces versus features in visual search, 1992, *Nature* 359, 231–233.)

1.2.4   Surface Shape as Primitives for Texture Segregation?

As mentioned earlier, texture segregation is another area in which researchers have generally thought that performance is determined by differences in receptive-field outputs in early visual processing. According to this conception, primitive shape differences are sensed by such postulated mechanisms, and texture boundaries are computed automatically by the filtering properties of early cortical neurons.

Yet, our surface hypothesis might apply here as well. Perhaps it is not primitive shape, as determined by early receptive-field mechanisms, but surface shape, determined after the process of surface formation and surface completion. To test this hypothesis, we arranged an experiment in which the observer is presented with a very brief visual display followed by a mask. The observer's task is to report whether the differently textured region is a rectangle oriented horizontally or vertically (He and Nakayama 1994b).

The texture displays are similar to that shown in Figure 1.25, where the textured central rectangle differs from its background by being either *Is* among *Ls* or vice versa. Here too the observer must report whether the

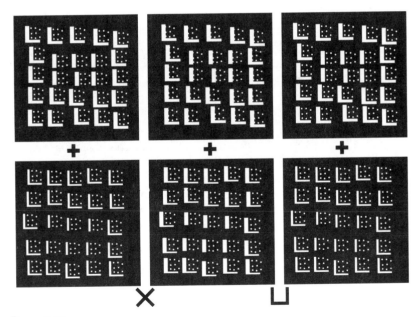

Figure 1.25
Stereograms showing texture segregation. Top row: texture elements are in front and texture segregation is easy. Bottom row: texture elements are in back, leading to amodal completion and more difficult texture segregation. (Reprinted by permission from Z. J. He and K. Nakayama, Perceiving textures: Beyond filtering, 1994, *Vision Research* 34, 151–162.)

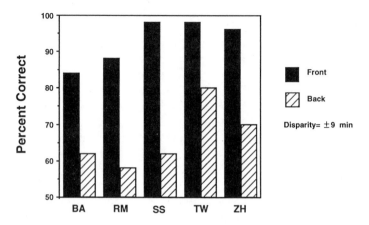

Figure 1.26
Percent correct identification of the orientation of a rectangular texture area defined by *Is* versus *Ls* (or vice versa). (Reproduced by permission from Z. J. He and K. Nakayama, Perceiving textures: Beyond filtering, 1994, *Vision Research 34*, 151–162.)

textured region is elongated horizontally or vertically. From the stereogram shown in this figure, it should be clear that it is much easier to discern the region of distinctive texture when the elements are in front than when they are in back. Compare the upper and lower stereogram. This difference is confirmed by the graph (Figure 1.26), which reports the percentage of correct scores of five observers for front versus the back cases.

In both visual search and visual-texture segregation experiments, we were able to change the surface representation so as to leave the image and, thus, the feature representation largely intact. This change in surface representation had a major effect. It was decisive in determining performance in very rapid visual tasks. This rapidity further underscores the importance of surface representation for immediate vision. It suggests that when we are confronted with an image under time constraints, we cannot respond to the shapes of the image fragments themselves. Our first impression is that of a surface representation.

1.2.5   Perception of Motion

Motion perception has generally been regarded as a fairly automatic and early visual function not dependent on higher-order visual input or top-down processing. As mentioned earlier, there are neurons selective to local motion in the striate and extrastriate cortices of primates, suggesting that at least some aspects of human motion perception are mediated by such cells (Hubel and Wiesel 1968; Britten et al. 1992). Yet there are a number of indications that motion perception cannot be determined simply by the outputs of motion-sensitive neurons with localized receptive fields. One of these is the aperture problem; another is the phenomenon of long-range apparent motion. We consider each of them in turn.

The Aperture Problem

Figure 1.27 illustrates the aperture problem. In Cases A and B, we show two directions of motion taken by different elongated surfaces textured with diagonal lines. Despite the large differences in motion, vertical for Case A and horizontal for Case B, the motion is indistinguishable when considered locally. When informed about these very different global motions through a circular aperture, our visual system defaults and sees neither horizontal or vertical motion but a diagonal motion of stripes, its direction being orthogonal to the orientation of the local oblique contours (equivalent to that depicted in Case C). This perception indicates that local measurements of motion (as accomplished by orientation-selective, motion-sensitive neurons) by themselves are insufficient to specify true motion direction. Wallach's famous barber pole illusion (1935) shows that if we change the shape of the moving stimulus aperture, perceived direction of motion changes dramatically. If the aperture is oriented horizontally (as

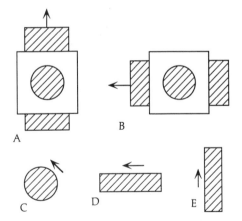

Figure 1.27
Moving oblique gratings viewed through various apertures. In (A) and (B) we show two very different motions of surfaces containing oblique lines, upward and leftward, respectively. Even though the true motion is different in each case, the local motion, as viewed through an aperture (as in C) and as it would be coded by motion-sensitive neurons, is the same, diagonal motion, up and to the left. (D and E) Wallach's barber pole illusion: perceived motion of the oblique lines is determined by the orientation of the elongated aperture.

in D), horizontal motion is seen to be moving horizontally. Similarly, vertical motion is seen in the vertical aperture (as in E).

Hildreth (1984), attempting to explain such perceptual phenomenon in terms of receptive-field-like entities, postulated a subsequent stage of analysis in which unambiguous motion of line terminators at the T-junctions (as in Figure 1.27) can propagate and overcome the ambiguity of such local motion (see also Nakayama and Silverman 1988; Yuille and Grzywacz 1988). According to this postulate, the "solution" to the aperture problem is an encapsulated one, "solved" exclusively within a motion module and operating only on a two-dimensional representation.

In line with our own understanding of visual surfaces, however, we analyze the problem very differently (Shimojo, Silverman, and Nakayama 1989). We ask whether the visual system regards the surface composed of stripes as continuing behind the aperture, in the same plane, or in front. From what we have said so far, one might expect that the moving stripes would be regarded as moving behind a rectangular aperture. This follows from the numerous T-junctions, which might indicate that the surface was behind and not bounded by the aperture. This depth cue, however, is in conflict with binocular disparity, which indicates that the diagonal stripes and the outline of the stripes are in the same depth plane.

Interestingly, if we look at the vertical barber pole illusion (as depicted in Case E, Figure 1.27) with only one eye; the bias toward motion along the aperture length is attenuated; that is, the illusion is weakened (Shimojo et al. 1989). This is not predicted by any receptive-field mechanism accompanied by velocity propagation from terminators. It can be explained at a surface level, however, when we realize that with monocular viewing the T-junctions denoting occlusion are no longer in conflict with the binocular cue of flatness. The surface itself is no longer seen as elongated but as boundless, appearing to extend beyond the aperture through which it is viewed. Not surprisingly, and for the same reason, the barber pole illusion is weakened further if we, by manipulating binocular disparity, arrange it so that the stripes are seen in back.

Figure 1.28 shows even more dramatically the importance of amodal surface completion behind occluders. In this experiment we changed the horizontal motion ordinarily seen in three horizontally oriented barber poles to vertical motion, simply by manipulating binocular disparity so that the configuration appears as a single, large vertical barber pole continuing behind nearer occluding stripes in front. This is accomplished by manipulating the binocular disparity of the two small stipled strips sandwiched between the three horizontal rows of oblique lines (Shimojo et al. 1989).

All these findings indicate that we cannot understand the perception of motion solely in terms of low-level motion signals. Even for the simple

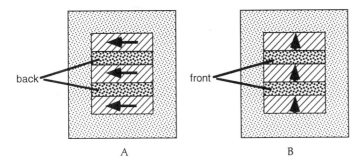

Figure 1.28
(A) Three small horizontal barber pole illusions showing movement to the left occurs if the stippled panels separating them are in back. (B) Putting the stippled panels in front, allows for the completion of all diagonal regions behind and leads to the perception of a large vertical rectangle. In this case, motion is seen as vertical. (Reprinted by permission from S. Shimojo, G. H. Silverman, and K. Nakayama, Occlusion and the solution to the aperture problem for motion, 1989, *Vision Research* 29, 619–626.)

coding of motion direction, the visual system needs information about the layout of surfaces in three-dimensional space.

Apparent Motion

The illusion of apparent motion was identified almost a hundred years ago by Gestalt psychologists (Wertheimer 1912). It occurs when stationary stimuli are flashed on and off in succession—one at time t_1, followed by the other at time t_2. This illusion is schematized in Figure 1.29.

It is interesting that the range of distances over which apparent motion can be seen is very large, spanning many degrees of visual angle. This poses problems for an explanation of motion perception in terms of motion-sensitive neurons in the visual cortex; such neurons are directionally selective but only over a very local area as small as a fraction of a degree in the striate cortex. Furthermore, the duration over which apparent motion is seen is long in terms of the measured properties of directionally selective cortical neurons. For these reasons, the processing of apparent motion has been classified separately from the processing of continuous motion and has been designated a long-range (as opposed to a short-range) motion process (Braddick 1974; Anstis 1980).

Still more interesting properties emerge with just the small addition of complexity to the usual apparent motion configuration. Ramachandran and Anstis (1983), for example, employed a 2 × 2 competitive-motion paradigm in which two pairs of stimuli occupying opposite corners of an imaginary rectangle flash alternately (see Figure 1.30). Note the potential ambiguity in this display. The element or token in, say, the upper-left

Figure 1.29
The simplest case of apparent motion. A time T_1 a small stationary square is flashed, followed at time T_2 by another flash.

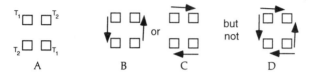

Figure 1.30
A. Bistable, competitive, apparent-motion paradigm where at time T_1 flashes at two stationary squares in opposite diagonal corners of an imaginary rectangle are followed at time T_2 by flashes at the complementary diagonal squares. Vertical (as in B) or horizontal motion (as in C) is perceived. Ambiguous motion (as in D) is not perceived.

corner flashing at time t_1 can be paired with other identical tokens flashing either at the upper-right or lower-left corners at time t_2. Despite this ambiguity, the perception of motion is pronounced. However it is also bistable. If the horizontal and the vertical distances are approximately the same, one sees either vertical or horizontal motion with approximately equal probability (as in Figure 1.30b, c). Surprisingly, we rarely, if ever, see a transitional motion perception in which, for example, the two targets split off to become the other two (as depicted in Figure 1.30d). This phenomenon illustrates what has been called the *correspondence problem* and its solution for apparent motion: Our visual system appears to make a binary decision, linking a token in frame 1 to a token in frame 2; there is no in-between solution or blending resolution.

What determines this correspondence? First, and most important in terms of establishing our experimental method, is the relative proximity between tokens. Correspondence is preferentially established between closer rather than distant tokens. Second is token shape. We consider each in turn.

The importance of token proximity can be easily demonstrated. If the relative vertical distance between tokens is decreased, motion will be predominantly vertical, whereas if the relative horizontal distance is decreased, horizontal motion will win out. Thus, if we keep vertical distance constant and gradually increase the horizontal distance in small steps, we can measure a motion-dominance function (see Figure 1.31) that summarizes the

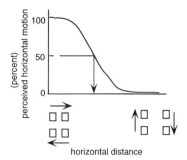

Figure 1.31
Motion-dominance function illustrating the proximity tendency for apparent motion. Keeping the vertical distance constant and increasing the horizontal distance decreases the likelihood of seeing horizontal motion. Horizontal affinity corresponds to the distance (marked by the arrow) where this function exceeds 50 percent (see text).

amount of horizontal motion seen as a function of horizontal distance. This function reflects the proximity tendency, which shows that matches are more likely to be made with nearest neighbors.

Less powerful but more pertinent for our immediate discussion is the role of shape in determining apparent-motion correspondence. Although shape matching is weak and can be easily overwhelmed by small differences in proximity, its existence is clearly revealed in a competitive-motion situation in which the various proximity tendencies between possible matches are more or less balanced. If, for example, we arrange our apparent-motion configuration so that one pair of identical shapes is presented sequentially in the top row alternately with a different pair of shapes in a bottom row, we see a preference for horizontal motion at intermediate horizontal distances, where the proximity tendency is more or less balanced for each possible match. In terms of the motion-dominance function seen in Figure 1.31, a bias toward matching identical tokens in the same horizontal row would shift the motion-dominance function to the right.

We have discussed the role of both shape and position in determining apparent motion correspondence strength, but we have not as yet linked these findings to the central theme of the chapter. In the context of visual surface representation, we need to define more precisely what is meant by *shape* and *position*. Is it shape as it might be defined in an image or image fragment or is it shape as defined after the processes of surface representation have been completed? Similarly, with position. Is it the position of the image patch narrowly defined, or after a surface representation has been established?

Shape Similarity in Apparent Motion

First, we deal with the issue of shape, showing that it is not image shape that determines correspondence but surface shape. To reveal its importance, we bias the competitive paradigm toward horizontal motion by increasing the relative similarity between potential horizontal matches, selecting the same shape between elements of the upper and lower pairs, respectively (see Figure 1.32). The upper row consists of oriented +45-degree bars, the lower row of oriented −45-degree bars. In each case, the pairs of white diagonal bars flash in opposite diagonal corners (as described earlier for Figure 1.30a), and flanking stationary nonflashing oblique rectangles are always present in all four positions.

Once again we use binocular disparity to manipulate the depth relations between the gray textured rectangle and the pairs of white bars. When the flashing white bars are in front no amodal completion between them can occur. They will be seen as two distinct diagonal bars and, because of the shape identity within a horizontal row, we expect to find the greatest horizontal affinity between the tokens, which should shift the motion-dominance function to the right. It should be very different, however, when binocular depth is reversed. Not only will the parallel bars be seen as behind but, more importantly, they will become part of a single surface

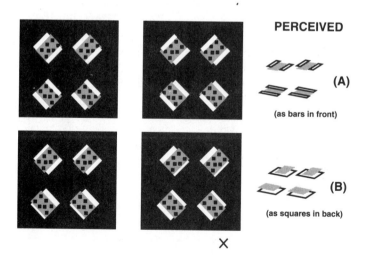

Figure 1.32
Stimulus showing that surface shape, not image shape, biases motion correspondence: flashing stimuli presented in the 2 × 2 paradigm. Note that the textured "occluders" are always present; only the parallel bars flash. (Stereograms use crossed fusion only.) (A) Parallel bars are coded in front and are therefore seen as parallel bars. (B) When parallel bars are coded in back, they are seen as parts of squares, completing behind.

completing *amodally* behind the rectangle, which now becomes an occluder. This shift, in turn, abolishes the preferential affinity between horizontal tokens because the white bars are now seen as part of completed squares in back.

This configuration should offer no opportunity for a shape bias and, as a consequence, less horizontal motion should be seen. The motion-dominance curve should shift to the left. Note that this particular expectation is not predicted by an image-based matching scheme, in which horizontal preference should be equal, whatever the depth relations. The results of this experiment are clear (see Figure 1.33). Greater horizontal motion bias is seen only for the bars-in-front case.

A final phenomenological observation confirms the critical role of surface encoding in a particularly revealing way. Because of the strength of the proximity tendency in relation to the weakness of the shape tendency, proximity can force vertical matches even if shape favors horizontal matches. For example, this can happen in the white-bars-in-front case, where the white bars are seen as distinct oriented bars. Because the orientation of the upper and lower tokens is different, the perceived-motion trajectory is no longer a simple vertical translation. The pair of bars is perceived both to rotate and to translate in the picture plan (as diagrammed in Figure 1.34a). If we think of what edges in each frame are

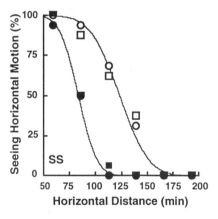

Figure 1.33
Surface shape, not image shape, biases motion correspondence (data from two observers). Open symbols refer to the case in which motion tokens are coded by binocular disparity to be in front. No amodal completion is expected; and, because the shapes match in the horizontal but not the vertical direction (as in Figure 1.32), there should be increased horizontal bias. Different times refer to two stimulus durations in the apparent-motion paradigm. (Reprinted by permission from Z. J. He and K. Nakayama, Surface shape not features determine motion correspondence, 1994, *Vision Research* 34, 2125–2136.)

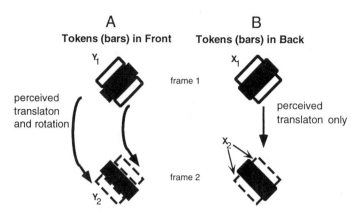

A
Tokens (bars) in Front

B
Tokens (bars) in Back

frame 1

perceived
translaton
and rotation

perceived
translaton only

frame 2

Figure 1.34
Phenomenology of rotational and vertical motion, showing that matches must be surface
based, not image based (see text).

matched from one frame to the other, it is clear that the upper-left edge
(labeled Y_1) of the oblique bar matches the lower-left edge of the ortho-
gonally oriented bar in the next frame (labeled Y_2).

Contrast this to the case in which vertical matches are made in the
bars-in-back case (as shown in Figure 1.34b) such that the bars are now
seen as part of a square surface. Here the perceived motion is that of a
simple vertical translation; no rotational motion component is seen. If we
analyze what edges are matched, the answer is telling. Note the image
edge X_1 labeled in Figure 1.34b for frame 1; there is *no* counterpart in the
image after the apparent motion in frame 2. Edge matching only occurs at
a level of surface representation between X_1, which has a visible counterpart, and X_2, much of which is an amodal contour hidden from view. This
phenomenon alone argues strongly for matching at a surface level.

Surface Position Changes Mediated via Amodal "Leakage"

So far we have shown the importance of surface shape relative to image
shape in determining correspondence in apparent motion. In this section,
we address the issue of surface position, as opposed to image position, by
selectively allowing a surface to amodally "leak" behind another (Shimojo
and Nakayama 1990). Consider the seemingly innocuous visual situation,
depicted in Figure 1.35a.

Frame 1 illustrates the same apparent-motion situation as before, but
with an added feature—a large stationary rectangle (marked with an X)
that can act as a potential occluder. We accentuate this role by altering its
depth so that is perceived to be in front of the flashing tokens (illustrated

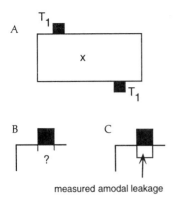

measured amodal leakage

Figure 1.35
Amodal leakage. (A) Motion tokens in back are flashed adjacent to a stationary rectangle (marked x) in front. (B) Hypothetical amodal leakage of a flashing target behind an occluder. (C) Measured estimate of amodal leakage obtained from the motion competition paradigm, expressed as the size of an equivalent visible surface. (See Shimojo, and Nakayama 1990.)

as black tabs in Figure 1.35a). Consider the encoding of the small token in the upper left. Because of stereoscopic depth plus the T-junction, the border between the small token and the larger rectangle "belongs" to the large rectangle. This means, of course, that the bottom boundary of this flashing tab is essentially unbounded and thus has the potential to complete behind the occluder. However, there is no similar unbounded surface nearby to which it can link (as in the conditions shown earlier, e.g., in Figure 1.1). Yet it seems unlikely to think that the lower boundary of the tab stops exactly at the occluder. Might not the visual system infer that the tab continued for some short distance behind the occluder? If so, how far would it extend (see Figure 1.35b)?

If we confine ourselves to simple phenomenology, such a question seems very imprecise and uncomfortably subjective. We would be hard-pressed to accept our own answers, let alone those of others. Fortunately, from the perspective of motion correspondence, a precise answer can be obtained. The motion-dominance function introduced in Figure 1.31 indicates that horizontal motion perceived 50 percent of the time denotes an indifference point, one at which horizontal and vertical motion affinity is equivalent. We can therefore calculate the relative horizontal distance that yields this indifference. Predictably, the motion-dominance function shifts toward vertical motion when the central closer rectangle (marked X) is placed just along the edges of the flashing tokens. Consequently, the indifferent point moves toward shorter horizontal motion. From the measured size of this shift, we can calculate the amount of amodal leakage behind the occluder and express it as the size of an equivalent visible surface that would cause such a shift, assuming a center of mass representation for a token position. This surface (as estimated from the data of six observers) is surprisingly large in relation to the size of the visible flashing tab. It is depicted as the white outlined area labeled amodal leakage in Figure 1.35c. (For further details, see Shimojo, Silverman, and Nakayama 1989.)

Taken together, these studies on motion indicate that apparent motion correspondence is dictated at a surface level of representation rather than one based on image shape or position. Why is this so? Our reasoning again rests on the computational problem posed by occlusion. Moving objects, no less than stationary objects, can be occluded by other objects. This means that motion-encoding schemes based on images alone are too unreliable; everyday perceptions of motion cannot be effectively mediated by the motion-sensitive neurons that respond to motion at the image level. As various parts of a surface become occluded or unoccluded, an image-based motion system would tend to sense spurious or nonrigid motion. For example, in examining the perception outlined in Figure 1.34b, an image-based matching system would perceive a rotary motion of individual bars; but because we see the two bars as part of a larger surface completing behind, such a spurious motion does not occur. Instead we see pure translational motion. The visual system codes, and we see, motion of a surface, not the motion of isolated image fragments.

1.2.6   Motion and Attention Dependent on Perceived Surfaces, not Three-Dimensional Geometry

In this final section reviewing experiments on surface perception, we address briefly a largely unexplored issue. Again, we challenge what we feel to be some implicitly, yet wrongly held views, those concerning the nature of space perception and spatial representation. Because we have a two-dimensional retina and because we live in a three-dimensional world, many have seen the problem of space perception as the recovery of the third dimension.

As such, conventional studies of visual space perception start with a spatial description of our environment inherited from geometry, in particular coordinate or Euclidian geometry. Perhaps it seems especially rigorous and scientific to think of space in terms of the XYZ Cartesian axes and of space perception as the recovery of the Z dimension—usually via binocular disparity—with X and Y being supplied by the retinal image. Distance, according to this view, is represented by the length of a straight line joining two environmentally localized points.

Yet there are reasons to think that this is not the manner in which spatial distance is encoded in the visual system. Perceptual psychologist J. J. Gibson (1966) argues that space is not perceived in this way but in terms of the surfaces that fill space. The most important and ecologically relevant surface is the ground plane. In Gibson's view, Euclidian distances between arbitrary points in three-dimensional space are not biologically relevant (see also Nakayama 1994). We see our world in terms of surfaces and plan our actions accordingly. Locomotion (except for flying animals or airplanes) is usually confined to surfaces.

To begin to understand how distance might be encoded in the visual system and to evaluate the role of surfaces, we have exploited the proximity tendency in apparent motion (He and Nakayama 1994a). You will recall that the motion-dominance functions shown in figures 1.31 and 1.33 reflect a strong tendency for the visual system to make matches between tokens having greater proximity, that is, shorter distances. But, as pointed out above, the exact definition of distance in defining proximity has not yet been fully elaborated. This is particularly true if we think of potential motion tokens as occupying positions on perceived surfaces, not as arbitrary points in three-dimensional space.

If simple distances in space are important, we would expect increased matches between horizontal tokens as we introduce binocular disparity between the upper and lower motion tokens (as in Figure 1.36). As binocular disparity increases, perceived three-dimensional distance between upper

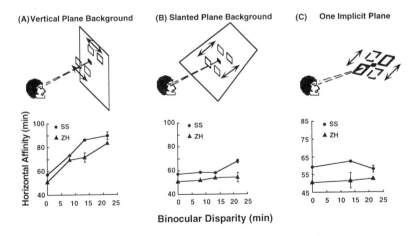

Figure 1.36
Apparent motion on different versus single plane receding in depth. (Reprinted by permission from Z. J. He and K. Nakayama, Apparent motion determined by surface layout not by disparity or 3-dimensional distance, 1994, *Nature* 367, 173–175.)

and lower tokens also increases; and by a proximity principle, motion matches should then be more prevalent between horizontal tokens. As Figure 1.36a illustrates, this is exactly what happens. The 50 percent point or indifferent point of the motion-dominance function (which we call the *horizontal affinity* because it reflects the strength of horizontal matches) increases with increasing binocular disparity. This experiment essentially replicates one originally reported by Green and Odum (1986), who demonstrated that matches were preferred between tokens having the same perceived depth.

Such an experiment does not, however, distinguish between this 3-D Cartesian view and one based on surfaces. Clearly, the outcome was predicted by a simple three-dimensional distance hypothesis. Yet, it is also the predicted outcome of a surface-binding hypothesis. If we hypothesize that perceived motion is preferentially bound to surfaces, it should be apparent that the two lower and the two upper tokens in Figure 1.36a define two implicit surfaces, which become increasingly distinguishable as binocular disparity is increased. If we also suppose that motion matches *within* a surface are preferred, we would also predict that horizontal matches would increase with increasing binocular disparity.

To differentiate a purely Cartesian depth hypothesis from our surface-binding hypothesis, we conducted two additional experiments. In each case, we varied the mean binocular separation between the upper and lower set of tokens, thus preserving the increase in perceived three-dimensional distances. However we also made subtle manipulations to accentuate the connections between the upper and lower tokens in terms of a surface representation. In the first case, we used exactly the same tokens as employed in the previous experiment, except that we added a stereoscopic receding plane composed of random dots upon which the tokens could "rest" (Figure 1.36b). In a second case, we increased the slant of each individual token so that if all four tokens were visible at the same time, they would be co-planar, lying in the same receding plane (Figure 1.36c).

We predicted that if motion is tied to surfaces rather than to three-dimensional depth per se, these two manipulations would greatly reduce the effect of binocular disparity—because such an increase would not be accompanied by a surface segregation. This is exactly what we found. The graphs in figures 1.36b and 1.36c show that binocular disparity in these situations does not increase the strength of horizontal matches. The results, therefore, emphasize the importance of surfaces. The preferential horizontal matches seen in the first experiment (shown in Figure 1.36a) were not due to increasing three-dimensional distance between vertically adjacent tokens. The same increase in three-dimensional distance had no effect when the increase in depth separation was accompanied by perception of a common surface.

Hypothesis: Motion is Tied to Surfaces Because Attention is Tied to Surfaces

In this section, we provide additional reasons to explain why motion is so closely tied to surfaces. We do so by advancing and testing a novel idea, namely that apparent motion is tied to surfaces because attention is also tied to surfaces. What is the basis of such a view? To start, let us return to the discussion of apparent motion and reexamine Figures 1.29 and 1.30. We noted there that apparent motion operates over a very large set of spatial intervals, comprising much larger distances than could be accounted for by known motion mechanisms in the striate and extrastriate cortex. Such neurons respond selectively to motion direction, but the distances between targets on successive frames are too small to explain apparent motion. For the central part of the visual field, they are approximately 0.5 degrees for striate cortex and from 2 to 4 degrees in area MT, the extrastriate cortical area specialized for motion. Apparent motion, however, can be seen over many tens of degrees.

Recently Cavanagh (1992) conducted an important set of experiments that strongly indicate that perceived motion is closely linked to attention. His findings show that if our attention is directed to one identifiable pattern and then to a similar one in a different position, we perceive apparent motion. Like the perception of motion obtained by following an actual moving target with eye movements, tracked attention provides the perceptual system with information about motion.

From Cavanagh's attentive motion, it is only a short step to the possibility that apparent motion is tied to surfaces because of its dependence on attention. We hypothesize that surfaces are also very important for the deployment of visual attention, arguing that attention cannot be arbitrarily directed to points or volumes in abstract space but is bound to perceived surfaces. Knowing that motion is so closely tied to attention helps explains why motion is also so closely bound to surfaces. To test such a view, however, we need to examine the deployment of attention in relation to surfaces more directly.

Our approach was to study directed focal attention in a cueing paradigm similar to that introduced by Posner (1980). The observer is presented with a cue at a site that is predictive of the target location in 80 percent of the trials (cue valid cases). The target appears at the uncued site on only 20 percent of the trials (cue invalid cases). We measured reaction times as a function of increasing binocular disparity separately for targets in the cue-valid and the cue-invalid trials. Our display was similar to that used in the apparent motion studies shown in Figure 1.36. Figure 1.37 shows the three stimulus conditions in which cued and uncued targets were presented: (A) in separate frontoparallel planes; (B) in separate frontoparallel planes resting on a common stereoscopic plane receding back;

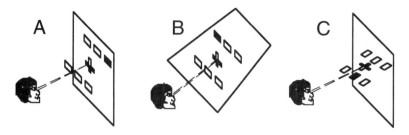

Figure 1.37
Cued attention to an upper or lower row examined as a function of stereoscopic depth separation. Three different configurations: (A) Tokens are frontoparallel seen against a background frontoparallel plane. (B) Tokens are frontoparallel and appear to rest on a receding plane, defined stereoscopically. (C) Tokens slant back according to the disparity difference in the upper and lower rows so that they form a single implicit plane, slanting back. Increased depth separation aids selective attention to a row only for case (A).

and (C) within a single stereoscopic plane receding back. In line with our attentional hypothesis, we predicted that only in condition A would there be an increased difference between cue-valid and cue-invalid cases as binocular disparity increased. In the other two conditions, we predicted, attentional focusing would not be as effective because attention would be automatically spread within surfaces (Case C) or spread evenly between separate surfaces lying on a common surface (Case B).

To begin the experiment, the observer fixated on a central cross flanked by an upper and lower row of three elements. At the start of each trial either upper or lower limb of the cross would brighten, thus pointing to the upper or lower row of gray squares. The observer was instructed to attend to this cued row. Approximately one second later all six squares were presented colored either red or green—with five squares in one color and the remaining, target square in a different color. The task of the observer was to release a button when he or she saw the odd colored target. On 80 percent of the trials (cue-valid), the odd colored target was in the row that had been cued. On 20 percent of the trials (cue-invalid case), the odd colored target was in the other row. The strength of attentional focusing was determined by noting the difference in reaction times between the cue-invalid and cue-valid cases.

As predicted, in Case A we found that increasing disparity aids the observer to maintain attention on the cued row. But, as in the case of apparent motion, the result does not distinguish between a pure depth hypothesis for attentional segregation and a surface-binding hypothesis. This was why we added conditions B and C. In these two situations, although there is the same increase in binocular disparity, the stimuli are more closely related to a common surface; they either rest on a common surface or comprise one. As the surface hypothesis predicts, under these conditions there is little difference in the ability of the observer to maintain attention as disparity increases.

This result shows that it is easier to confine attention to distinguishable surfaces than to confine one's attention to particular regions within a surface. This new idea, confirmed here, provides a mediating explanation for why apparent motion is confined to surfaces. Because attention is confined to surfaces, and because apparent motion is dictated by the mobility of attention (Cavanagh 1992), apparent motion is preferentially tied to surfaces.

1.2.7   The Perceptual and Phenomenological Primacy of Surfaces: A Critical Explicit Link?

In the first part of this chapter relying primarily on phenomenological observations, we described how small changes in binocular disparity can have dramatic effects on surface completion. In this second section we have basically confirmed these phenomenological observations by using objective methods that show very strong evidence of the processes underlying surface perception. Surface shape, not image shape, determines whether we see texture as segregated, whether single targets pop out of a display of distractors, whether motion is seen to conform to the aperture it is enclosed in, whether motion correspondence will occur, and so on. Surface properties rather than image properties are decisive. It appears that all higher visual processes must have, as a data format, a surface representation. We think it justified, therefore, to consider surface representation an indispensable link between low-level and higher-level vision.

Our perception cannot be conceptualized as a simple combination of image properties without understanding what specific visual entities must be coded. Because of occlusion, one of the prime candidates for explicit encoding is visual surfaces, stable, enduring aspects of the world that provide appropriate inputs for higher-order visual functions.

1.3   Possible Mechanisms of Visual Surface Representation

So far, we have mainly stressed the functional aspects of a surface representation, emphasizing the importance of surfaces in mediating very rapid visual processes and underscoring the idea that surface representation is a relatively primitive bottom-up process. Yet it is also one that appears to be governed by functional interactions not obviously related to the known properties of neuronal receptive fields. The coding of surfaces is better understood in terms of more macroscopic concepts: border ownership, depth, modal and amodal completion, and so on. To help bridge the gap between what appear to be qualitatively distinct levels of processing, we would like to specify an anatomical locus, a probable cortical site where visual surface representation might begin.

1.3.1   Surface Representation May Begin as Early as the Striate Cortex (V1)

If we look at the diagram of the known extent of the visual brain, as shown in Figure 1.2, we note the large number of topographic maps, all of which are specialized for seeing. Where in this complex hierarchy of projections might surface representation begin?

A number of converging lines of evidence suggests that it must begin fairly early. Because surface representation seems to require little in the way of object-specific knowledge, it is likely to be antecedent to cortical areas in which object knowledge is stored. Therefore, it would probably not be postponed until, say, the infero-temporal cortex, the "what" or object-recognition system to Ungerleider and Mishkin (1982). In fact, we have argued elsewhere that it must be prior to object recognition to be of any use (Nakayama et al. 1989) as it must parse images into the appropriate surface units upon which object recognition can act.

The need for a visual surface representation in mediating rapid visual processes was outlines in Part 1.2. It plays a critical role in the perception of motion, the segregation of texture, and the processes required for rapid visual search. That such a broad range of functions not requiring object knowledge are so critically dependent on a surface representation argues strongly for an early rather than a late anatomical site for such processing.

The strongest piece of evidence that at least some part of surface representation must begin very early is the phenomenon of DaVinci stereopsis (Nakayama and Shimojo 1990; Anderson 1994). In our discussion of this phenomenon we noted that the placement of subjective contours and surfaces was critically dependent on which eye received the unpaired information. We found that right-eye-only points elicited subjective contours to their immediate left, whereas left-eye-only stimulation elicited subjective contours and surfaces to their immediate right (see Figures 1.16, 1.17). This indicates that critical aspects of surface perception are determined by very unusual sorts of information, of a class not generally available to us as conscious perceivers. To appreciate this fact, look around you with the right eye covered, then the left. No obvious difference exists in our perceptions unless there is some gross interocular anomaly. At a conscious level of perception, explicit eye-of-origin information is unavailable to us as perceivers. Interestingly, this information also appears to be hidden from most of the higher visual system as well. Cortical neurons from V2 and beyond respond more or less equally to stimulation delivered to one eye or the other (Maunsell and Newsome 1987; Burkhalter and Van Essen 1986). These neurons do not, therefore, carry eye-of-origin information. Only earlier, in the striate cortex, where the inputs from the two eyes are physically segregated into ocular-dominance columns, is explicit eye-of-origin information preserved (Hubel and Wiesel 1968). This suggests that cells in the striate cortex are the only ones available to signal the presence of subjective contours from unpaired points and inform us of whether an occluding contour lies to the left or the right of a given dot. The implications of this line of thought are potentially far-reaching. They mean that at least some aspects of surface representation must begin very early and must rely on information coming directly from cortical area V1, the striate cortex.

This very early cortical site for the beginnings of surface processing is also broadly consistent with findings that neurons in cortical area V2, the next stage of visual processing after V1, are responsive to subjective contours (von der Heydt et al. 1984, 1989). In these studies, receptive fields cells in area V2 were localized and an orientation preference determined (see Figure 1.38a for the position of a receptive field). It is significant that a stimulus similar to that shown in Figure 1.38b also excites V2 cells. These patterns elicit perception of a subjective contour yet have no luminance boundaries within the measured receptive field. As an important control, von Heydt and his colleagues showed that very small changes in the configuration (as shown in Figure 1.38c) abolish both the impression of an illusory contour and the neuronal response of these cells.

These two independent sources of evidence point to an explicit surface representation that is likely to begin somewhere in the neighborhood of cortical areas V1 and V2. In the broader picture of visual processing outlined in Figure 1.2, this suggests that visual surface representation is strategically placed just before the branching points of different functional visual streams—the "what" and "where" systems of Ungerleider and Mishkin (1982) or, alternatively, Goodale's framework for conscious perception versus visuomotor action (Chapter 5, this volume). Although we acknowledge the need for additional evidence, the view that surface processing is occurring at such early anatomical stages is appealing for a number of reasons.

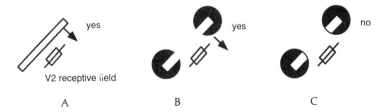

A   B   C

Figure 1.38
Schematic diagram showing cells responsive to illusory contours.

First, it is broadly consistent with the multiple higher-order functions just mentioned. All such functions are likely to need a visual surface representation as an input data format. This is evident in the area of object recognition (see Nakayama et al. 1990; Biederman 1987; Biederman, Chapter 4, this volume), but it is also likely for visuomotor control. Goodale's patient DF, for example, can reach for and grasp visual objects appropriately, even though she cannot report on their identity and shows little or no conscious awareness of their spatial properties. We suggest that DF may have preserved mechanisms of surface representation, adequate for the visuomotor processing. Second, surface processing at an early stage is consistent with the very broad range of findings reported in Part 1.2. We have shown that many primitive visual tasks, such as motion perception, visual search, and visual textural segregation depend on a surface representation. This dependency suggests that visual surface representation must be one of the earliest visual functions beyond simple coding of image properties by low-level cortical receptive fields. Third, such a view invites us to think more mechanistically. Areas V1 and V2 are some of the most completely characterized portions of the cortical visual system; their inputs and outputs are more clearly identified than any other cortical visual structure. Furthermore, in comparison to other cortical areas, the receptive-field properties of cells here have been well characterized.

Can we, therefore, begin to envision how these cells might account for the surface properties described here? Perhaps. Yet, impressed as we are with this evidence regarding cortical localization, it does not follow that a reductionistic understanding of surfaces in terms of neural circuitry is imminent. Localization is only the very first step in understanding a function mechanistically, that is, in terms of specific classes of neural connections. Several of the most difficult and challenging questions lie ahead.

First, there is no indication of how amodal completion of surfaces and contours behind occluders is encoded in the firing patterns of visual neurons. So far, the only connection we know of between neuronal properties and surface perception is through the phenomenon of modal completion, that is, subjective contours. Although this provides striking confirmation of our early visual system's ability to make important inferences about surfaces, subjective contours represent only a small fraction of the occasions on which we need surface-completion phenomena in our daily lives. Except for silhouettes and cases of very low illumination, the real boundaries of objects are almost always accompanied by physical "visible" luminance changes in the image. Such zero-contrast boundaries become even rarer for longer contours. Amodal completion, the completion of boundaries behind occluders, is much more common. There is rarely a scene in which the need for such completion is absent. Furthermore, the image distance over which such completions are necessary are often substantial, subtending many degrees of visual angle. Thus amodal completion, one of the most important aspects of surface completion, has as yet no known neurophysiological counterpart or correlate.

The second issue is border ownership. In thinking about surface completion, this has been an important concept; it determines whether or not surface fragments group and, if so, whether they do so in front of or in back of other surfaces. An attempt to account for at least part of surface completion in terms of end-stopped and binocular-disparity-specific cells was suggested in an earlier paper (Shimojo, Silverman and Nakayama 1989); this finding, however, accounts for only cases in which surfaces are covered or bounded by lines, not those created by general textures. It also is not explicit about how a boundary gets assigned to one region or another. As yet, we are lacking in a plausible neuronal explanation of how border belongingness is attached to a given image region.

What we are saying is that an as-yet-unbridged conceptual gap lies between the coding of image properties and the coding of surfaces. So, despite the success in relating some aspects of surface representation to, say, the striate cortex (through DaVinci stereopsis) and the important discovery that V2 neurons respond to subjective contours, a satisfactory scientific explanation of the coding of surfaces in terms of specific neural properties and neural circuitry remains elusive.

How then should we proceed? How can we begin to understand how image properties as measured by neuronal receptive fields are related to the more inferred representation of surfaces? Given the difficulty of the problem, it might be advantageous to step back, to think more broadly about how surface perception might emerge through development and the process of perceptual learning.

1.3.2   A View from Developmental Neurobiology: The Critical Role of Associative Learning

We mention recent work on visual development because it provides a strong argument for the importance of learning and plasticity at a cellular level. This work, which has emerged over the past ten to fifteen years, demonstrates the profound influence of neural activity in shaping neuronal connections, both prenatally and postnatally. Even the gross features of central nervous system topography, such as the lamination of geniculate nucleus and the ocular-dominance structure of visual cortex, are determined by activity-dependent cellular learning mechanisms. The interplay of Hebbian learning and the statistical pattern of correlation between neighboring afferent inputs from the two eyes to higher centers accounts for much of the observed gross structure and connections at a millimeter scale (see Kandel and Jessell 1991). The firing patterns of retinal ganglion cells coding the same general direction of the visual field are correlated within the same eye but not between the two eyes. Research has shown that this correlation is decisive in forming the selective associations between the particular connections at the lateral geniculate nucleus and striate cortex, leading to the characteristic pattern of eye-of-origin lamination of the lateral geniculate nucleus and the ocular-dominance columns of the striate cortex (Stryker and Harris 1986; Shatz 1990). The results imply that, aside from the topographic maps established by the mechanisms of neuronal growth and guidance (Jessell 1991), the exact connections a given neuron makes with its neighbors is profoundly dependent on experience, that is, the past history of its inputs from other neurons. Similar mechanisms are also likely to be responsible for the formation of binocular connections needed for stereopsis (Hubel and Wiesel 1965), motion sensitivity (Daw and Wyatt 1976), and even the refined retino-topographic map itself (Schmidt 1985). The pervasiveness of learning at a cellular level to fashion the most dominant, well-documented connections of the visual cortex indicates that we need a similar understanding of the role of learning for other visual functions that develop through visual experience.

We also need, however, a conceptual framework for understanding which aspects of visual experience may be relevant—a means of identifying the visual and environmental events that must be functionally associated that is analogous to the statistical correlation of ganglion-cell discharges that occur prenatally. Because one of the most important challenges in understanding the visual coding of surfaces is establishing a relationship between image-based and surface-based representations, we start here. We hypothesize that the visual experience of the young, mobile observer, sampling images of surfaces from varied vantage points, provides the defining context within which to understand the learning of a visual surface representation.

1.3.3   Surface Transparency, a Proposed Example of the Associative Learning between an Image and a Surface

In Nakayama and Shimojo (1992), we developed a theoretical framework to explain the learning of a surface representation. Here, we condense this argument by resting our case on a single example, showing how analysis of the association between an image and a surface representation can explain an otherwise bizarre perceptual phenomenon, the emergence of perceived transparency in stereograms.

In the stereogram shown in Figure 1.39a, we created a stimulus made up of four repeats of a simple pattern of bipartite bars set against a black background frame. Each bipartite bar is divided into a gray and white region. Stereoscopic information is sparse, consisting of only two discrete

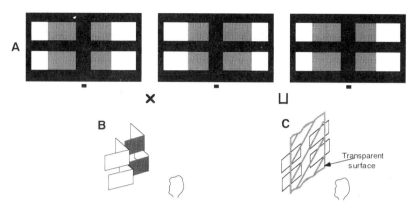

Figure 1.39
(A) Stereogram showing transparency. The gray-white border is coded in front; all other features are seen in back, in the plane of the frame. (B) Folded cards, the expected perception given depth interpolation. (C) Perceived transparency seen for the configuration in (A). (Modified by permission from K. Nakayama and S. Shimojo, Experiencing and perceiving visual surfaces, 1992, *Science* 257, 1357–1363.)

disparity values, front and back. The vertical contour dividing each of the bars is stereoscopically coded as in front. Every other contour, including that of the frame, is coded in back. It should be clear that only vertically oriented contours can supply binocular depth information in this stereogram. All other depth values, such as those along the horizontal contours or within the interior of the figure, are indeterminate. This occurs because there is no image variation in the horizontal direction from which to make a binocular match in the two eyes. Although the ordinary rules of stereopsis do not explicitly predict the depth of these indeterminate regions, we can justifiably assume that the perceived depth of these regions can be obtained by interpolation. The perception of sparse, yet textured stereograms with curved surfaces is consistent with this view. When observing such stereograms viewers see continuous curved surfaces interpolating appropriately in local regions where no texture exists (Julesz 1971; Ninio 1981).

Such an interpolation would predict a set of perceived surfaces like those in Figure 1.39b, a set of folded cards whose the convex edges face the observer. This configuration represents the simplest form of linear interpolation between points of defined binocular disparity. Surprisingly, this simplest interpretation is rarely, if ever, seen. Instead of seeing surfaces slanting in depth and connected at a fold, our perception is qualitatively different (Figure 1.39c). We see two disconnected sets of surfaces, one in front of the other, each frontoparallel with respect to the observer. More striking, we also see a *material* change. The closer surface appears as transparent and partially occluding a white surface in back (Figure 1.39c). The perceived transparency is so potent that it makes our visual system see a filled surface over the whole transparent region perceived. Thus the gray transparent material actually appears to invade the black region and to be enclosed by a subjective contour that bounds the spreading. This is even more clearly seen in studies using colored stimuli, in which red areas spread into the otherwise black regions (Nakayama and Shimojo 1990; 1992). The importance of depth in eliciting transparency is clearly demonstrated when we view the stereogram in the reverse configuration, with the gray-white contour coded in back. Here we see no transparency, no color spreading, and no subjective contours. Instead, we perceive an opaque surface that is behind and seen through rectangular apertures.

Why, returning to the perception of transparency, does the visual system opt for what seems like an unusual interpretation, seeing a large global transparent surface instead of the folded cards? Why does the visual system avoid what would seem to arise naturally from depth interpolation? Our answer considers the question in terms of perceptual learning. We argue that through learning some critical feature of the binocular image shown in Figure 1.39a becomes associated with a transparent surface.

The essence of the general argument is simple. Those images most strongly associated with given surfaces determine which surfaces are perceived. What remains, then, is to develop a principled approach for estimating the association between images and surfaces. We take as a starting point the work on aspect graphs, a concept popular in machine vision pioneered by the mathematical insights of Koenderink and van Doorn (1976). In their seminal paper, Koenderink and van Doorn outline the characteristic pattern of topological stability and change of images sampled during shifts in viewer position.

Following their lead, we consider all of the various images (views) that can be associated with a given surface configuration as an observer takes all possible positions around a given surface. If we assume essentially random motions of an observer with respect to surfaces, then the determination of the probabilistic association between images and surfaces becomes an exercise in solid geometry. We need to estimate the volume in space from which a given images can be sampled. To illustrate this analysis, we consider the sampling of images from a familiar set of surfaces, a cube, from various positions in space.

To simplify the analysis we looked at potential vantage points in terms of a set of regions on a "viewing sphere" (see Figure 1.40a). The totality of such spheres of varying radii constitute all the possible vantage points that can be taken relative to a surface. It should be clear that a number of possible topologically defined classes of image can be sampled from posi-

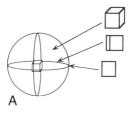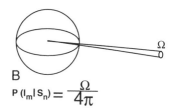

$$P\,(I_m|S_n) = \frac{\Omega}{4\pi}$$

Figure 1.40
(A) Cube viewed from different positions on a "viewing sphere." Note that three topological classes of image can be sampled: one-faced, two-faced, and three-faced. One-faced and two-faced images can be sampled only from very restricted positions on the sphere, from points at the intersections of the circles and on the circles, respectively. Three-faced images can be sampled from all other positions on the viewing sphere. (B) Diagram indicating that the probability of seeing a given view from a given surface is related to the ratio of two solid angles.

tions on this sphere. From most vantage points, we see the image in the usual three-quarter, generic view, with three faces visible. We can also see it, however, from unusual vantage points so that we see accidental views in which just one or two faces are visible. Thus the vantage points that can give rise to such accidental views are very much more restricted than those giving rise to generic views. An image sample of just one face, is seen from just six vantage points or discrete loci on the sphere, as defined by the intersections of the circles. The number of vantage points yielding an image sampling from two faces is somewhat larger but still very limited, along a line defined by the three circles. From all other positions on the sphere, we obtain the generic view in which the image has three faces.

If we assume random locomotion around the cube, the probability of an association between a surface representation and a given image reduces to the quotient of two solid angles. More formally, the conditional probability of obtaining a particular image I_m, given surface S_n, can be approximated by the following ratio:

$$p(I_m|S_n) = \Omega/4\pi \tag{1}$$

where Ω is the solid angle over which we can sample a particular image class, and 4π is the solid angle comprising the total set of vantage points from which the surface can be sampled. From this analysis, it should be clear that the probable association between image I_m, three faces, approaches unity as the distance from the cube increases. The probability of the other accidental images, two faces and one face, approaches zero.

In general, we can conceive of the totality visual experience (images) and sets of possible surfaces as depicted by the matrix in Figure 1.41. Each cell in the matrix represents the value of $p(I_m|S_n)$ that, as outlined above,

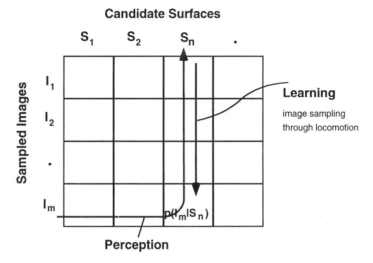

Figure 1.41
Generalized associative matrix denoting the probability of obtaining an image I_m, given surface S_n. We suggest that through locomotion—which places the viewer at random positions with respect to given surfaces—various images are sampled with determinate probabilities. This is shown as the conditional probability of sampling image I_m given surface S_n. We assume that these probabilities can be learned and represented in the connection strengths between an image and a surface representation. Learning of these probabilities through experience is expected to proceed along the downward arrow. Assuming that associative learning between various surfaces and images has occurred, the act of perception, denoted by the bent arrow connecting a particular I_m to a particular S_n, is hypothesized to depend the on strongest connection strength (conditional probability) for a given row, that is, for a given image.

can be plausibly estimated from geometry. It summarizes the visual experience of a mobile observer in terms of the images sampled from a surface. We hypothesize that these probabilities can be encoded in the nervous system as simple connection strengths between representations of images and representations of surfaces. Given these assumptions, the task for perception is clear. When confronted with an image, I_m, it must come up with the perceived surface representation, S_n, most closely associated with the image. In terms of the matrix, for any given row the perceptual system must find the cell having the highest associative strength [i.e., the highest $p(I_m S_n)$], which in turn defines the perceived surface. In Figure 1.41, therefore, we can envision the route from the image to perception. It is depicted as the bent arrow, starting from the image to the strongest connection for the image, thus pointing to the associated surface representation.

Let us apply this analysis to the cube, and consider what alternative surface interpretations might plausibly be evoked by various images of a

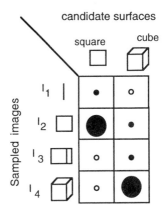

Figure 1.42
Associative matrix for cubes and squares with the same representation as in Figure 1.41, except that the probabilities of sampling are denoted by symbols (see text).

cube. For this purpose, we consider the exhaustive set of images that could arise from a cube and a square described in Figure 1.42. We illustrate the image sampling probabilities schematically: high probabilities or generic views are represented by the symbol ●; low probabilities or accidental views by ●; and zero probabilities by o.

According to this scheme, when presented with, say, image I_4, we see the cube because it corresponds to the column having the largest associative strength. Interesting, when presented with image I_2, we do not see a cube, even though seeing a cube is compatible with this image class. Instead we see a square because of the greater associative strength between I_2 and the square.

We can now turn to the case of the stereogram shown in Figure 1.43 to explain why we see a transparent surface instead of the folded cards. First, consider again the relative rate of sampling of the various images from each surface type. In the case of the folded cards, the two classes of image that could be sampled, I_x (straight) and I_y (bent) are shown in Figure 1.43a. This diagram demonstrates that image I_y can be sampled from many vantage points; it is thus an generic view of a fold. The image presented in stereogram I_x is sampled only from very special vantage points, where the observer is at exactly the same height as the configuration. As such, I_x is an accidental view of the folded card and has very low probability. It is quite otherwise when sampling the same image I_x from a transparent occluding surface (Figure 1.43b). Any changes in the viewer's position, including motions up and down, preserves the same image. Image I_x is a generic view of this surface. Thus, for each of the associative strengths outlined in our image/surface matrix for the cube, we can see an analogous

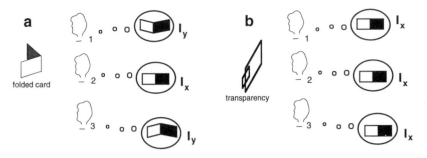

Figure 1.43
Generic and accidental views of candidate surfaces. (A) The sampling of image I_x from folded cards, which can occur only under a restricted set of vantage points at just the right height. (B) In contrast, the sampling of image I_x from the transparent surface can occur over a wide range of observer viewpoints.

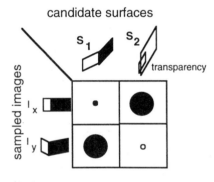

Figure 1.44
Associative matrix for folded cards and transparent surfaces.

situation for the folded card and the transparency. When confronted with image I_x, the visual system has the greatest associative strength in relation to the transparent surface, not the folded card. Consequently, image I_x as shown in Figure 1.44, elicits the perception of transparency, not a folded card.

Local Mechanisms of Inference: From Image Fragment to Surface Properties

So far, our analysis relating images to surfaces has consisted of extended images and surfaces (cubes, squares, folded cards, transparent surfaces, and so on). Although pedagogically clear, these examples are too complex to provide a credible way in which to envision how local image features might be used to build the fragments of a local surface representation. Yet such a simpler process might be expected, or at least hoped for. Early cortical neurons, by virtue of their retinotopically organized receptive fields, analyze the image locally, providing information about the patterning of the image in a small, limited region of the visual field. It seems plausible that surface representation too might be built up by an inferential and mechanistic associative process similar to that outlined but at a more local level that links image fragments to surface properties.

How might such a local process occur? First, we can analyze the perceived transparency in Figure 1.39a at a slightly more microscopic level. A reasonable clue is the existence of critical T-junctions. Recall, however, that ordinary T-junctions accompany occluding contours, with the top of the T occluding the stem of the T. But our configuration does not contain the usual T-junction. Rather, the stem of the T, as dictated by binocular disparity, is coded as closer than the top of the T. This configuration is incompatible with the usual case, in which the top of the T occludes the stem. As we cannot appeal to the properties of ordinary T-junctions, therefore, we must analyze this configuration at a more essential level.

Figure 1.45, represents two stereoscopic image fragments, I_1 and I_2, from the image-sampling matrix outlined earlier. In both cases the vertical contour is coded binocularly as in front. It should be clear that while image fragment I_1 could have arisen from the folded convex corner, that likelihood is very low and requires an accidental vantage point. On the other hand, image fragment I_2, which does not contain any accidental collinear

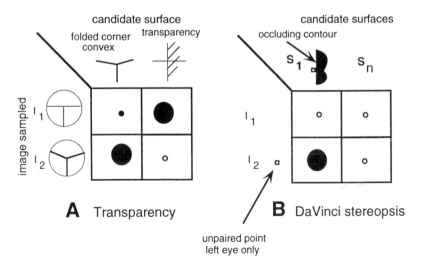

Figure 1.45
From image fragment to surface properties. Associative matrices explain transparency junctions (A) and subjective occluding contours from unpaired points (B).

lines, could have arisen from the corner from a very wide range of vantage points. Employing reasoning more or less identical to that used for whole surfaces, we can predict which image configurations will lead to the appropriate local surface properties; the system needs simply to make the connection of the image fragment to the most likely surface property. Thus, even at a very local level, when confronted with an image, say fragment I_1, the system, thanks to associative learning, is equipped to signal transparency, a surface property.

A similar analysis can be extended to DaVinci stereopsis. Here, we should note that image I_2, an unpaired left-eye-only point, is always paired with a closer occluding edge to its immediate right. Such a configuration containing unpaired points serves to reinforce the occluding status of the edge, that it is not, for example, a surface marking, i.e., paint on a surface. It is interesting to consider what happens when single unpaired dots are presented alone, as in Figure 1.16, illustrating DaVinci stereopsis. Here, we interpret the appearance of subjective contours as a ghostly pale reflection of this process of associative pairing.

1.3.4   The Generic View Principle: An As-If Heuristic

In the preceding section we sketched out the beginnings of a low-level mechanistic approach to surface perception, arguing that associative pairing of image fragments to a surface representation provides a plausible framework within which to understand relations between image data and a surface representation.

In this final section, we continue this argument, but more broadly. Our line of thinking has three parts. First, we state what is emerging across many disciplines as an important principle of visual analysis, the generic-view principle. Second, we argue that this principle can be conceived of as an automatic and passive consequence of the type of perceptual learning we have outlined. Third, we show that the generic-view principle has broad explanatory power, helping us understand at a deeper level some of the macroscopic concepts of surface representation we initially formulated as ad hoc rules.

Simply stated, the generic view principle asserts that *when faced with more than one surface interpretation of an image, the visual system assumes it is viewing the scene from a generic, not an accidental, vantage point* (Nakayama and Shimojo 1992).

Thinking back over our foregoing discussions of associative learning, we can see that the generic-view principle is a more formal and succinct assertion of the more extended arguments we have already made. By treating the probabilistic assertions we have made as discrete entities, the principle dichotomizes images or views into two categories. Stated in this form, the principle is not new. It has been one of the core assumptions of machine-vision algorithms (Guzman 1969), as well as a key insight for Biederman's (1987) theory of human object recognition. However, with the exception of Rock's (1984) important work, it has rarely been formulated and applied as a general explanation to perceptual phenomena, to the human perception of depth and surface layout in simple scenes, both binocular and monocular (Nakayama and Shimojo 1990; 1992).

One of the advantages of the generic-view principle is that it codifies an approach to thinking about perception, providing a simple as-if rule stripped of any mechanism. As such it becomes a justifiable shorthand by which to apply the ideas just outlined without having to invoke a detailed discussion of probabilities, associations, and so forth. It is also becoming a more widely accepted view, independent of our own attempt to explain it in terms of perceptual learning (see, e.g., Freeman 1994; Albert and Hoffman 1995). To show some of its broad explanatory power, we pick two well-known but surprisingly misunderstood phenomenon: (1) the so-called impossible objects, and (2) the so-called figure-ground phenomenon.

First, consider what have been called impossible objects, objects that, shown pictorially, do not correspond to any real object we have ever encountered or could imagine. In Figure 1.46, we show the famous Penrose triangle (Penrose and Penrose 1958). At first glance, it appears to be a line drawing of a three-dimensional object, but it becomes almost immediately obvious that there is an inconsistency. We cannot conceive of how the various parts fit together as a real three-dimensional object.

An important face regarding this drawing is, however, underappreciated. The Penrose figure does *not* depict an impossible object but a real, *misperceived* object. Furthermore, it is misperceived in a way that provides strong support for the distinct level of surface representation we have

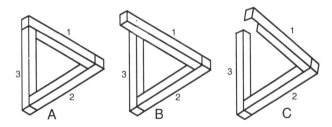

Figure 1.46
(A) The Penrose "impossible" triangle. (Redrawn by permission from L. S. Penrose and R. Penrose, Impossible triangles: A special type of illusion, 1958, *British Journal of Psychology*, 49, 31.) (B and C) Demonstration that the Penrose triangle is in fact a physically realizable object consisting of three bars that when skillfully notched (as in C) and viewed from an accidental vantage point, is perceived as an impossible triangle (A).

been advocating, an autonomous level of representation independent of a coding of known objects.

Gregory (1970) and later Ernst (1992) constructed such real objects, showing that, viewed from a specific accidental vantage point, they look like the Penrose "impossible" triangle. Each researcher produced a physical object, the tri-bar, and photographed it from various angles. The tri-bar is on display at various science museums, including the Exploratorium in San Francisco. It consists of an extended wooden figure in three dimensions with three arms or bars. For the sake of clarity, we show it first in a somewhat unfinished form (Figure 1.46b). The three arms are joined at approximately right angles and two of them extend in space. Arms 1, 2, and 3 are arranged so that each one is farther away from the viewer. Next, by using a precise woodworking techniques, the closer arm (1) was cut down to size and notched to form the three-dimensional object seen in Figure 1.46c. It is important for readers to understand that we are still discussing the same three-dimensional object and positioning it so that the arms have the same depth ordering as before. Readers must now imagine themselves as just slightly changing their vantage point, moving upward and to the right around the object, imagining that they are now viewing the tri-bar from an accidental vantage point that results in the image shown as shown in Figure 1.46a. Because of the skillful cut shown in Figure 1.46c, which permits the exact optical alignment of the near and far ends of the tri-bar from this vantage point, viewers will sample an image and see the so-called impossible triangle.

The results are the same and equally dramatic for viewers moving around the real tri-bar. Although they know full well that it is a three-dimensional object and not a triangle, when in the critical position, viewers see an impossible triangle. They cannot visualize the object in front of their eyes as anything else. Why do they see the impossible triangle and not bar 1 in front of bar 2, that is, the tri-bar in its correct spatial configuration?

We suggest that what is happening is perhaps the strongest, most powerful, and most dramatic example of the generic-view principle at work. Recall the principle again: When faced with more than one surface interpretation of an image, the visual system assumes it is viewing the scene from a generic, not an accidental, vantage point. Note that the new T-junction formed by the accidental alignment of arm 3 with the notch in arm 1 strongly indicates that that the lower surface of arm 1 is occluded by the front surface of arm 3 (Figure 1.46a). In accord with the generic-view principle, our visual system assumes that we are viewing this T-junction from a generic vantage point and thus recover the expected set of surfaces. What is so striking is that the generic-view principle is so strong at a local level that it recovers a surface representation of an image that is literally

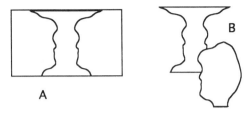

Figure 1.47
Face-vase revisited. (A) The classical face-vase illusion. (B) Same face, partially occluding the vase. Both figures are seen simultaneously with no difficulty despite the partial occlusion of the vase.

impossible from the point of view of object knowledge. Our surface processing is so powerful and autonomous that it generates an object we can't even conceive of. Furthermore, even when we have just seen and touched the tri-bar, our recent experience with and knowledge of the real object is of no help in resisting the generic-view interpretation.

This single example is the strongest piece of evidence in favor of a distinct level of visual representation independent of object knowledge or object representation. It provides the generic-view principle with a most unusual opportunity to demonstrate its predictive power.

The generic-view principle also shows its explanatory power in understanding another important perceptual demonstration, the figure-ground reversing configuration. When we presented the famous Rubin face-vase image in an earlier discussion, we invoked the need for an autonomous process at a pre-object-recognition level. Various portions of even unfamiliar figures can flip so that a surface that was the figure becomes the ground. The term *figure-ground*, in fact, gives the impression that there is a *figure* existing prior to recognition, a hypothetical proto-object.

It is possible that such a level does exist. Yet we should not be blind to the potential role of even more primitive and local processes of surface formation to explain the figure-ground reversal. In fact, we argue that local processes of border ownership and surface completion are more explanatory than processes at a hypothetical figural level. The generic-view principle again clarifies the important role of these more primitive processes.

To remind ourselves that we cannot so easily appeal to a prototypical figure to explain the figure-versus-ground reversal, in Figure 1.47b we show the same two components, but in the more usual, real-world arrangement. Here the face and the vase share a common border in the more common everyday case in which one partially occludes the other. In distinction to the ceaseless competition between the two in the Rubin configuration, we easily see the face and vase as concurrent figures, even when part of one figure, the vase, completes amodally behind the face. Our claim is that the critical and obvious event in figure-ground reversal is not the reversal of objects or figures but the local reversal of border ownership. Each time we experience the perceptual flip, the common border between the two image regions changes. The reader should confirm this for selected instances of figure-ground reversals, recalling our Rule 1, which states that different surface patches cannot share a border and that the visual system must decide which region owns any common border between surface regions.

Here again, the generic-view principle provides strong explanatory power. Consider two separate surfaces at varying distances from an observer. What is the probability that a boundary between the two surfaces will be viewed in such a way that the edge of one surface coincides exactly with the edge of another? Even if they were to have the same shape (which is already unlikely), the chance that we would be just at the correct vantage point to see them as aligned is vanishingly small. Thus, the probability of two objects aligning to form a common border owned by both surfaces is essentially zero. The visual system does not assume an accidental alignment of surface boundaries in an image. This implies that borders between different surface regions cannot be shared, a strong rationale for the border ownership principle (Rule 1.)

1.4   Concluding Comments

We live in a three-dimensional world, full of objects resting on various surfaces. As visual creatures we rely on reflected light to obtain information from the world around us. Such reflections, of course, arise only from the boundaries of objects, the interface between various states of matter, usually solids and gases (air) but also water and air, water and solids. (Gibson 1966).

Thus, in general, surfaces constitute the only visually accessible aspect of our world. We cannot, for example, obtain visual information from the interior parts of ordinary objects. Yet even the surfaces of objects are not fully accessible to us as observers. Surfaces occlude other surfaces. Furthermore, "the amount of occlusion varies greatly, depending on seemingly arbitrary factors—the relative positions of the distant surface, the closer surface, the viewing eye. Yet, various aspects of visual perception remain remarkably unimpaired. Because animals, including ourselves, seem to see so well under such conditions and since this fact of occlusion is always with us, it would seem that many problems associated with occlusion would have been solved by visual systems throughout the course of evolution" (Nakayama and Shimojo 1990).

We argue that because of occlusion, the visual system is forced to adopt certain strategies for dealing with optical information. Most important, we claim, the visual system must develop as a prerequisite for further analysis a representation of the scene as a set of surfaces. Image-based representations, although indispensable first steps in capturing optical information, are insufficient bases for higher-level vision.

We have pursued a number of goals in the present chapter. First, we wished to establish the existence of a surface representation and to argue that it is a legitimate domain of inquiry and a definite stage of visual processing vital to other higher-order processing. Second, we sought to show some of the properties of this representation and to indicate that it seems to have rules distinct from those of neuronal receptive fields but also distinct from processes involved in the coding of familiar objects. Finally, and at a more mechanistic level, we were concerned with the issue of implementation, the possible manner by which surface properties might be derived from image information. Our goal here was to emphasize the powerful role of learning and to outline a possible low-level associative mechanism linking image samples to surface representations.

Appendix: Free Fusion of Stereograms without Optical Aids

Ever since the invention of the stereoscope by Wheatstone (1838), it has been recognized that binocular vision is important for depth perception. By finding a way for an observer to fuse two pictures, Wheatstone demonstrated that our brains are able to synthesize the perception of three dimensions from two flat images.

To appreciate how stereopsis works, we need first to understand binocular image sampling from visual scenes. Figure 1.48a shows a very simple diagram of physically realizable objects consisting of two rods, the left one closer than the right. By comparing left- and right-eye views, it is evident that the angular distance between the two rods is different in the two views, being smaller on the left. Wheatstone discovered that this difference, called binocular disparity, is sufficient to elicit the perception of depth and demonstrated it with his stereoscope (Figure 1.48b), which consists of two mirrors (labeled m and n) that supply the eyes separate images. An example of two such separate images are the center and right images in Figure 1.48c. The perceived depth relations of the two rods, which appear in front of the observer, is labeled by points 1 and 2 in Figure 1.48b.

More recently, large numbers of observers have been taught to fuse two separate images without a stereoscope by training them to misalign their eyes so that each views separate pictures. This has spawned a popular

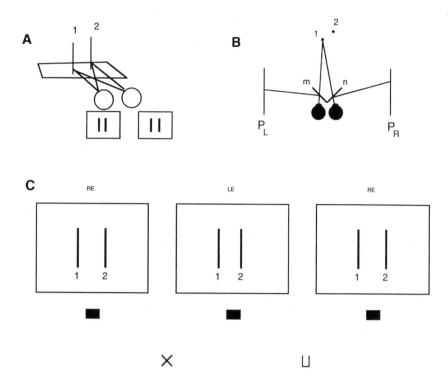

Figure 1.48
(A) Binocular viewing of a real-world scene showing two vertical rods, the left of which is closer than the right. Note that in the image of the two rods, the angular separation in the left eye is smaller than that on the right. (B) Schematic diagram of Wheatstone's mirror stereoscope, illustrating the position of the perceived rods (numbered 1 and 2) when viewing pictures P_L and P_R of the rods taken from two different vantage points. (C) Separate left-eye and right-eye images of the rods. Note that this is a three-panel stereogram in which the image to the left eye is presented in the middle and the image for the right eye is presented twice, at the extreme right and left. For this stereogram, the observer should cross fuse the left and center image or parallel fuse the center and right image (as described in Figure 1.49 and in text).

new art form in which observers peer into large pictures with repetitive patterns called autostereograms. To their surprise, they see otherwise invisible figures emerging in depth. A number of popular books using such displays is available. We recommend that readers who have difficulty fusing the stereograms in this chapter purchase several of these books for additional instruction and practice. Learning how to fuse stereograms sometimes takes a bit of practice. Like learning to ride a bicycle or to swim, some pick it up immediately, while others have a harder time. Once mastered, however, it is almost impossible to forget. If you know someone

A. NORMAL VIEWING   **B**. PARALLEL FUSION   **C**. CROSSED FUSION

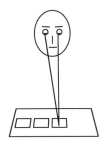   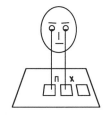   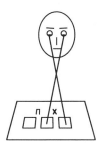

Figure 1.49
(A) Normal viewing of a book containing stereograms. (B) Parallel fusion of a stereogram, showing the viewing of the center and right images. (C) Cross fusion of the stereogram, showing the viewing of the left and center images.

who has mastered the technique, you may find it easier to learn with his or her help.

To help you understand the various techniques, we show the usual method of viewing a page (see Figure 1.49a). The eyes are not parallel but converge on the point of interest on the page. Each eye sees the same panel, and no stereopsis is obtained because there is no fusion of disparate images. There are two ways to misalign the eyes to obtain fusion. First, as shown in Figure 1.49b, we can diverge them abnormally in such way that the eye alignment is roughly parallel (parallel fusion). Alternatively, we can overconverge our eyes to obtain crossed fusion (Figure 1.49c).

When using the method of parallel fusion the observer imagines he or she is peering through the picture to a distant point beyond the picture. Doubling and blurring of vision is a good sign. Try not to focus the eyes immediately (to remove the blur) and concentrate on obtaining the perception of three images; the center image will eventually become the fused three-dimensional image. Eventually, the blur should disappear. To view these patterns (and to try the crossed fusion technique), experiment with holding the book closer and farther away or with removing or putting on your glasses. It is also useful to make sure the picture is viewed straight on (not at an oblique angle), at eye level, and in even illumination. Right-eye and left-eye images should be of the same brightness, and there should be no shadows on the page. For parallel fusion, most observers find that starting with close observation works better than viewing from a great distance.

Figure 1.49c shows the method for crossed fusion. This technique can usually be started by holding the book somewhat farther away than for parallel fusion. Holding the tip of a pencil in front of the eyes, close each eye alternately and move the pencil tip so that it lines up with a visible feature on "crossed" pictures. The pencil tip should be inserted just where the lines of sight shown in Figure 1.49c cross. Thus, the right eye should be viewing an image to the left of the image presented to the left eye. It is most appropriate to try this on the two left-most images of the three-paneled stereogram shown as Figure 1.49c.

By now the reader has probably noticed that, with few exceptions, the stereograms in this chapter consist of three, instead of two pictures. This feature allows the observer to fuse them by either the crossed or parallel fusion method. For crossed fusion, the symbol **X** placed between the left and center image serves as a reminder that these are the two images to be fused. For parallel fusion, the center and right images should be fused. These two images are marked with the symbol **U**.

At times the text of this chapter asks the reader to view the reversed stereogram, or reversed-eye configuration. Under these circumstances, he or she should look at the wrong pattern, the one marked by the **U** for crossed fusion or **X** for uncrossed fusion.

For readers who have yet to master one of these two techniques, there is one additional method of viewing the stereograms in this chapter. One can purchase a very inexpensive stereo-viewer, which consists of two lenses (and prisms) that aid in attaining parallel fusion. Here, the observer should put the viewer over the image pair marked **U** and otherwise follow instructions.

Suggestions for Further Reading

The present chapter presents, in the spirit of this volume, a case history of scientific inquiry. Much of the research effort has rested on the technique of binocular stereopsis. Although studies in the past have used binocular vision, they have used it to understand the encoding of depth. Our endeavour, on the other hand, has been mainly concerned with the consequences rather than the causes of depth. We employ binocular vision only as a powerful tool to create depth. In interpreting our results, we were fortunate to be able to draw on a rich tradition of well-written, important books on perception and vision. In particular we recommend the work of J. J. Gibson (1950; 1966; 1979) whose series of books outlines an evolving theoretical context for understanding visual perception. Strongly influenced by Gibson is David Marr (1982), who attempts to synthesize information from many fields, including visual psychophysics, visual physiology, and computer vision. Although the details of Marr's work need revision, his broad conception of the enterprise of understanding vision remains as a guideline for future work. Also important, is the book by Kanizsa (1979), who provides a particularly clear example of perceptual phenomenology in the service of a deeper inquiry into the representation of surfaces. Additional insight regarding perception and inference can be obtained by referring to Rock (1984), Gregory (1970), and Hochberg (1978). We also recommend the work of Julesz (1971), a pioneer in the study of binocular stereopsis whose book is full or stimulating stereoscopic demonstrations and comments. For a wider exposure to current approaches to vision and perception, we also recommend basic textbooks, including those by Blake and Sekuler (1994), and Bruce and Green (1992).

Problems

1.1   What is wrong with problem-solving approach to the study of vision that presents visual perception as a problem to be solved by reasoning?

1.2   What is wrong with a physiological reductionistic approach to the study of visual perception? Comment on ways in which these problems may be overcome.

1.3   Argue for or against the statement that we do not "see" light but the surfaces of objects.

1.4   If you were a neurophysiologist, how might you find out whether neurons are coding attributes of surfaces?

References

Albert, Marc K., and Donald D. Hoffman (1995). Genericity in spatial vision. In D. Luce, ed., *Geometric representations of perceptual phenomena: Articles in honor of Tarow Indow's 70th birthday*, 95–112. New York: L. Erlbaum.

Allman, J. M., and J. H. Kaas (1974). The organization of the second visual area (VII) in the owl monkey: A second-order transformation of the visual hemifield. *Brain Research* 76, 247–265.

Anderson, B. L. (1994). The role of partial occlusion in stereopsis. *Nature* 367, 365–368.

Anstis, S. (1980). The perception of apparent movement. *Philosophical Transactions of the Royal Society of London* B290, 153–168.

Barlow, H. B. (1953). Summation and inhibition in the frog's retina. *J. Physiol.*, 119, 69–88.

Barlow, H. B. (1972). Single units and sensation: A neuron doctrine for perceptual psychology? *Perception* 1, 371–394.

Barlow, H. B., C. Blakemore, and J. D. Pettigrew (1967). The neural mechanism of binocular depth discrimination. *Journal of Physiology* 193, 327–342.

Barlow, H. B., and W. R. Levick (1965). The mechanism of directionally selective units in rabbit's retina. *Journal of Physiology* 178, 477–504.

Beck, J., K. Prazdny, and A. Rosenfeld (1983). A theory of textural segmentation. In J. Beck and A. Rosenfeld, eds., *Human and machine vision*. New York: Academic Press.

Biederman, I. (1987). Recognition by components: A theory of human image understanding. *Psychological Review* 94, 115–117.

Blake, R., and R. Sekuler (1994). *Perception*. New York: McGraw-Hill.

Boring, E. G. (1942). *Sensation and perception in the history of experimental psychology*. New York: Appleton-Century-Crofts.

Braddick, O. (1974). A short-range process in apparent motion. *Vision Research* 14, 519–528.

Bregman, A. L. (1981). Asking the "what for" question in auditory perception. In M. Kubovy and J. R. Pomerantz, eds., *Perceptual organization*. Hillsdale, NJ: L. Erlbaum.

Britten K. H., M. N. Shadlen, W. T. Newsome, and J. A. Movshon (1992). The analysis of visual motion: a comparison of neuronal and psychophysical performance. *Journal of Neuroscience* 12, 4745–4765.

Bruce, V., and P. R. Green (1990). *Visual perception: Physiology, psychology, ecology*. Hillsdale, NJ: L. Erlbaum.

Burkhalter, A., D. J. Felleman, W. T. Newsome, and D. C. Van Essen (1986). Anatomical and physiological asymmetries related to visual areas V3 and VP in macaque extrastriate cortex. *Vision Res.*, 26, 63.

Cavanagh, P. (1992). Attention-based motion perception. *Science* 257, 1563–1565.

Crick, F., and C. Koch (1992). The problem of consciousness. *Scientific American* 267, 152–159.

Daw, N. W., and H. J. Wyatt (1976). Kittens reared in a uni-directional environment: Evidence for a critical period. *Journal of Physiology* 257, 155–170.

DeValois, R. L., H. Morgan, and D. M. Snodderly (1974). Psychophysical studies of monkey vision: III. Spatial luminance contrast sensitivity tests of macaque and human observers. *Vision Research* 14, 75−81.

Ernst, W. (1992). *The eye beguiled*. Cologne: Benedikt Taschen.

Fogel, I., and D. Sagi (1989). Gabor filters as texture discriminator. *Biological Cybernetics* 61, 103−113.

Freeman, W. T. (1994). The generic viewpoint assumption in a framework for visual perception. *Nature* 368, 542−545.

Gibson, J. J. (1950). *Perception of the visual world*. Boston: Houghton Mifflin.

Gibson, J. J. (1966). *The senses considered as perceptual systems*. Boston: Houghton Mifflin.

Gibson, J. J. (1979). *The ecological approach to visual perception*. Boston: Houghton Mifflin.

Green, M. (1986). What determines correspondence strength in apparent motion? *Vision Research* 26, 599−607.

Green M., and J. V. Odom (1986). Correspondence matching in apparent motion: Evidence for three-dimensional spatial representation. *Science* 233, 1427−1429.

Gregory, R. L. (1970). *The intelligent eye*. New York: McGraw Hill.

Gregory, R. L. (1972). Cognitive contours. *Nature* 238, 51−52.

Guzman, A. (1969). Decomposition of a visual scene into three-dimensional bodies. In A. Graselli, ed., *Automatic interpretation and classification of images*. New York: Academic Press.

He, Z. J., and K. Nakayama (1992). Surfaces versus features in visual search. *Nature* 359, 231−233.

He, Z. J., and K. Nakayama (1994a). Apparent motion determined by surface layout not by disparity or 3-dimensional distance. *Nature* 367, 173−175.

He, Z. J., and K. Nakayama (1994b). Perceiving textures: Beyond filtering. *Vision Research* 34, 151−162.

He, Z. J., and K. Nakayama (1994c). Surface shape not features determines apparent motion correspondence. *Vision Research* 34, 2125−2136.

He, Z. J., and K. Nakayama. Deployment and spread of attention to perceived surfaces in 3-dimensional space. Proceedings National Academy of Science. (in press).

Helmholtz, H. von (1867). *Treatise on physiological optics*, Vol. III. Trans. from the 3rd German edition, J. P. C. Southall, ed. New York: Dover Publications, 1962. First published in the Handbuch der physiologischen Optik, 1867, Voss.

Hildreth, E. C. (1984). *The measurement of visual motion*. Cambridge, MA: MIT Press.

Hochberg, J. (1978). *Perception*. Engelwood Cliffs, NJ: Prentice-Hall.

Hubel, D. H., and T. N. Wiesel (1959). Receptive fields of single neurones in the cat's striate cortex. *J. Physiol.*, 148, 574−591.

Hubel, D. H., and T. N. Wiesel (1965). Receptive fields and functional architecture in two nonstriate visual areas (18 and 19) in the cat's visual cortex. *J. Neurophysiol.*, 148, 229−289.

Hubel, D. H., and T. N. Wiesel (1968). Receptive fields and functional architecture of monkey striate cortex. *Journal of Physiology* 195, 215−243.

Jessell, T. M. (1991). Cell migration and axon guidance. In E. R. Kandel, J. H. Schwartz, and T. M. Jessell, eds., *Principles of neural science*. New York: Elsevier.

Julesz, B. (1971). *Foundations of cyclopean perception*. Chicago: University of Chicago Press.

Julesz, B. (1986). Texton gradients: The texton theory revisited. *Biological Cybernetics* 54, 245−251.

Kandel, E. R., and T. Jessell (1991). Early experience and the fine tuning of synaptic connections. In E. R. Kandel, J. H. Schwartz, and T. M. Jessell, eds., *Principles of neural science*. New York: Elsevier.

Kanizsa, G. (1979). *Organization in vision: Essays on gestalt perception*. New York: Praeger.

Kellman, P. J., and T. F. Shipley (1991). A theory of visual interpolation in object perception. *Cognitive Psychology* 23, 141−221.

Koch, C., and S. Ullman (1985). Shifts in selective visual attention: Towards the underlying neural circuitry. *Human Neurobiology* 4, 219–227.

Koenderink, J. J., and A. J. van Doorn (1976). The singularities of the visual mapping. *Biol. Cybernetics*, 24, 51–59.

Lettvin, J. Y., H. R. Maturana, W. S. McCulloch, and W. H. Pitts (1959). What the frog's eye tells the frog's brain. *Proceedings of the Institute of Radio Engineers* 47, 1940–1951.

Malik, J., and P. Perona (1990). Preattentive texture discrimination with early vision mechanisms. *Journal of the Optical Society of America* A7, 923–932.

Marr, D. (1982). *Vision*. New York: W. H. Freeman.

Maunsell, J. H. R., and W. T. Newsome (1987). Visual processing in monkey extrastriate cortex. *Annual Review in Neuroscience* 10, 363–401.

Metelli, F. (1974). The perception of transparency. *Scientific American* 230, 90–98.

Michotte, A. (1964). *La perception de la causalité*. Louvain: Publications Universitaires.

Nakayama, K. (1985). Biological image motion processing. A review. *Vision Research* 25, 625–660.

Nakayama, K. (1990). The iconic bottleneck and the tenuous link between early visual processing and perception. In C. Blakemore, ed., *Vision: Coding and efficiency*. Cambridge, Eng.: Cambridge University Press.

Nakayama, K. (1994) James J. Gibson—An appreciation. *Psychological Review* 101, 329–335.

Nakayama, K., S. Shimojo, and G. H. Silverman (1989). Stereoscopic depth: Its relation to image segmentation, grouping and the recognition of occluded objects. *Perception* 18, 55–68.

Nakayama, K., and S. Shimojo (1990). DaVinci stereopsis: Depth and subjective contours from unpaired monocular points. *Vision Research* 30, 1811–1825.

Nakayama, K., S. Shimojo, and V. S. Ramachandran (1990). Transparency: Relation to depth, subjective contours and color spreading. *Perception* 19, 497–513.

Nakayama, K., and S. Shimojo (1990). Toward a neural understanding of visual surface representation. In T. Sejnowski, E. R. Kandel, C. F. Stevens and J. D. Watson, eds., *The Brain* 55, 911–924. Cold Spring Harbor, NY: Cold Spring Harbor Symposium on Quantitative Biology.

Nakayama, K., and S. Shimojo (1992). Experiencing and perceiving visual surfaces. *Science* 257, 1357–1363.

Nakayama, K., and G. H. Silverman (1988). The aperture problem II: Spatial integration of velocity information along contours. *Vision Research* 28, 747–753.

Ninio, J. (1981). Random-curve stereograms: A flexible tool for the study of binocular vision. *Perception* 10, 403–410.

Penrose, L. S. and R. Penrose (1958). Impossible objects: A special type of illusion. *British Journal of Psychology* 49, 31.

Poggio, G. F., and B. Fischer (1977). Binocular interaction and depth sensitivity in striate and prestriate cortex of behaving rhesus monkey. *Journal of Neurophysiology* 40, 1392–1405.

Posner, M. I., C. R. R. Snyder, and B. J. Davidson (1980). Attention and the detection of signals. *Journal of Experimental Psychology: General* 109, 160–174.

Ramachandran, V. S., and S. M. Anstis (1983). Perceptual organization in moving patterns. *Nature* 304, 529–531.

Rock, I. (1984). *The logic of perception*. Cambridge, MA: MIT Press.

Rubin, E. (1921). *Visuall wahrgenommene Figuren*. Copenhagen: Gylden Kalske Boghandel.

Sagi, D. (1990). Detection of an orientation singularity in gabor textures: Effect of signal density and spatial-frequency. *Vision Research* 30, 1377–1388.

Schmidt, J. T., and D. L. Edwards (1983). Activity sharpens the map during the regeneration of the retinotectal projection in goldfish. *Brain Res.*, 269, 29–39.

Shatz, C. J. (1990). Impulse activity and the patterning of connections during CNS development. *Neuron* 5, 745–756.

Shimojo, S., G. H. Silverman, and K. Nakayama (1989). Occlusion and the solution to the aperture problem for motion. *Vision Research* 29, 619–626.

Shimojo, S., and K. Nakayama (1990). Amodal presence of partially occluded surfaces determines apparent motion. *Perception* 19, 285–299.

Shimojo, S., and K. Nakayama (1990). Real world occlusion constraints and binocular rivalry interaction. *Vision Research* 30, 69–80.

Schmidt, J. T. (1985). Formation of retinotopic connections: Selective stabilization by an activity-dependent mechanism. *Cellular Molecular Neurobiology* 5, 65–84.

Stryker, M. P., and W. A. Harris (1986). Binocular impulse blockade prevents the formation of ocular dominance columns in cat visual cortex. *Journal of Neuroscience* 6, 2117–2133.

Treisman, A. (1982). Perceptual grouping and attention in visual search for features and for objects. *Journal of Experimental Psychology: Human Perception and Performance* 8, 194–214.

Ungerleider, L. G., and M. Mishkin (1982). Two cortical visual systems. In D. J. Ingle, M. A. Goodale, and R. J. W. Mansfield, eds., *Analysis of visual behavior.* Cambridge, MA: MIT Press.

von der Heydt, R., E. Peterhans, and G. Baumgartner (1984). Illusory contours and cortical neuron responses. *Science* 224, 1260–1261.

von der Heydt, R., E. Peterhans, and G. Baumgartner (1989). Mechanisms of contour perception in monkey visual cortex: I. Lines of pattern discontinuity. *Journal of Neuroscience* 9, 1731–1748.

Van Essen, D. C., D. J. Felleman, E. A. DeYoe, J. Olavarria, and J. Knierim (1990). Modular and hierarchical organization of extrastriate visual cortex in the macaque monkey. *Cold Spring Harbor Symposium of Quantitative Biology* 55, 679–696.

Wallach, H. (1935). Uber visuell wahrgenommene Bewegungrichtung. *Psychologische Forschung* 20, 325–380.

Wertheimer, M. (1912). Experimentelle Studien über das Sehen von Bewegung. *Zeitschrift für Psychologie* 61, 161–265.

Wheatstone, C. (1838). On some remarkable, and hitherto unobserved, phenomena of binocular vision. *Philosophical Transactions of the Royal Society of London* B128, 371.

Yuille, A. L., and N. M. Grzywacz (1988). A computational theory for the perception of coherent visual motion. *Nature* 333, 71–74.

Zeki, S. (1978). Functional specialization in the visual cortex of the rhesus monkey. *Nature* 274, 423–428.

Chapter 2
Attention and Visual Perception: Analyzing Divided Attention
Harold Pashler

Attention is one of the most active areas of experimental research in cognitive science. This chapter focuses on an aspect of attention that has been studied intensively over the past twenty-five years or so: capacity limitations on our ability to perceive multiple visual inputs. The basic questions are simple but far-reaching. How much visual information can we take in at one time? What can we do with this information? Do we recognize objects one at a time, or can we recognize a large number simultaneously? This chapter describes several different methods that investigators have used to address these questions, along with some of the most important conclusions to emerge from this research.

2.1   Attention in Ordinary Language and Psychology

Attention is a common word in ordinary language as well as the name of a field of study. Other research areas in cognitive psychology, such as memory or problem solving, have also borrowed their names from ordinary language. More often than these other terms, however, the term attention tends to be used not only as a name for a set of phenomena, but also as a theoretical construct for explaining these phenomena. As various writers have noted (e.g., Broadbent 1982), smuggling terms like attention into scientific discourse from ordinary language can cause serious confusion. One way it can do so is by leading people to assume that the prescientific notions that go along with the word must be correct. For this reason, we shall use the term sparingly in this chapter, pending an examination of the empirical evidence.

Of all the phenomena associated with the concept of attention, the most conspicuous are *selectivity* and *capacity limits*, both of which are important aspects of mental life that must be dealt with by any psychology. Our momentary experience of the world depends not just on what stimuli are presented to our sensory apparatus, but on which we choose to become aware of. Selection is most obvious when it is difficult, as for example when we strain to listen to one voice in a crowd while ignoring other, perhaps louder voices. However, much or all of our experience reflects selection, since there are always more stimuli impinging on our sensory systems than we happen to notice at any moment. The fact that people can generally control which stimuli have the most pronounced effects on them bedeviled early stimulus-response theorists who hoped to find laws linking objectively defined stimuli and responses to them. For the most part, modern cognitive scientists take it for granted that unobservable internal structures and processes are important. They thus find attention a source of puzzlement rather than of consternation.

Capacity limitations are another undeniable feature of human mental life. Certain limitations on human perception and performance are dictated by the nature of the stimuli and of the peripheral sensory apparatuses. One example of this constraint is our limited acuity for stimuli presented in the periphery of the visual field. Others follow from the structure of our bodies and motor systems: for example, we have difficulty typing and drinking coffee at the same time. In many cases, however, tasks that we could perform individually are impossible to perform at the same time, quite apart from any sensory or physical limitations. For example, few people can profitably read a novel while listening to a symphony, and only the reckless try to tune their car radio while driving in busy traffic. These sorts of capacity limits are generally termed *attentional limitations*.

The concept of attention is part of what might be called a folk-psychological theory—that is, an informal set of propositions people rely on to explain their own and other people's daily experience and behavior. The folk-psychological theory of attention postulates an internal substance or process (attention) and uses it to explain both selectivity and capacity limits. The informal theory goes something like this: Attention can be devoted to stimuli, to memories, and to activities (and perhaps other things as well). When this happens, less attention (or sometimes no attention) remains available for other experiences or activities. Attention is also thought to be precondition of conscious awareness; if one doesn't attend to a stimulus, one remains oblivious to it. Paying attention is effortful, sometimes aversive (hence "paying"), but doing so yields a return: performing just about any sensory, cognitive, or motor activity improves it (except, perhaps, for automatic ones). A task that is given the benefit of full attention is accomplished more quickly and efficiently (except, again, certain automatic activities). Attention is ordinarily under the control of the "self," or the "will," although external stimuli or intruding thoughts can sometimes grab attention despite the best efforts of the will to prevent this from happening.

This intuitive theory is so compelling that one is tempted to think it self-evidently true. William James seems to have that idea in mind with his famous dictum that "everyone knows what attention is" (James 1890). If the dictum is valid, it would seem to make sense for attention researchers to take the basic notion of attention as a given and to move on to working out the details. The earliest psychologists who wrote on attention in the late nineteenth century set about to explore attention in essentially this way. Their main focus was on how attention affects people's conscious experience of external stimulation. They debated such questions as whether attending to a stimulus increases its subjective intensity or clarity or only changes its salience. The issue of divided attention was also discussed extensively. Here, too, the existence of something called attention was taken as a given, and writers debated whether this substance was ever truly divided. Oswald Kulpe, for example, was skeptical about the idea that people could genuinely divide their attention, while James argued that people could attend to more than one thing at a given instant, but only when they apprehended all of them as part of a single object (James 1890). Others argued that people could attend to as many as five or six different things at once. For the most part, these writers relied on introspective evidence, although a few studied people's ability to count a collection of objects or to name briefly presented stimuli (e.g., Cattell 1885).

Most present-day attention researchers work in the tradition sometimes called *information-processing psychology*. Their avowed goal is not to characterize conscious experience per se, but rather to trace the flow of information among different representational systems in the mind/brain. For the most part, they place little stock in introspection as a means of achieving this goal, relying instead on recording observations of human behavior in laboratory settings. These behaviors often involve subjects' reports of stimulus objects presented to them. The reports are taken as phenomena to be explained—by postulating a sequence of underlying information-processing operations resulting in the overt report. For the most part, however, subjects are not asked to analyze their underlying mental processes (for example, where their attention was directed).

While introspection as a methodology is universally dismissed in present-day cognitive science, the basic notion of conscious experience is still generally accepted. For example, contemporary researchers often distinguish between what observers experience at the moment they look at a stimulus and what they might report about the stimulus immediately thereafter. There is much less agreement, however, about the criteria to use in making this distinction (for examples of debates that have hinged on these criteria, see Treisman and Schmidt 1982; Fagot and Pashler 1995; Cheesman and Merikle 1986). In this regard, information-processing psychologists rarely adhere to the rigorous precepts of psychophysicists, such as Brindley (1970), who argue that subjective observations can be meaningful only when they show that observers cannot discriminate between physically distinct stimuli. (An example of an effect that does satisfy this standard of rigor is metamerism in color vision, in which blends composed of different wavelengths of light appear to have the same color as light of a pure wavelength). While some of the inferences used in studies of attention could be recast to fit this strict standard, many could not.

Even though modern attention researchers are usually dubious about introspection as a way to gain access to psychological processes, they often seem to assume that the folk-psychological theory of attention is at least partly right. For example, researchers often speak as if any task that is difficult in some way *consumes* attention. For example, when subjects hold digits in memory while doing some other task, this other task is often said to be performed under conditions of *divided attention*. Similarly, it is sometimes suggested that attention can be allocated to mental contents as diverse as positions in visual space or semantic categories in long-term memory. These ways of speaking make sense in terms of the folk-psychological theory of attention, but obviously involve assumptions that might or might not be valid.

2.2   Capacity Limits in Visual Perception

The present work focuses exclusively on one aspect of divided attention: people's ability to perceive multiple visual stimuli at the same time. This question can be posed without assuming the intuitive notion of attention described above and, indeed, without even using the term attention. For the reasons described above, this practice has much to recommend it and will be followed—with occasional lapses—in the following several sections. If the intuitive notion of attention is valid, its virtues should emerge when empirical results are considered; in that case it will have proven most instructive to have put the concept to the test.

We begin with the following question: If a person is presented with several visual stimuli and attempts to perceive them all, what information can be derived from them in parallel? If the stimuli belong to familiar categories, for example, can they be categorized simultaneously? Alternatively, do we have to scan the items, recognizing first one and then the next? Or perhaps people can perceive multiple objects simultaneously, but not as quickly or efficiently as when fewer items are present. This idea is usually called *capacity-limited parallel processing* and designates *serial processing* as the most extreme form of capacity limitation. The first several sections of this chapter consider these questions and describe two approaches used to answer them. Later sections examine how capacity limits in perception of multiple objects relate to capacity limits in thought and action. Here the broad conception of attention described in the preceding paragraphs is considered. The final section explores the relationship of capacity limits to selectivity and control and makes some observations about the intuitive notion of attention in light of the results discussed in earlier sections.

What exactly is meant by *perception* or *recognition*? Attention researchers generally use these terms to refer to deriving an internal representation that makes explicit the fact that an object belongs to some learned category, such as a particular letter, word, or type of object. Thus, a letter is not recognized by the retina just because its identity could be recovered from the pattern of retinal activity, because that representation is not explicit or usable without further processing. (Unfortunately, the term recognition is also used in memory research to refer to the judgment that something has been encountered before; that meaning will not be relevant here.) Visual perception involves a great deal more than just recognition, however. For example, our perceptual apparatus informs us about the color, position, and texture of different surfaces, perhaps the most likely three-dimensional arrangement of an object's parts, the sizes of objects and their parts, and so on. These different representations seem to be computed by at least partly independent brain areas (see Chapters 3 and 5). Partly for historical reasons, though, attention theorists have focused heavily on the recognition of familiar stimuli. Many of the issues considered here could also be raised about other perceptual operations (for a rare example in which this has been done, see Epstein, Babler, and Bownds 1992).

The question of whether visual object recognition is subject to capacity limits sounds straightforward enough. Probably the most obvious strategy would be to have observers name the objects they see in a brief display; the objects might be letters, digits, words, or pictures. Can an observer report the identity of all the elements in the display equally well regardless of how many items are present? If performance breaks down as the number is increased, one might conclude that capacity limits are present and that recognizing some of the objects depletes the capacity available for recognizing others. If it does not break down, one might infer that recognition is parallel.

The problem with this strategy is that reporting what was in the display not only involves identifying the objects, but also storing the results long enough to report them. Even though the layperson does not usually think of memory as a limiting factor over the time scale of a few seconds, it often is. Sperling's classic studies of the perception of brief displays provide one demonstration of this (Sperling 1960). When he asked subjects to report as many letters as they could from a display, he found they could report only about four or five. This was the case whether the display was presented for just one millisecond or for as long as one second (a factor of a thousand difference). It is also true whether the display contains six letters or many more. The main cause of this strange invariance is the fact that in order to report the information about what is present, the person must store the information in a limited-capacity memory system that is capable of holding only four or five items; getting even a few items into memory can be problematic.[1]

If information must get into short-term memory and be kept there to be reported, then report tasks are a problematic way of looking at recognition per se. To get around these limitations, investigators, beginning in the 1960s, started using visual monitoring or search tasks. In a monitoring task, a subject searches a display for a target and indicates whether or not he or she believes it to be present. (In the forced-choice version of the task, the subject indicates which of several possible targets is present.) Monitoring tasks greatly reduce the amount of information that must be stored in memory (to perhaps one bit). Estes and Taylor found that even with displays so brief that people could only report four or five letters, their detection performance often showed they must have processed more than four or five letters. The term *processed* here refers to whatever operations are used to discriminate targets from nontargets (Estes and Taylor 1964). A compelling illustration of the same point is provided by Molly Potter's studies involving rapid, serial visual presentation (RSVP) of pictures (Potter 1976). When her observers tried to determine whether a target picture (e.g., a girl sitting in a bed) was present in a sequence of sixteen pictures exposed at a rate of 167 milliseconds (msec) per picture, they detected over 70 percent of them. On the other hand, when the pictures were presented first and *then* subjects were told what targets to report, their performance was much worse. It seems, then, that an observer can rapidly analyze a sequence of pictures when the purpose is to find something, even when the pictures are exposed so briefly that he or she cannot store more than a tiny fraction of what is seen. The subjective experience of looking at RSVP displays is consistent with this. At moderately fast presentation rates, people often report feeling that they comprehend each picture but somehow lose this information the moment the next picture is presented. Indeed, the quick-cut exposure of scenes favored by movie and television directors, would be hard to explain if it produced only bewilderment in a viewer.

The fact that a person's visual system can process a great deal of information in a detection task—far beyond what the person can report—

1. The interpretation of these classic studies has often been questioned and sometimes misunderstood; for a recent discussion of the nature of the short-term memory system used in Sperling's whole-report task, see Pashler, in press).

still does not tell us whether object recognition operates in parallel and whether it is subject to capacity limits. It does, however, suggest a possible way of answering that question: measuring how the number of objects in a display affects observers' accuracy of detection. Estes examined this issue with displays of letters and found that accuracy declined as the number of characters in the display (*display set size*) increased (Estes and Taylor 1964). This might sound like proof of capacity limits, but there is a further methodological hurdle. To understand it, we consider a concept psychophysicists refer to as the *ideal observer*. This ideal observer is a hypothetical observer subject to no capacity limits, that is, capable of processing any given item equally well, regardless of how many total items are presented. The performance of this ideal observer can be formalized as follows. Let the probability that the ideal observer mistakes a given nontarget for a target be F, and let the probability that the observer mistakes a target for a nontarget be M. Both M and F are independent of display set size N. (While the observer is ideal in the sense that it is not subject to capacity limits, it is not flawless or omniscient, as reflected in the fact that both F and M are greater than zero.)

What happens to the ideal observer's accuracy when no target is present and N is varied? One might think it would remain constant, but in fact the ideal observer becomes more and more likely to report a target falsely as N increases. The exact amount of increase depends on exactly how the ideal observer makes decisions. The simplest and most discrete decision process would consist of reporting the presence of a target if *any* channel appears to contain one. Under these assumptions, the probability of a false alarm is $1 - (1 - F)^N$. (As N grows, $(1 - F)^N$ shrinks, and $1 - (1 - F)^N$ grows.) If the decision process works in some other way (it need not be discrete, for example), the false alarm rate would follow a different equation but would still increase with N (unless some rather exotic conditions hold[2]). This statistically based increase in error rates produced by increases in N will be termed *statistical decision noise*.

The full implications of statistical decision noise are often neglected. For example, some of the best-known studies of visual capacity limitations have plotted increases in response times (RTs) to detect correctly various kinds of targets, as a function of N (e.g., Treisman and Gelade, 1980). The logic of these studies is intuitively very natural: if every additional item in the display increases RTs by the same amount, then one presumes that people are scanning the display one item at a time. Error rates are disregarded in this formulation, which seems reasonable since the actual error

2. The exotic condition is a so-called high-threshold model according to which the probability of mistaking a nontarget for a target is always zero, although the probability of detecting a target may be below 1.0. Since people generally *do* make false alarms, this model would have to suppose that some pure guessing occurs.

rates in these kinds of studies are very low and differences among conditions are often insignificant. It would seem, therefore, that the issue of statistical decision noise can safely be ignored in this kind of experiment. This is not the case, however. Subjects are capable of trading speed for accuracy in essentially any task—taking a little more time and consequently making fewer errors. The existence of speed-accuracy trade-offs raises some troubling possibilities. For example, when subjects are confronted with a larger display, they may compensate for statistical decision noise by taking more time to process them. If subjects actually achieve the same error rates at different display set sizes, they must be doing something of this kind. However, the typical RT experiment provides little information either way about how display set size affects error rates, because accuracy is generally very close to ceiling (Wickelgren 1977). It would require careful analysis of the joint pattern of RTs and error rates— probably combined with an unusual degree of speed stress—to unravel these issues. Given these factors, one needs to be cautious in interpreting experiments that simply look at the effects of display set size on RTs. When RTs are unaffected by display set size it is a pretty good indication of parallel processing. Enormous slopes make sequential processing seem very plausible, but the technique suffers from serious logical limitations.

2.3   A Better Method: Comparing Simultaneous and Successive Displays

The preceding section makes it clear that diagnosing capacity limits is not quite as easy as it sounds. Fortunately, however, several strategies can provide more leverage on the problem than one gets by simply looking at error rates or RTs as a function of display set size. The first method was developed by Eriksen and Spencer (1969) and Shiffrin and Gardner (1972). Here the display set size is held constant (thereby keeping statistical decision noise constant), and the timing of the presentations is varied (see Figure 2.1). Two conditions are compared. In the simultaneous condition, all items are presented at the same time and followed by masks. In the successive condition, the N items are presented k at a time. Each item is always followed by a mask after a fixed delay, which is the same in the two conditions. If there are capacity limits, the successive condition is bound to be easier; here capacity can be shared among fewer items at any given moment. If processing is sequential—the most extreme form of capacity limit—the cost should be very substantial.

The method requires several assumptions. One is that in the successive condition, information extracted from the first display is not forgotten by the time a response is chosen. If it were, accuracy would be worse when

Successive Condition

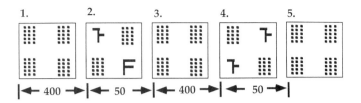

Simultaneous Condition

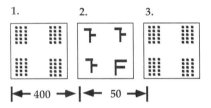

Figure 2.1
Method used in Shiffrin and Gardner (1972) experiment comparing simultaneous and successive exposures of four characters. The subject's task is to decide if a T or an F is present. In both conditions, stimuli are presented for T msec and then masked; however, in the successive condition only two stimuli are shown at any given time, whereas in the simultaneous condition all four are present simultaneously.

the target occurs in the first display rather than in the second. This outcome has not been observed (e.g., Shiffrin and Gardner 1972). Second, the method assumes that masks terminate perceptual processing fairly abruptly; if observers have hundreds of milliseconds beyond the onset of the masks to continue working on the display, the simultaneous condition would be effectively equivalent to the successive condition. Various results, including those described below, provide some justification for this assumption.

The simultaneous versus successive method has been used extensively with search tasks involving alphanumeric characters. The general finding is that detection performance is identical, or close to identical, in the two conditions. This implies that these displays can be processed with virtually no capacity limits (Shiffrin and Gardner 1972; Duncan 1980b). This finding holds even when the task involves finding the highest digit in a display of digits (Pashler and Badgio 1987). The highest-digit task is of interest because, while people might detect a single prespecified target character without actually identifying distractors, it seems unlikely they could find the highest digit without computing the identities of all the digits, since any unidentified digit might turn out to be higher than the apparent highest.[3]

People are not always as accurate in detecting targets in simultaneous displays as they are with successive displays, however. The simultaneous condition falls behind when subjects are required to make more complex and numerous visual discriminations. For example, when the displays are composed of two words, the successive condition enjoys a very large advantage over the simultaneous condition. This holds true whether the task involves detecting a particular word (Duncan 1987) or detecting an unspecified member of a semantic category (Shiu and Pashler, in preparation). The fact that an advantage for successive presentations emerges when more complex discriminations are required is intuitively reasonable. It implies, however, that perceptual capacity limits are fundamentally different from verbal short-term memory (STM) capacity limits. STM seems able to hold a certain number of familiar items; whether they are words or letters doesn't make much difference.

Another condition in which successive displays enjoy an advantage over simultaneous displays is when two targets are presented and must be detected separately. Duncan carried out a series of studies involving detection of variable numbers of targets in brief displays (Duncan 1980b). In a typical experiment, subjects monitored displays of four masked letters rather like those shown in Figure 2.1. In one condition, though, subjects made a separate detection response for each of the diagonals. Since subjects responded to information in each diagonal separately, each diagonal can be referred to as a separate information *channel*. Doing this makes it easier to discuss experiments of this sort, so we will use the term henceforth. When a single target is present in one of the channels and no other target is present, subjects are almost as accurate in the simultaneous condition as in the successive condition. When two targets are present, however, one in each channel, performance is significantly worse for simultaneous presentations. The probability of detecting a target in one channel (say, channel A) depended on which happened in the other (channel B). If the subject successfully detected a target on channel B (a "hit") or falsely reported there to be a target there ("false alarm"), there was a large decrement in detection on channel A.

It turns out that detecting a target in any brief presentation on one channel generally reduces the ability to detect a target on another channel. What makes this tendency especially interesting is that it happens in the

3. Of course, one could postulate a search process that searches first for the highest potential target, then for the next highest, and so on. Given the ease and rapidity with which people perform this task upon first attempting it, this algorithm seems a little implausible.

very tasks in which people can monitor two channels simultaneously without loss (that is, when there is only one target present). Duncan showed that the effect occurs even for detection of relatively simple visual features (Duncan 1985).

Although the focus in this volume is on visual perception, it is worth noting that the results just described generalize very nicely to auditory detection performance. A representative study was conducted (Puleo and Pastore 1978) in which subjects listened for a target (defined by frequency) in one ear and a target of a different frequency in the other ear. The subject knew which ear might contain a target at which frequency. In the monaural listening condition, targets were presented only in the ear the subject monitored. In the selective attention condition, targets could occur in either ear, but the subject was instructed to detect targets in only one ear. In the divided attention condition, targets could occur in either ear, and subjects had to report on each ear separately.

When target frequencies in the two ears were similar, simultaneous targets were perceptually fused and listeners heard a single tone. (Not surprisingly, this made it hard to detect both targets.) Targets widely separated in frequency did not fuse, however. Here, detection was as good in the selective attention condition as in the monaural condition, implying that subjects could prevent target detection if they chose. In the divided attention condition, however, performance depended on what was played in the opposite ear. When no target was played, detection was comparable in selective and monaural conditions. When the other ear received a target, however, performance was substantially worse. Thus, Puleo and Pastore's results closely parallel Duncan's later findings with visual detection. Like the visual effect, the auditory double-detection decrement cannot be due to sensory interactions; a contralateral target interferes even when it is a *gap* rather than a tone (e.g., Gilliom and Mills 1974, cited by Pohlmann and Sorkin 1976). The auditory double-detection problem is also robust. Neville Moray, for example, had people monitor continuous inputs (rather than discrete trials, as in the Puleo and Pastore studies), listening for targets ranging in complexity from tones to spoken words (Ostry, Moray, and Marks 1976). Again dual-channel monitoring was about as good as single-channel monitoring unless a target was present on the other channel, in which case a large decrement occurred (Moray 1975).

We see, then, that people can often monitor several visual or auditory channels for a target in parallel without showing any capacity limitations. However, detecting a target on one channel regularly impairs detection on another channel. Unfortunately, no one seems to have asked the obvious question of whether detecting an auditory target impairs detection of a visual target, and vice versa, so it is not clear whether the underlying limitation is restricted to a given sensory modality. There is one final intriguing parallel between the auditory and visual double-detection decrements. The effect is essentially abolished when the two targets are attributes of the same "object"—for example, orientation and texture of a line (Duncan 1984) or the loudness and timbre of a tone (Moore and Massaro 1973). The reasons for these close parallels between visual and auditory attention effects are not well understood. However, it will be suggested below that they are more likely to reflect similarly organized but separate attentional processes in different modalities than a single common attentional system serving both visual and auditory inputs.

2.4   A Converging Approach Using Reaction Times

As described earlier, many detection studies that require people to respond quickly about the presence or absence of a target analyze the reaction times in order to characterize perceptual capacity limitations. When the task is to detect a visual target that differs markedly from the background elements in a simple feature like color or orientation, RTs are barely affected by the number of (dissimilar) distractors (Green and Anderson 1956; Treisman and Gelade 1980). This essentially rules out the possibility that featural discriminations are carried out sequentially across the whole display. The parallel search may be carried out using mechanisms that normally function to segment the scene by locating discontinuities in visual features (see Chapters 2, 4, and 5).

When observers must detect a letter or digit target, on the other hand, results are more equivocal. Every additional distractor typically adds 20 or 30 msec to the total RT on negative trials. In some cases the slopes for positive (target-present) trials are about half those for negative trials, while in other cases the ratio is closer to 1:1. A 1:2 ratio for positive:negative slopes suggests a serial, self-terminating search, that is, a search that stops when a target has been found. This is because a self-terminating search will, on average, stop halfway through the array. A serial, self-terminating search might involve scanning clumps composed of more than one item at a time rather than just individuals, which results in a 1:2 slope ratio (Pashler 1987).

The more general question, however, is whether moderate positive slopes imply that character recognition is being carried out sequentially, or at least is subject to capacity limitations. Logically speaking, the slopes are consistent with that possibility; they are also consistent with several alternatives. Perceptual processing might, for example, be parallel, and other, later parts of the task might take longer when display set sizes increase. Even if all the items in the display are analyzed in parallel, a comparison or decision process might still be necessary to generate a response, which might take longer for displays composed of more items. A second alternative, already noted above, is that while perceptual analysis might operate in parallel, subjects may try to equalize their accuracy over different display set sizes to overcome the effects of statistical decision noise. This might be done by taking more time for larger display set sizes than for smaller ones.

Can one use RTs to sort out whether perceptual processing is serial or parallel? One fairly promising technique for doing so involves manipulating not just display set size, but also the difficulty (and hence the duration required) of perceptual analysis of each item in the display. There are several ways of prolonging perceptual analysis—for example, by reducing the intensity of the characters or degrading them with dots. In the

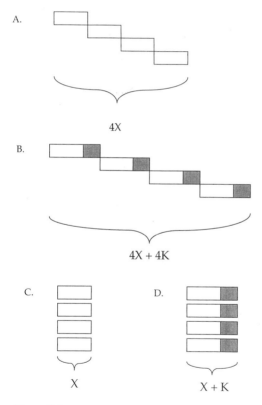

Figure 2.2
Predicted effects of adding extra time (shaded portion, lasting k msec) to duration of perceptual analysis. (A and B) Serial Perception Model; (C and D) Parallel Perception Model. (B) and (D) represent cases in which the stimuli are degraded (adding the extra time indicated by shading). In (B) Serial Model, RTs are increased by 4*k, whereas in (D) Parallel Model, they are only increased by k. Thus, the models make very discrepant predictions.

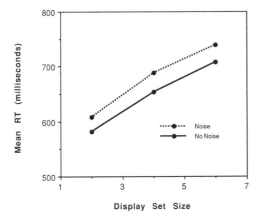

Figure 2.3
Data from experiment in which subjects name the highest digit in an array of two, four, or six digits, either all degraded or all undegraded. Degradation produces a constant increase (as predicted by the model shown in Figure 2.2, Panels C and D). (Data reproduced by permission from H. Pashler and P. C. Badgio, Visual attention and stimulus identification, 1985, *Journal of Experimental Psychology: Human Perception and Performance* 11, 105–121.)

experiments described below, the entire display was either undegraded (e.g., all letters very clear) or degraded (e.g, tiny dots superimposed on all the letters). The key test is how degradation effects on RTs vary according to display set size. As shown in Figure 2.2, parallel and serial hypotheses make completely different predictions. If perceptual analysis is sequential, slowing perception by X msec increases RTs by an amount equal to the product of X and the display set size. Thus, if, for example, degrading the display adds 20 msec to the time it takes to process each element in the display, degrading a display of four items will add a total of 80 msec to the RT, while degrading a display of eight will add 160 msec. If, on the other hand, perceptual analysis operates in parallel, the 20 msec slowing should be added into the RT regardless of display set size.[4] Thus, the hypothesis of sequential perceptual analysis predicts that effects of display set size and visual degradation will combine in a multiplicative manner, while the parallel hypothesis predicts that they will combine in an additive fashion.

In fact, the results show the additive pattern predicted by the parallel hypothesis (Pashler and Badgio 1985). Figure 2.3 shows the results of a speeded-up task in which—rather than detecting a prespecified letter—

4. Strictly speaking, a slight (and perhaps undetectable) interaction is to be expected whenever a manipulation increases the *variance* of processing times (see Pashler and Badgio 1985).

subjects named the highest digit in a display of digits. Here too the effects combined additively.

In summary, while reaction times often increase somewhat with display size in speeded-up detection tasks involving moderate numbers of alphanumeric characters, these slopes probably represent the effects of display set size on postperceptual processes, rather than sequential perceptual analysis. The results are therefore compatible with the conclusions described in Section 2.3: moderate numbers of fairly simple objects like letters and digits can be identified in parallel. We argued in that section that recognition of more complex objects like words is probably constrained to operating sequentially (or at least is severely capacity limited). The second method described above—analyzing display set size and visual degradation effects on search RTs—has not been tried with stimuli as complex as words. Doing so would provide a useful converging test.

2.5   Interpreting Perceptual Capacity Limits

So far we have focused on the question of how much visual analysis a person can carry out in parallel while trying to perceive several inputs concurrently. Since people can process a great deal more than they can ever report, visual search turns out to be more useful for this purpose than tasks requiring overt naming of stimuli. The problem of decision noise makes it trickier than one might have expected, however, to determine whether processing is capacity limited or not. This is because accuracy falls when there are more stimuli to be processed, even if perceptual analysis is not subject to any sort of capacity limitations. We described two methods that (at least partially) overcome this problem. One involves comparing simultaneous and successive displays, and the other involves looking at the joint effects of display set size and visual degradation on RTs. The results of both methods suggest that people can identify a number of characters at the same time without capacity limits. The accuracy experiments further suggest that this is not true of more complicated stimuli such as words; here the analysis is clearly subject to capacity limitations and may operate sequentially. We turn now to the broader implications of these conclusions.

The conclusion that parallel analysis of characters is possible in the context of a search task raises several questions. The virtue of search tasks is that they get around the need to hold more than a tiny bit of information in memory. However, they also have a disadvantage: when people search for a target of some kind, they may be performing perceptual analysis that is specialized for search. Each item in the display may not be identified but merely processed in a such a way as to figure out whether it is a target or not. This is plausible but not compelling. With a high degree of practice, subjects do seem to construct specialized detection skills limited in their generalizability to other stimulus sets (Shiffrin, Dumais, and Schneider 1981). On the other hand, our perceptual systems do not have unlimited flexibility; even when one might expect that top-down information would be used to tune perceptual machinery, this does not always happen (Johnston and Hale 1984). There is reason to suspect that visual search may depend on genuine object identification. For one thing, the highest-digit detection task produced results quite similar to those found with ordinary search tasks, as noted above. Furthermore, people often do a good job searching rapid sequential displays for such unknown members of a target semantic category as animal names (or, in Potter's RSVP experiments, pictures of animals).

It would seem reasonable, therefore, to conclude tentatively that search does reflect parallel object recognition, and that object recognition is subject to moderate capacity limits that are exceeded by a few complicated objects but not by a few simple ones. There is much more to be done to figure out the generalizability of these conclusions, however. For example, are there capacity limits in recognizing nonlinguistic stimuli like faces, tables, and chairs? Is recognizing an object like a table or a face comparable to recognizing a letter or a word in terms of its capacity demands? As the technology for presenting brief images of scenes becomes widely accessible, we will undoubtedly learn much more about the capacity limits affecting a broader range of visual tasks and stimuli.

A much deeper question is what causes the perceptual capacity limits that have been documented. At the present time little is known about this issue. Some contemporary neural-net models of visual processing predict a gradual breakdown of processing as visual load increases (Mozer 1991). These models generally predict display-set-size effects that are highly dependent on similarity (worse performance with more similar objects). The evidence for this is mixed. The most extreme manipulation of similarity is modality. Treisman and Davies showed that perceptual analysis in two different modalities produces less mutual interference than processing within a modality (Treisman and Davies 1973). On the other hand, Duncan recently found no evidence that the similarity of two classification tasks within the visual modality affected the degree of mutual interference (Duncan 1993). Clearly, it will take further research to unravel the causes of the interference. From a functional, psychological perspective, we may have to be content with a description of how and when performance breaks down and the ways in which output of perceptual analyses become available for different purposes. Answers to the deeper questions of *why* these processing limitations exist may only become possible when the neural circuitry is better understood.

Another intriguing issue is what kind of internal representations people form when they look at a scene composed of multiple objects. From the studies described earlier we concluded that when people choose to do so (as in a search task), they can identify multiple objects at the same time—subject to the capacity limits described above. But beyond allowing a person to locate the search target, what purposes does this identification process serve? Does it produce a mental representation of the identities of the objects in the scene? If so, can the identity information be accessed at will? Is the knowledge of the identity of objects tied to a spatial representation of the scene as a whole, like a map in a military command center with labels pinned to it? Since we seem to have the experience of seeing a whole scene at a glance, it might seem that there must be such a representation.

There are two kinds of studies that indicate, however, that this may not be the case. The first set of studies investigates people's ability to detect changes in a scene they have been looking at for a period of time. Consider what happens when subjects look at a display for a second or two, then it disappears for a short interstimulus interval (ISI) and then reappears. If asked to say whether anything in the display has been altered, can they do so accurately? George McConkie and his coworkers explored this issue in situations where the observer makes a saccadic eye movement during the ISI (McConkie 1990). He finds that people are remarkably insensitive to changes in either text or natural scenes, except for a small portion they choose to monitor, presumably using visual short-term memory (see also Irwin 1991). This finding has a remarkable implication: our subjective sense that we see the world as stable when we move our eyes may be illusory, because our brains do not actually check out the stability (a possibility first raised by Mackay, 1962).

Of course, it is possible that the changes are missed simply because of the eye movement: a maplike representation of the scene is built up during the initial fixation, but when the eye moves the representation is flushed and a new one is created when the eye comes to rest. To test this proposition, one can ask what happens when the display simply flickers off and on, with no eye movement being made. If the ISI is very brief—say, a few milliseconds—observers see flicker or motion where the change is introduced and have no trouble using that as a cue to detect the change (e.g., Phillips 1974; 1983). When the ISI is lengthened to, say, 100 milliseconds —still a very brief offset—an observer sees flickering in all parts of the scene. Because the flicker is everywhere, it provides no clue to where changes might have taken place. Does an internal maplike representation of what was in the display allow one to perform well? It seems not: without the flicker as a cue, observers are very poor at detecting changes. This result has been observed with displays composed of unfamiliar matrices of light and dark squares (Phillips 1974) as well as with displays of objects like letters and digits (Pashler 1988). More recently, Ling-Po Shiu and I have begun looking at change-detection by using photographs of real-world scenes. Here, too, observers often miss gross changes (e.g., the removal of a sailboat from a harbor front scene, or a change in the color of the pants worn by a person standing in the foreground). It seems, then, that while attending to the whole scene gives us the impression that we "take it all in at once," this may be in some sense illusory; at least, we don't form a rich internal representation of what objects are present in the scene and where each one is located. Or, if we do form such a representation, we don't seem able to use it to detect changes. Is it possible that people actually do form such a representation but lose it as soon as the display disappears? Or might they simply lack the ability to use their scene representation to detect change? These ideas sound reasonable enough, but they raise metaphysically troubling questions such as whether it makes any sense to say that someone knows that an object is in a certain position if the person finds nothing amiss when when it is removed.

It may be possible to get around some of these metaphysical questions by means of another method of testing, one that is similar in some ways to the RT experiments described above. An observer looks at a scene for a short while ("preview"); then a visual probe is presented next to one of the objects. The subject's job is to name the probed object, responding as quickly as possible. The task is similar to Sperling's (1960) classic partial report experiments, except that the display remains present for a short time after the probe appears, and the response is speeded up. If people really construct a mental representation of what objects are present in the scene and where each one is located, the subject should be able to retrieve the identity of the probed element from memory when the probe appears (after preview). As they should not have to identify the element "from scratch," RTs should not depend on the difficulty of identifying this item, since any slowing on that account will have been "absorbed" by the preview time. Experiments using displays of letters and various kinds of degradation did not, however, find any such absorption (Pashler 1984). (The conditions of this study are shown in Figure 2.4 and the results in Figure 2.5.) Apparently, then, subjects do have to start from scratch when the probe appears.

In summary, although we seem to be able to analyze more than one object at a time and recover their identities in parallel, we don't seem to be able to store the outputs of these analyses and access them for whatever purposes we might have (such as retrieving an identity when given its location). There are various possible interpretations of this finding. One is that we simply lack any way to hold onto the results beyond the moment they are retrieved. This could also explain the failures of change detection.

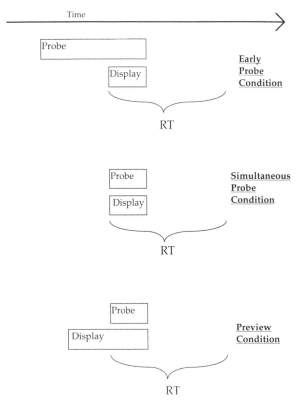

Figure 2.4
Timing of the three conditions (early probe, simultaneous probe and preview) in the probe-and-preview experiment described in text. In all cases, subject names the probed letters as rapidly as possible. RTs are recorded from the first instant that both probe and display are present on the screen as the subject has, potentially, the information to respond from this time forward.

However, a mere memory limitation does not explain the RT results, since subjects are looking at the display when the probe appears. Another interpretation would say that location information is lost when multiple objects are analyzed in parallel. This possibility fits in nicely with certain computational models of visual object recognition (e.g., Mozer 1991). There are, however, psychophysical observations that seem to conflict with it. When people detect the presence of even a simple object in a search task they seem to be able to report accurately the location at which the detection occurred (Green 1991; Johnston and Pashler 1991). This suggests yet another formulation: our knowledge of identities and locations might be more like a set of one-way pointers from identities to locations, rather than

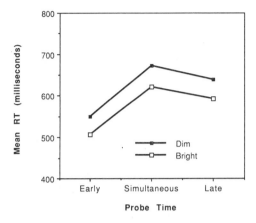

Figure 2.5
Typical results from experiments of the kind depicted in Figure 2.4. Reducing the intensity of the characters slows responses even when the probe is late (giving the subject the opportunity to preview the entire display). These data are from an unpublished experiment conducted by the author in 1991 and basically replicated in Pashler (1984), Experiment 3.

a set of bidirectional linkages (like a map). Thus, given the identity of an object, a person might be able to recover the location of the object in that location but, given the location, might not be able to recover the identity. Getting that information would require starting "from scratch" and re-processing the item alone. Another possibility is that the same neural machinery is used to analyze the whole scene and to process just a single element in it, but at a higher resolution.

Obviously, more experimentation is needed before one can assess the merits of these proposals. One conclusion seems warranted, however: visual awareness of a scene may reflect more impoverished knowledge structures than one might suspect on the basis of casual introspection. Experimental results confirm that people can take in a lot of information in parallel, but what they can do with that information seems to be limited in ways that are counterintuitive. Of course, our intuitions are based mostly on metaphors involving the objects in the physical world (e.g., maps) and, recently, digital computers. Given that the brain is constructed so differently from these things, it should probably not surprise us to learn that the fate of visual information does not conform to our initial guesses.

2.6   Perceptual versus Central Processing Limitations

So far we have discussed perceptual aspects of divided attention; that is, trying to take in many different stimuli at once and figure out what they

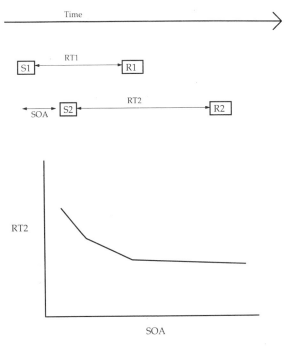

Figure 2.6
The Psychological Refractory Period (PRP) effect. Two stimuli are presented (S1 and S2) and subject makes a rapid response to each. The responses to the second stimulus are lengthened as the interval between the stimuli (stimulus onset asynchrony or SOA) becomes shorter.

are. People also have limitations in carrying out more central kinds of cognitive operations, such as thinking, deciding, and acting. Clear and striking evidence of these limitations comes from a very simple experimental situation. If a person must make a speeded-up response to each of two stimuli presented close together in time, the response to the second stimulus is almost invariably slowed. (Occasionally the response to the first is slowed as well.) The closer the stimuli in time, the greater the interference (Figure 2.6). The slowing of the second RT is usually referred to as the psychological refractory period (or PRP) effect. The PRP effect arises in a very wide range of circumstances: for example, when the two stimuli are in different modalities (Davis 1959), or when responses use diverse modalities like eye movements (Pashler, Carrier, and Hoffman 1993) or even foot movements (Osman and Moore 1993).

The chief cause of the PRP effect seems to be a central bottleneck that prevents people from selecting two responses at the same time (Welford 1952; Pashler 1993). The fundamental limitation (depicted in Figure 2.7)

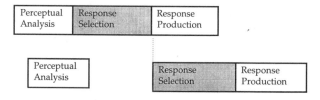

Figure 2.7
The response selection bottleneck account of the PRP effect. The selection of the response to the second stimulus waits for the selection of the response to the first stimulus to be completed, while other stages operate in parallel with each other. This represents the case in which perceptual interference is absent (e.g., where S1 is a tone and S2 is a letter); if both S1 and S2 are visual, perceptual interference can occur.

is cognitive, not "motoric." Deciding what to do in Task 2, rather than actually producing the response, is delayed by the corresponding stages in Task 1. What is the evidence for this claim? One kind of evidence comes from experiments manipulating the duration of the response production on the first task; this slows the response to Task 1 but barely delays the second at all (Pashler and Christian, submitted). Another key piece of evidence is that slowing the cognitive stages of Task 2 delays the second response by a constant amount, regardless of the interval between the stimuli (McCann and Johnston 1992). If the PRP effect were caused by a bottleneck in the producing responses, cognitive effects would decrease as the interval between tasks is shortened. However, this is not to say that there are no limits on response production; there does, in fact, seem to be an additional limitation preventing people from making two manual responses at the same time (even if one involves the left hand and the other the right hand). In practice, this probably makes no difference in a typical PRP experiment in which subjects make a single button-push response to each task because the cognitive bottleneck insures that the two responses would not overlap anyway (Pashler 1994b).

The folk-psychological theory of attention described at the beginning of this chapter attributes problems in dual-task performance to limited availability of attention. Indeed, subjects in PRP experiments often say their responses were slowed because they had trouble paying attention to the two tasks at once. Does the bottleneck that results in the PRP effect have the same underlying cause as the capacity limitations in visual detection tasks? Several lines of evidence argue that it does not. Recall that when people must perform two simultaneous and difficult visual discriminations (e.g., reading two words), accuracy is impaired (section 2.3). The degree of impairment is the same, whether the person must respond to each stimulus immediately or at his or her leisure (Pashler 1989). Thus, response selection conflicts cannot be the fundamental cause of this interference, as they are of the PRP effect. Similarly, recall that people have difficulty detecting two simultaneous targets, as noted by Duncan (see section 2.3). This occurs even when the person only needs to make a single response to indicate the presence of both targets (for example, counting them; Duncan 1980b). By contrast, the PRP effect depends on being forced to choose two actions. The dual-detection problem is avoided, however, when the subject reports on two attributes of the same object (e.g., the orientation and texture of a briefly presented line; Duncan 1984). By contrast, the PRP effect is the same whether both responses are to different attributes of the same object or to different attributes of different objects (Fagot and Pashler 1995). For example, Fagot and I found the usual PRP effect when subjects both named a character and made a button-push response to its color. Finally, as noted earlier, the PRP effect arises whether stimuli are presented in the same sensory modality or in different modalities (for example, a light and a tone). By contrast, perceptual capacity limits seem to be much less severe when stimuli are presented in different modalities (Treisman and Davies 1973).

All of these results argue that the central attentional limits found in the PRP effect have a different source than the perceptual capacity limits documented in detection tasks. There seem to be two sources of interference at work. First, there is a perceptual limitation that is most acute when we try to perceive multiple objects in the same sensory modality; this limitation is indifferent to the difficulty of (or even the need for) response choice and production. Second, there is a postperceptual bottleneck: when a person retrieves a response to one stimulus (or engages in memory retrieval; Carrier and Pashler, in press), he or she cannot retrieve anything else at the same time. Either or both of these conditions might arise in any given dual-task situation.[5] The final section explores these issues in a broader context, returning to the commonsense notion of attention with which we began.

2.7   Conclusions

The results described above distinguish between several different sorts of limits arising in divided-attention experiments. What overall picture of human information-processing limitations can account for these diverse effects? Clearly, supposing that there is a single resource or mechanism called attention that gets divided up in these different contexts will not get us very far. On the other hand, we can identify some grains of truth in the

5. These are not the only factors; others include the manual-response limitation described earlier and a general problem in attaining and maintaining preparation for any task (Pashler, in press).

intuitive conception of attention. One is the linkage between capacity limits and selectivity that is implicit in the ordinary notion of attending. Capacity limits do indeed arise from processing that is—for the most part—subject to voluntary control. This applies to both the central bottleneck and the perceptual capacity limits. For example, there is no PRP effect when a secondary stimulus is presented but the person chooses to ignore it (Pashler and Johnston 1989).[6] In the perceptual domain, when one ignores a portion of a visual display the contents do not tax perceptual capacity limits (Duncan 1979). Thus, selective control and capacity limits go hand in hand with regard to both central and perceptual capacity limits.

A second grain of truth in the ordinary notion of attention is that selecting a stimulus for one purpose often makes information derived from it available for other purposes. For example, when one selects an object in order to respond to its color, information about its form can be stored in short-term memory with minimal interference, while it is difficult to store the form of a different object at the same time (Pashler 1994a). What is selected for one purpose is thereby made available for other purposes (and what is not selected for one purpose is thereby unavailable for other purposes). Were it otherwise, the notion of attending might be useless indeed.

If the folk-psychological notion of attention is not supported, however, is there some alternative conception that fits with the results described earlier? At this point, the most that can be ventured is a very tentative sketch drawn with broad strokes. The following paragraphs offer a brief version of one such sketch, which assumes that selective control works by excluding rather than enhancing particular inputs. There is some evidence for this assumption, but this is not the place to describe it.

Associated with each sensory modality is a set of filtering mechanisms that shut out unwanted stimuli. In the case of vision, information is excluded on the basis of location—although there must be some interaction between the exclusion of a location and the segregation of figure and ground, since people can attend to one of several spatially overlapping objects (Neisser and Becklen 1975). This filtering process operates "early" —that is, prior to object recognition but probably after some initial featural analysis necessary for figure/ground segregation. Stimuli that are allowed passage by this mechanism may activate multiple codes corresponding to various levels of description of the stimulus. This activation may also store the results of all analyses of the object in visual short-term memory, where they are at least partially insulated from being masked by subsequent visual input. The machinery that carries out these analyses can

6. This statement applies when Task 2 is a choice task, but not when it is a simple RT task (see Pashler, in press).

be overloaded: if too many stimuli are admitted, accuracy suffers. In one sense, then, stimuli that pass through the filtering mechanism could be said to compete for perceptual "capacity," while those blocked do not. When an object activates a description that matches with a central cognitive activity (e.g., if it is a target in a search task, for example), the filtering mechanisms constrict around the location of the object. Obviously this notion of matching is very vague; there is little data on how central processes determine detection responses. We do not know, for example, whether a person can think about a sheep *without* searching for sheep. In any case, this constriction makes it difficult to detect two simultaneous targets, even when the discrimination between targets and distractors is so easy that enough capacity is available to search the entire display without capacity limits (Duncan 1985).

Central access to the outputs of these analyses seems far less complete than one might expect, even given the existence of capacity and acuity limits. Although spatial location may be implicitly coded for certain purposes, our conscious awareness of a scene does not seem to generate a maplike representation of the identities of objects in the scene and their locations. At least, if it does, we do not seem able to use the representation to see if anything is changed in a scene, nor can we look up the identity of the object present in a given location.

The perceptual limitations described in these few paragraphs are, of course, only half the story of divided attention. In addition, there is a central bottleneck that arises quite separately from these perceptual capacity limits. This bottleneck is seen most clearly in the PRP effect. Whenever an action is selected or information is retrieved from memory, other operations of the same type are delayed. These delays are mysterious from a computational standpoint, since looking up a few responses seems like a trivial operation, compared to the remarkable computational feats achieved by our perceptual and motor-control system.[7]

What, then, about attention—the concept we banished from the discussion back in section 2.1? Can attention be equated with any of the mechanisms proposed in the preceding paragraphs? One might refer to the machinery that "gates" stimuli as attention and say that an object is attended to when it is allowed access to perceptual machinery. This fits in reasonably well with ordinary usage, although it provides no justification for some common ways of talking. For example, if attention is identified with perceptual gating, it is hard to justify speaking of holding onto a

---

7. Alan Allport (1989), argues that the key problem of divided attention is preventing incompatible actions, but central interference seems to arise even when tasks involve decision making without any overt responses. The key problem seems to be memory retrieval, not action per se.

memory load while doing some other task as a divided-attention manipulation. Nor should one speak of "attending to a semantic category," "attending to a color," or many other common locutions. Since these are such common ways of talking, it may be best not to use the term *attention* for any of the mechanisms revealed by studies of performance or physiology; instead we might simply regard attention as a pretheoretical concept that is being replaced by empirically based concepts.

In summary, it seems increasingly clear—and perhaps hardly surprising—that the limitations we describe as involving attention in fact reflect the operation of a number of distinct control mechanisms and performance limitations. The ordinary conception of attention, taken for granted by many psychologists as well as laypersons, contains some elements of truth but seems to underestimate the functional separation between perceptual and cognitive limitations. The fine-grained study of performance in divided attention tasks—the examination of concurrent perceptual, cognitive, and motor activities—is a relatively new enterprise. Nonetheless, it seems to be revealing some fundamental limitations about what is humanly possible and offering up several clues to essential features of human beings' cognitive/neural architecture.

Suggestions for Further Readings

The book edited by R. Parasuraman and D. R. Davies (1984) contains a useful collection of articles on many different aspects of attention—including important topics in attention not discussed here, such as selective attention, set, and vigilance—although many recent developments are not included. A very clearly written analysis of capacity limitations and how they can be demonstrated is by John Duncan (1980a). For those interested in learning more about decision noise and how it can be modeled in the context of threshold perception, Norma Graham (1989) is lucid and definitive. A large literature describes the use of reaction times to analyze visual capacity limits; see, for example, Treisman and Gelade (1980) and Wolfe, Cave, and Franzel (1989), both of which focus on perceptual combination of features rather than recognition of familiar forms like letters. James Townsend and F. Gregory Ashby (1983) describe a wide variety of formal methods for discriminating parallel from serial processing. For a brief description of how limitations in dual-task performance have been analyzed using the psychological refractory effect, see Pashler (1993); for a more detailed review of studies of dual-task interference, see Pashler (1994b).

Problems

2.1   Consider the task of determining whether a display contains a barely detectable dark gray square (presented against a black background). In the *location-uncertain* condition, the observer knows that the square might occur in any of four possible positions. In a *location-certain* condition, the observer is told in advance where the square will occur if it is presented. If detection performance is better in the location-certain condition than in the location-uncertain condition, does that imply that people can allocate some sort of attentional capacity to a specified location and that this attentional capacity improves the processing of faint sensory input?

2.2   Consider the following hypothetical experiment. The subject sees a display of digits and simultaneously hears a spoken digit. The task is to verify that the display contains at least one token of each of the digits from one to the spoken digit. For example, if subjects hear FIVE their task is to verify that there is at least one 1, one 2, one 3, one 4 and one 5 in the display. Consider only the RTs on positive trials—that is, the trials in which subjects correctly respond YES—and, specifically, the effects of two variables: (1) what the spoken digit is; and (2) the visual quality of the display (degraded or normal). Suppose it should turn out that there is a *multiplicative interaction* between the spoken digit and visual quality: that is, RTs are slower when the display is degraded, and the amount of slowing is in direct proportion to the particular spoken digit (e.g., three times as much slowing to respond to THREE than ONE). Offer a straightforward interpretation of this finding and relate it to some of the ideas described in the text.

2.3   Casual observation suggests that people can drive while having a conversation but that doing so becomes risky if the traffic is heavy or the road is narrow. How would you reconcile this general impression with the findings about central and perceptual capacity limits described in the text?

References

Brindley, G. S. (1970). *Physiology of the retina and visual pathway* (2nd ed.). London: Edward Arnold.

Broadbent, D. E. (1982). Task combination and the selective intake of information. *Acta Psychologica 50*, 253–290.

Cattell, J. M. (1885/1947). The intertia of the eye and brain. In A. T. Poffenberger, ed., *James McKeen Cattell, 1860–1944, Man of Science*. Lancaster, PA: Science Press.

Cheesman, J., and P. M. Merikle (1986). Distinguishing conscious from unconscious perceptual processes. *Canadian Journal of Psychology 40*, 343–367.

Davis, R. (1959). The role of "attention" in the psychological refractory period. *Quarterly Journal of Experimental Psychology 11*, 211–220.

Dennett, D. C. (1991). *Consciousness Explained*. Boston: Little, Brown.

Duncan, J. (1979). Divided attention: The whole is more than the sum of its parts. *Journal of Experimental Psychology: Human Perception and Performance*. 5, 216–228.

Duncan, J. (1980a). The demonstration of capacity limitation. *Cognitive Psychology 12*, 75–96.

Duncan, J. (1980b). The locus of interference in the perception of simultaneous stimuli. *Psychological Review 87*, 272–300.

Duncan, J. (1984). Selective attention and the organization of visual information. *Journal of Experimental Psychology: General 113*, 501–517.

Duncan, J. (1985). Visual search and visual attention. In M. I. Posner and O. S. M. Mavin, eds., *Attention and Performance XI*, 85–105. Hillsdale, NJ: L. Erlbaum.

Duncan, J. (1987). Attention and reading: Wholes and parts in shape recognition—A tutorial review. In M. Coltheart, ed., *Attention and Performance XII: The Psychology of Reading*, 39–61. London: L. Erlbaum.

Duncan, J. (1993). Similarity between concurrent visual discriminations: Dimensions and objects. *Perception and Psychophysics 54*, 425–430.

Epstein, W., T. Babler, and S. Bownds (1992). Attentional demands of processing shape in three-dimensional space: Evidence from visual search and precuing paradigms. *Journal of Experimental Psychology: Human Perception and Performance 18*, 503–511.

Eriksen, C. W., and T. Spencer (1969). Rate of information processing in visual perception: Some results and methodological considerations. *Journal of Experimental Psychology: Monograph 79*, 1–16.

Estes, W. K., and H. A. Taylor (1964). A detection method and probabilistic models for assessing information processing from brief visual displays. *Proceedings of the National Academy of Science* 52, 446–454.

Fagot, C., and H. Pashler (1995). Repetition blindness: Perception or memory failure? *Journal of Experimental Psychology: Human Perception and Performance* 21, 275–292.

Graham, N. (1989). *Visual pattern analyzers*. New York: Oxford University Press.

Green, M. (1991). Visual search, visual streams, and visual architectures. *Perception and Psychophysics* 50, 388–403.

Green, N. F., and L. K. Anderson (1956). Color coding in a visual search task. *Journal of Experimental Psychology* 51, 19–24.

Irwin, D. E. (1991). Information integration across saccadic eye movements. *Cognitive Psychology* 23, 420–456.

James, W. (1890/1950). *Principles of Psychology*, vol. 1. New York: Dover.

Johnston, J. C., and B. L. Hale (1984). The influence of prior context on word identification: Bias and sensitivity effects. In H. Bouma and D. G. Bouwhuis, eds., *Attention and Performance X: The Control of Language Processes*, 243–256. Hillsdale, NJ: L. Erlbaum.

Johnston, J. C., and H. E. Pashler (1991). Close binding of identity and location in visual feature perception. *Journal of Experimental Psychology: Human Perception and Performance* 16, 843–856.

Kahneman, D., and A. Treisman (1984). Changing views of attention and automaticity. In R. Paraduraman and D. R. Davies, eds., *Varieties of Attention*, 29–62. New York: Academic Press.

Mackay, D. M. (1962). Theoretical models of space perception. In C. A. Muses, ed., *Aspects of the theory of artificial intelligence*. New York: Plenum.

McCann, R. S., and J. C. Johnston (1992). Locus of the single-channel bottleneck in dual-task interference. *Journal of Experimental Psychology: Human Perception and Performance* 18, 471–484.

McConkie, G. (1990). Where vision and cognition meet. Paper presented at the H. F. S. P. Workshop on Object and Scene Perception, Leuven, Belguim.

Moore, J. J., and D. W. Massaro (1973). Attention and processing capacity in auditory recognition. *Journal of Experimental Psychology* 99, 49–54.

Moray, N. (1975). A data base for theories of selective listening. In P. M. A. Rabbitt and S. Dornic, eds., *Attention and Performance V*, 119–135. New York: Academic Press.

Mozer, M. C. (1991). *The Perception of Multiple Objects: A Connectionist Approach*. Cambridge, MA: MIT Press.

Neely, J. H. (1977). Semantic priming and retrieval from lexical memory: Roles of inhibitionless spreading activation and limited-capacity attention. *Journal of Experimental Psychology: General* 106, 226–254.

Neisser, U., and R. Becklen (1975). Selective looking: Attending to visually specified events. *Cognitive Psychology* 7, 480–494.

Osman, A., and C. M. Moore (1993). The locus of dual-task interference—Psychological refractory effects on movement-related brain potentials. *Journal of Experimental Psychology: Human Perception and Performance* 19, 1292–1312.

Ostry, D., N. Moray, and G. Marks (1976). Attention, practice and semantic targets. *Journal of Experimental Psychology: Human Perception and Performance* 2, 326–336.

Parasuraman, R. and D. R. Davies eds., (1984). *Varieties of Attention*. Orlando: Academic Press.

Pashler, H. (1984). Evidence against late selection: Stimulus quality effects in previewed displays. *Journal of Experimental Psychology: Human Perception and Performance* 10, 429–448.

Pashler, H. (1987). Detecting conjunctions of color and form: Reassessing the serial search hypothesis. *Perception and Psychophysics* 41, 191–201.

Pashler, H. (1988). Familiarity and visual change detection. *Perception and Psychophysics* 44, 369–378.

Pashler, H. (1989). Dissociations and dependencies between speed and accuracy: Evidence for a two-component theory of divided attention in simple tasks. *Cognitive Psychology* 21, 469–514.

Pashler, H. (1993). Doing two things at the same time. *American Scientist* 81, 48–55.

Pashler, H. (1994a). Divided attention: Storing and classifying briefly presented objects. *Psychonomic Bulletin and Review* 1, 115–118.

Pashler, H. (1995b). Dual-Task interference in simple tasks: Data and theory. *Psychological Bulletin.*

Pashler, H. (in press). Structures, processes, and the flow of information. In E. Bjork and R. Bjork, eds., *Handbook of Perception and Cognition*, vol. 10, *Memory.* Orlando, FL: Academic Press.

Pashler, H., and P. C. Badgio (1985). Visual attention and stimulus identification. *Journal of Experimental Psychology: Human Perception and Performance* 11, 105–121.

Pashler, H., and P. Badgio (1987). Attentional issues in the identification of alphanumeric characters. In M. Coltheart, ed., *Attention and Performance XII*, 63–81. Hillsdale, NJ: L. Erlbaum.

Pashler, H., M. Carrier, and J. Hoffman (1993). Saccadic eye movements and dual-task interference. *Quarterly Journal of Experimental Psychology* 46A, 51–82.

Pashler, H., and J. C. Johnston (1989). Chronometric evidence for central postponement in temporally overlapping tasks. *Quarterly Journal of Experimental Psychology* 41(n1-A), 19–45.

Phillips, W. A. (1974). On the distinction between sensory storage and short-term visual memory. *Perception and Psychophysics* 16, 283–290.

Phillips, W. A. (1983). Short-term visual memory. *Philosophical Transactions of the Royal Society*, London, B302, 295–309.

Pohlmann, L. D., and R. D. Sorkin (1976). Simultaneous three-channel signal detection: Performance and criterion as a function of order of report. *Perception and Psychophysics* 20, 179–186.

Potter, M. C. (1976). Short-term conceptual memory for pictures. *Journal of Experimental Psychology: Learning, Memory and Cognition* 2, 509–522.

Puleo, J. S., and R. E. Pastore (1978). Critical-band effects in two-channel auditory signal detection. *Journal of Experimental Psychology: Human Perception and Performance* 4, 153–163.

Shiffrin, R. M., and G. T. Gardner (1972). Visual processing capacity and attentional control. *Journal of Experimental Psychology* 93, 78–82.

Shiffrin, R. M., S. T. Dumais, and W. Schneider (1981). Characteristics of automatism. In J. B. Long and A. D. Baddeley, eds., *Attention and Performance IX*, 223–238. Hillsdale, NJ: L. Erlbaum.

Sperling, G. (1960). The information available in brief visual presentations. *Psychological Monographs: General and Applied*, Whole No. 498, 1–29.

Townsend, J., and F. Gregory Ashby (1983). *Stochastic modeling of elementary psychological processes.* Cambridge, Eng.: Cambridge University Press.

Treisman, A., and A. Davies (1973). Dividing attention to ear and eye. In S. Kornblum, ed., *Attention and Performance IV*, 101–117. New York: Academic Press.

Treisman, A., and G. Gelade (1980). A feature integration theory of attention. *Cognitive Psychology* 12, 97–136.

Treisman, A., and H. Schmidt (1982). Illusory conjunctions in the perception of objects. *Cognitive Psychology* 14, 107–141.

Welford, A. T. (1952). The "psychological refractory period" and the timing of high speed performance—A review and a theory. *British Journal of Psychology* 43, 2–19.

Wickelgren, W. A. (1977). Speed-accuracy tradeoff and information processing dynamics. *Acta Psychologica* 41, 67–85.

Wolfe, J. M., K. R. Cave, and S. L. Franzel (1989). Guided search: An alternative to the feature integration model for visual search. *Journal of Experimental Psychology: Human Perception and Performance* 15, 419–433.

Yantis, S., and J. C. Johnston (1990). On the locus of visual selection: Evidence from focused attention tasks. *Journal of Experimental Psychology: Human Perception and Performance* 16, 135–149.

Chapter 3

Dissociable Systems for Visual Recognition: A Cognitive Neuropsychology Approach

Martha J. Farah

Parsimony is a guiding principle in cognitive science as in other sciences. Of course, it is not an infallible principle; nature is sometimes more complex than we expect it to be. In the study of visual pattern recognition, most people's initial assumption is that we have a single, general purpose system for recognizing all the different types of stimuli in our visual world. After all, there is no obvious reason why a system that can recognize a face should not also be able to recognize, say, an armchair or an airplane, and a single, general purpose system has the advantage of parsimony. However, in this case the assumption of parsimony appears to be wrong. As the studies reviewed in this chapter will demonstrate, the recognition of faces and common objects appear to be functions of distinct subsystems with separate neural substrates and different ways of representing shape.

The idea that face recognition is special is not new. In addition to the neuropsychological support I discuss in this chapter, evidence from normal subjects suggests that face recognition is different from other types of object recognition. For example, infants are born with a preference for gazing at faces rather than at other objects. At just thirty minutes of age, they will track a moving face farther than other moving patterns of comparable contrast, complexity, and so on (see Morton and Johnson 1991, for a review of this and other studies of infant face perception). The "face inversion effect" to be discussed in more detail later, provides another indication that face recognition is special. Whereas most objects are only a bit harder to recognize upside down than right side up, inversion makes faces dramatically harder for normal adult subjects to recognize. (See Valentine 1988, for a review of this research.)

The writing of this chapter was supported by ONR grant N00014-93-I0621, NIMH grant R01 MH48274, NINDS grant R01 NS34030, Alzheimer's Association/Hearst Corporation Research Grant PRG-93-153, a University of Pennsylvania Research Foundation grant and an NSF STC grant to the Institute for Research in Cognitive Science at the University of Pennsylvania. I would like to thank Jim Tanaka for his collaboration in developing the concept of holistic representation for faces, and my other co-authors on the papers cited herein for vital input into this research.

These findings from normal subjects indicate two clear differences between face recognition and the recognition of other objects: face recognition has earlier developmental precursors and is more orientation sensitive than other types of object recognition.

These differences need not, however, imply that different systems are involved. How might one individuate different systems? In this chapter I will use three commonsensical criteria. To be considered different, two systems must: (1) be functionally independent, such that either can operate without the other; (2) be physically distinct (which is not necessarily redundant with the first criterion—two programs running on the same computer can be functionally independent); and (3) process information in different ways, so that it is not merely a physical duplicate of another. By these criteria, the foregoing data on face tracking and inversion effect do not tell us whether face and object recognition are accomplished by different systems. Faces could be the first type of shape a general purpose system represents. Similarly, faces might require a special orientation-sensitive type of shape representation derived within a physically unitary and functionally indivisible system.

In this chapter, the hypothesis that face recognition depends upon a specialized system will be tested with data from both brain-damaged and normal subjects. Although disease and injury do not normally confine their damage to functionally defined subdivisions of the brain, occasionally a person does sustain a brain injury with relatively selective effects on one cognitive system. Such individuals have much to teach us about the functional architecture of the mind. They have been called "experiments of nature" because their brain damage can be viewed as an (unfortunate) experimental manipulation that eliminates one component of the cognitive architecture and allows us to observe the results. Normal subjects figure in this research in two ways: as control subjects who provide baseline performance data against which to measure patients' impairments, and as subjects of interest in their own right. In the latter case, some hypotheses that arise from the context of neuropsychological research can be more conveniently tested with normal subjects.

3.1   The Visual Agnosias: Impairments of Visual Recognition

The most relevant neuropsychological impairment for present purposes is *visual agnosia*. The term *agnosia* refers to an impairment of object recognition that is not attributable to a loss of general intellectual ability or to an impairment in such elementary visual perceptual processes as brightness, acuity, depth, and color (see Farah 1990, for a detailed overview). Thus by definition, agnosics retain full knowledge of the nonvisual aspects objects and can recognize them by touch, hearing characteristic sounds, or listening to a verbal definitions. They can also perceive at least some of their visual properties.

In *associative agnosia*, perception can be remarkably well preserved, to the extent that the person may be able to draw a good copy of a drawing or object he or she cannot recognize. Indeed, the term was coined in the nineteenth century because it seemed that perception was intact in these cases and that the problem must therefore lie in associating perception with knowledge of the objects. Our understanding of vision has now progressed to the point where we can identify different levels of visual representation—from those early and intermediate representations that make explicit the edges and surfaces of an image to higher-level representations that make explicit the more stable shape properties of the distal object (see Chapter 4). Associative visual agnosia is currently viewed by most neuropsychologists as an impairment at the highest levels of visual representation, rather than as an inability to associate normal visual representations with other types of knowledge. According to this view, the ability of associative agnosics to draw the object results from their use of lower-level visual representations, whereas recognition requires higher-level representations. The observation that associative visual agnosics tend to make visual errors—that is, that they mistake objects for visually similar objects—is also at least suggestive of an impairment in visual perception (see Farah 1990).

Associative visual agnosia does not always seem to affect the recognition of all types of stimuli equally. The selectivity observed in some cases of agnosia lends support to the hypothesis that there are specialized systems for recognizing particular types of stimuli. The best-known example of this is *prosopagnosia*, the inability to recognize faces after brain damage.

Prosopagnosics cannot recognize familiar people by their faces alone and must rely on other cues for recognition, such as a person's voice or distinctive clothing or hairstyles. The disorder can be so severe that the patient will not even recognized close friends and family members. One prosopagnosic described sitting in his club and wondering why another member was staring at him so intently. When he asked a steward to investigate, he learned that he had been looking at himself in a mirror (Pallis 1955)! Although many prosopagnosics also experience some degree of difficulty recognizing objects other than faces, in other cases the deficit appears to be strikingly selective for faces (e.g., DeRenzi 1986).

3.2   Prosopagnosia: Damage to a Specialized Recognition System?

The most straightforward interpretation of prosopagnosia is that the highest levels of visual representation are subdivided into specialized systems, and prosopagnosics have lost the specific system that is necessary for recognizing faces but not essential—or at least less necessary—for recognizing other types of objects. However, it is possible that faces and other types of objects are recognized using a single, general purpose recognition system but that faces are simply the most difficult type of object to recognize. Prosopagnosia could then be explained as a mild form of agnosia in which the impairment is detectable only on the most taxing form of recognition task. This account has the appeal of parsimony in that it requires only a single type of visual recognition system. Perhaps for this reason, it has gained considerable popularity (see e.g., Damasio, Damasio, and Van Hoesen 1982).

To determine whether prosopagnosia is truly selective for faces, and hence whether the human brain has specialized mechanisms for recognizing faces, we must therefore assess the prosopagnosic performance on faces and nonface objects *relative* to the difficulty of these stimuli. One technical difficulty encountered here is that normal subjects invariably perform both face and nonface recognition tasks nearly perfectly. The resultant ceiling effect thus masks any differences in difficulty that might exist between tasks and makes it pointless to test normal subjects in the kinds of recognition tasks traditionally administered to prosopagnosic patients.

With this problem in mind, Karen Klein, Karen Levinson, and I looked for a visual recognition task that would allow us to manipulate task difficulty for normal subjects, with the goal of setting normal performance at a moderate level (Farah, Klein, and Levinson, in press). The performance of a prosopagnosic subject on face and nonface stimuli could then be assessed relative to normal performance on the same tasks, answering the question of whether the subject was disproportionately impaired at face recognition.

Our subject, LH, was a forty-year-old man who has been prosopagnosic since an automobile accident in college. He is profoundly prosopagnosic, unable to recognize reliably his wife, children, or even himself in a group photograph. Yet he is highly intelligent, has no difficulty recognizing printed words, and only minimal difficulty recognizing objects. Although he has a degree of impairment with recognizing objects in drawings, this appears less severe than his impairment with faces.

We employed a recognition memory paradigm in which subjects first studied a set of photographs of faces and nonface objects, then performed an old/new judgment on a larger set of photographs, half of which were old. In a first experiment, we compared the recognition of faces to the recognition of a variety of nonface objects, which were paired with very similar foils, (as shown in Figure 3.1). In this experiment, we succeeded in equating the difficulty levels of the two sets of stimuli for a set of normal undergraduate subjects at approximately 85 percent correct. LH was given additional study time with the stimuli to ensure that he would

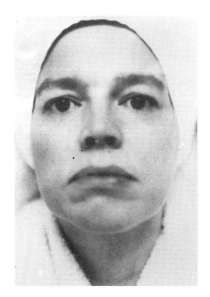

(a)

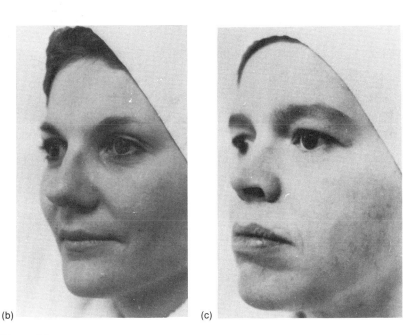

(b)   (c)

Figure 3.1
Examples of faces and objects used in recognition memory study with normal subjects and a prosopagnosic subject. The top item in each triple was studied, and the bottom two items were test items.

(d)

(e)

(f)

Figure 3.1 (cont.)

perform above a chance level. LH showed a significantly larger performance disparity for the two stimulus sets than the normal subjects, achieving only 62 percent correct for faces and 92 percent correct for objects.

In a second experiment, we attempted to test a particular version of the hypothesis that face recognition is just harder than object recognition, a view that has recently been promoted by Damasio and his colleagues (1982). Accordingly to this account, it is the fact that faces are highly similar exemplars all belonging to the same category (namely *face*) that makes them particularly taxing. We tested this hypothesis by comparing recognition of exemplars of the category face with an equivalent number of highly similar exemplars from a single nonface category, namely eyeglass frames. Examples of the stimuli are shown in Figure 3.2.

The faces and eyeglass frames were divided evenly into sets of old items, which appeared in both the study and the test phases of the experiment, and sets of new items, which appeared only in the test phase. Similar-looking eyeglass frames were separated into old and new sets to make the task more challenging; for example, there were both old and new horn-rims, and old and new aviator-style frames. As before, LH was disproportionately impaired at face recognition relative to nonface recognition compared to normal subjects. In this experiment, normal subjects found face recognition considerably easier than eyeglass frame recognition. Normal undergraduates achieved, on average, 87 percent correct responses on faces and 67 percent correct on eyeglass frames. A second group of normal subjects matched in age and education level with LH showed the same disparity; they achieved, on average, 85 percent correct on faces and 69 percent correct on eyeglass frames. LH showed significantly less face superiority in this task than normal subjects, achieving 64 percent correct for faces and 63 percent correct for eyeglass frames. Like the first experiment, this one also suggests that LH's impairment in face recognition cannot be attributed to a general problem with object recognition. The results also suggest that the problem does not lie with the recognition of specific exemplars from any visually homogeneous category but is specific to faces.

A final experiment was undertaken to address the specificity of LH's face-recognition impairment. In essence, the first two experiments compared LH's performance with faces and his performance with stimuli that are similar to faces—in their recognition difficulty and their membership in a visually homogeneous category—but are not processed by the hypothesized face-specific recognition mechanism. Stating the experimental design in this way suggests the ideal nonface comparison stimulus: upside-down faces. As mentioned earlier, inverting a face makes it much harder for normal subjects to recognize. On the basis of the face-inversion effect, it is generally assumed that if a specialized face-recognition mechanism exists,

Figure 3.2
Examples of faces and eyeglass frames used in recognition memory study with normal subjects, prosopagnosic subject, and object-agnosic subject.

it is specialized for the processing of upright faces. Inverted faces, therefore, constitute ideal comparison stimuli: They are equivalent to upright faces in virtually all physical stimulus parameters, including complexity and inter-item similarity, but they do not engage (or engage to a lesser extent) the hypothesized face-specific processing mechanisms.

My colleagues and I reasoned that if LH's underlying impairment was not face-specific, he would show a normal face-inversion effect (Farah, Wilson, Drain, and Tanaka, in press). In other words, he would perform normally with upright faces relative to his performance on inverted faces. In contrast, if he had suffered damage to neural tissue implementing a specialized face-recognition system, he would show an absent or attenuated face-inversion effect. That is, he would be disproportionately impaired with upright faces relative to his performance on the comparison stimuli, inverted faces.

LH and normal subjects were tested in a sequential matching task, in which an unfamiliar face was presented, followed by a brief interstimulus interval, followed by a second face, to which the subject responded "same" or "different." The first and second faces of a trial were always in the same orientation, and upright and inverted trials were randomly intermixed. As expected, normal subjects performed better with the upright than with the inverted faces, replicating the usual face inversion effect: 94 percent versus 82 percent correct, respectively.

LH's results were more surprising. He was significantly more accurate with inverted faces, achieving 58 percent correct for upright and 72 percent correct for inverted faces! This outcome was not among the alternatives we had considered. We had assumed that if he had an impaired face processor, it would simply not be used in this task and he would, therefore, show an absent or attenuated face-inversion effect. Instead, it appears, he has an impaired face-specific processor, which is engaged by the upright but not by the inverted faces, and used even though it is impaired and thus disadvantageous. This result was confirmed in additional studies, which invariably showed either statistically significant or nonsignificant trends in the same direction.

Two major conclusions follow from LH's "inverted inversion effect." First, LH's prosopagnosia results from damage to a specialized face-recognition mechanism. Inverted faces are the perfect control stimulus for equating faces and nonface objects for such factors as complexity and inter-item similarity. LH's disproportionate impairment on upright relative to inverted faces is therefore strong evidence that an impairment of face-specific processing mechanisms underlies his prosopagnosia.

A second, and unexpected, finding was that LH's specialized face-perception system was contributing to his performance, even though it was impaired and clearly maladaptive. This demonstrates the involuntary nature of the specialized face system and provides very direct neuro-psychological support for Fodor's (1983) characterization of special-purpose perceptual systems ("modules") mandatorily engaged by their inputs.

The general conclusion of these three studies with LH is that proso-pagnosia represents the selective loss of visual mechanisms needed for face recognition, and not needed (or less necessary) for other types of object recognition. There is, therefore, specialization within the visual recognition system in which faces are recognized differently than other objects.

3.3   Selective Impairment of New Face Learning

Prosopagnosics such as LH are equally impaired at learning new faces and recognizing previously familiar faces, as we would expect from damage to the substrates of face representation. My colleagues and I recently encountered someone with an even more selective impairment. CT is impaired at learning new faces, but his ability to recognize previously familiar faces and to learn other nonface visual objects is relatively intact. (Tippett, Miller, and Farah, in preparation). This pattern of performance is consistent with a disconnection between intact face representations and an intact medial-temporal memory system. As such, it provides additional evidence that the neural substrates of face representation are distinct from the representation of other objects, as they can be selectively disconnected from the substrates of new learning.

CT's face perception was normal on a variety of measures, including the face-inversion task used with LH. His overall level of performance was also good relative to normal subjects, and he showed a normal inversion effect. His learning of verbal material and even visual material other than faces is also normal. However, when given the face- and eyeglass-learning task, he performed about as well as LH, achieving 58 percent correct for faces and 63 percent correct for eyeglasses. Additional evidence of his inability to learn faces comes from his identification of famous faces. For people who were famous prior to his head injury, CT performed within the range of eight age-matched control subjects on a forced choice famous/not famous task; whereas for more recently famous individuals he performed at chance level. One celebrity allowed us to make an especially interesting comparison between premorbid and current face recognition. In the case of Michael Jackson, the singer's extension plastic surgery following CT's injury provides us with a "within-celebrity" comparison of face recognition. Despite the greater popularity and media exposure of Michael Jackson in recent years, CT recognized an older picture of the celebrity and failed to recognize an up-to-date photograph.

3.4   Object Agnosia with Preserved Face Recognition: Further Clues to the Functional Architecture of Visual Recognition

Some associative agnosics appear to have more difficulty with object recognition than with face recognition, presenting us with the mirror image of the prosopagnosic's impaired and spared abilities. This pattern of impairment is interesting for two reasons. First, it offers further disconfirmation of the hypothesis that prosopagnosia is just a mild disorder of a general purpose object recognition system, with faces simply being harder to recognize than other objects. If this were true, how could a person do better with faces than with other objects? Second, it distinguishes two possible relationships that might hold between the specialized face system and the nonface object system. As illustrated in Figure 3.3, the two systems could be arranged in parallel, so that a stimulus would be recognized by one or the other. Alternatively, the two systems could be arranged in series, so that all stimuli would first be processed by the one system, with faces then receiving further processing by the other system.

Given the intuition that face recognition requires processing that is somehow more elaborate or demanding than object recognition, which presumably motivated the alternative accounts described in the last section, one might expect the latter, serial arrangement to hold. According to

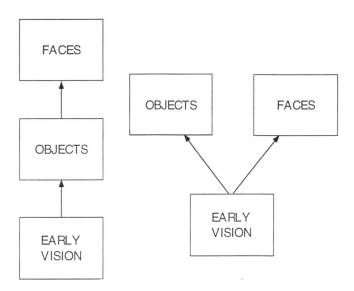

Figure 3.3
Schematic diagram of two different ways in which face recognition could be distinct from object recognition.

this view, there is a specialized face system but it is not functionally independent of the object system; it requires input from the object system and performs further processing on that input. In other words, face recognition involves normal object recognition plus some additional processing. This arrangement contrasts with the first one, according to which earlier visual processes deliver their products to two parallel, independent systems, one required to recognize faces and the other objects.

If there are indeed associative visual agnosics with relatively intact face recognition, then the systems subserving object and face recognition must be arranged in parallel. Marlene Behrmann and I recently set out to confirm experimentally the clinical observation that recognition of faces can be disproportionately spared. We used the same faces and eyeglasses experiment earlier administered to LH.

The subject in this experiment was CK, a thirty-five-year-old man who sustained a head injury in an automobile accident. An MRI showed bilateral thinning of the occipital lobes but no other focal abnormality. CK is agnosic for objects and printed words. His pattern of performance differs from normality in the direction opposite to LH's: He is 98 percent correct for faces and only 48 percent correct for eyeglasses. He shows a larger superiority of faces over eyeglasses than expected on the basis of either of the sets of normative data collected for the experiments with LH. This result, taken together with the earlier findings from LH, implies that the systems specialized for face and object recognition are functionally independent. Put more precisely, there are two systems—one more important for face recognition than for nonface object recognition and another system (or set of systems) more important for nonface object recognition than for face recognition. And they are arranged in parallel.

3.5   Functional Differences between Face and Nonface Processing

Having concluded that there are at least two specialized subsystems underlying visual recognition, let us now turn to the question of what these specialized systems might be specialized for, in terms of the kinds of visual information processing they carry out. Before addressing this question, it would be helpful to review a bit of what we know about nonface object recognition.

A recent review of published cases of associative visual agnosia suggests that face, object, and printed-word recognition are all pairwise dissociable, but that not all possible three-way combinations of impaired and spared face, object, and word recognition are possible (Farah 1991). Object recognition, in particular, was found to be impaired only if either face

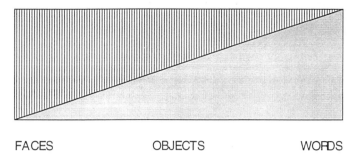

FACES   OBJECTS   WORDS

Figure 3.4
Graphic representation of the relations between two hypothesized types of pattern recognition. (See text.)

recognition or word recognition was also impaired.[1] This leads to the following hypothesis concerning the number of specialized recognition systems and their domains of applicability: There are two systems, one of which is essential for face recognition, useful for common object recognition, and not at all needed for printed word recognition, and the other of which is essential for printed-word recognition, useful for common object recognition, and not at all needed for face recognition. Figure 3.4 illustrates the inferred contributions of these two systems to face, object, and word recognition. The two hypothesized systems can account for the disproportionate face-recognition impairment of subjects like LH (damage to the first system), and the disproportionate object- (and word) recognition impairment of subjects like CK (a more severe degree of damage, to the second system). In contrast, there is no way to damage the two systems to produce an impairment in common object recognition alone, which explains the apparent absence of such cases in the literature.

This two-system interpretation of the patterns of co-occurrence of the different types of associative visual agnosia offers a clue to the nature of the nonface object recognition system. Whatever type of visual information processing this system performs, this processing is more taxed by printed words than by common objects.

1. As discussed in the original review, one case report mentioned "mild object agnosia" with no accompanying reading or face recognition difficulties in a table, but referred to the same subject as nonagnosic in the text. The information in the table represents the only violation of the pattern found by me in the literature. Rumiati, Humphreys, Riddoch and Bateman (1994) recently reported that they found another violation of this pattern. However, their subject does not appear to be a visual agnosic: His errors in naming objects are overwhelmingly semantic (e.g., "cup" for saucer) rather than visual (e.g., "record" for saucer), and he has similar problems interpreting object *names* as well as object pictures. It is not clear why such a pattern of performance would be interpreted in terms of an impairment in visual object recognition.

One salient property of words, as a type of visual pattern, is that they are composed of numerous individually recognizable parts, namely letters. In fact, research with normal subjects has shown that printed words are recognized by first recognizing their letters. For example, Johnston and McClelland (1974) found that tachistoscopic word recognition was significantly more disrupted by a mask made up of letters than by one made up of letter fragments. This finding is consistent with the idea that a necessary stage in word recognition is the explicit recognition of component letters.[2] There is also evidence that the underlying impairment in subjects who have lost the ability to recognize printed words consists of an inability to recognize multiple shapes. Such individuals typically resort to reading letter by letter, as if they can only recognize one part of the word at a time. For a fuller discussion of the role of visual perception in acquired impairments of printed-word recognition, see Farah and Wallace (1991).

Like words, most objects can also be subdivided into component parts. In fact, as reviewed by Biederman in Chapter 4, many current theories of object recognition hypothesize some form of structural description that is a representation of object shape in terms of parts, which are themselves explicitly represented as shapes in their own right. The more extensive the part decomposition, the more parts there will be in an object's representation, but the simpler those parts will be. The less the part decomposition, the fewer parts there will be in an object's representation, but the more complex those parts will be.

The conjecture being put forth here is that word recognition involves extensive part decomposition and, therefore, requires the ability to represent a large number of parts; face recognition, on the other hand, is holistic in that it involves virtually no part decomposition, and hence requires the ability to represent complex parts. Common objects are represented using a mixture of the two types of representation.

3.6   Face Recognition and Holistic Shape Representation: Empirical Tests

The patterns of co-occurrence among disorders of face, object, and word recognition suggest the existence of two complementary systems of shape representation. Consideration of the types of representations underlying

2. The word superiority effect, by which letters embedded in words are perceived better than words presented in nonwords or alone, might appear to imply that words are perceived holistically, without decomposition into letters. However, its implications are weaker than this. It implies only that, in addition to individual letter representations, word or letter-cluster representations are also activated, and that the activation states of the latter representations influence those of the former.

word recognition led, above, to a conjecture about those underlying face recognition. Specifically, if the system that is essential for recognizing words is specialized for the representation of numerous but relatively simple parts, then the system that is essential for recognizing faces might be specialized for the representation of complex but relatively few parts. In collaboration with James Tanaka and others, I carried out several tests of this hypothesis.

In one set of studies, we reasoned as follows (Tanaka and Farah 1993): To the extent that some portion of a pattern is explicitly represented as a part for purposes of recognition, then when that portion is presented later in isolation, subjects should be able to identify it as a portion of a familiar pattern. In contrast, if a portion of a pattern does not correspond to the way the subject's visual system parses the whole pattern, then that portion presented in isolation is less likely to be recognized. Tanaka and I taught subjects to identify a set of faces, along with a set of nonface objects, and then assessed their ability to recognize both the whole patterns and their parts. Examples of study and test stimuli are shown in Figure 3.5. Relative to the recognition of houses, face recognition showed a greater disadvantage for parts relative to wholes: Subjects achieved, on average, 81 percent and 79 percent accuracy for parts of houses and whole houses, respectively, and 65 percent and 77 percent for parts of faces and whole faces, respectively. This is what we would expect if the representations underlying face recognition do not explicitly represent parts or do so to a lesser degree than nonface objects. Similar results were obtained with inverted faces and scrambled faces as the nonface comparison stimuli.

My collaborators and I recently adapted Johnston and McClelland's masking paradigm (mentioned earlier in connection with word recognition) to a new test of the hypothesis that face perception involves less part decomposition than the perception of such other stimuli as words, houses, or inverted faces (Farah, Wilson, Drain, and Tanaka, 1995). Recall that Johnston and McClelland found that word perception was more disrupted by a mask composed of letters, compared with one composed of letter fragments; they inferred that a necessary stage in word recognition is letter recognition. In our first experiment, word and face perception were assessed in a sequential same/different matching task in which the first stimulus (word or face) was presented only briefly and was followed by a mask. We used two kinds of masks. In the "part mask" condition, either letters or facial features were presented in spatial arrangements that did not make real words or faces. In the "whole mask" condition, a word or a face was used to mask the first stimulus. We predicted that if faces are perceived holistically, without explicit representations of their parts, the part masks should not be very disruptive of face perception compared to the disruption caused by a whole face mask. With word perception, on

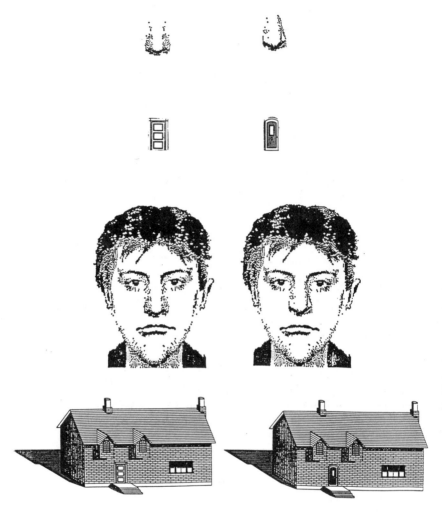

Figure 3.5
Examples of face and house stimuli used in memory study with normal subjects.

the other hand, part masks should be effective. This is what we found: Whereas subjects correctly judged 78 percent of the faces with a part mask, their accuracy dropped to 73 percent with a whole mask. In contrast, their performance for words with parts and whole masking was 78 percent and 77 percent, respectively. In subsequent experiments, we found that the difference between part and whole masks is found only for upright faces; inverted faces show no difference. We also found that the perception of houses showed an intermediate degree of sensitivity to part masks. These results accord well with the hypothesis of a parts-based system and a holistic system, used together for the recognition of objects such as houses and used separately for the perceptions of words and faces, respectively.

In the final experiment to be described, we bring the research back to prosopagnosia and the neural bases of face recognition. The neuropsychological results described earlier imply that there is some neurologically distinct subsystem that is more important for face recognition than for other kinds of object recognition. The results of the experiments described imply that normal subjects perceive faces more holistically than they perceive other kinds of objects. Taken together, these findings suggest that the face-recognition system damaged in prosopagnosia is a one of relatively holistic representation. The final experiment tests this hypothesis directly.

Tanaka, Drain, and I compared the relative advantage of whole faces over face parts for normal subjects and for the prosopagnosic LH. Our initial plan was to administer the same task Tanaka and I used with the normal subjects to LH; but despite intensive effort, LH could not learn to recognize a set of faces. We therefore switched to a short-term memory paradigm in which a face was presented for study, followed by a blank interval, followed by a second presentation of a face. The subject's task was to say whether the first and second faces were same or different. There were two different conditions for presentation of the first face: it was either "exploded" into four separate frames containing the head, eyes, nose, and mouth (in their proper relative spatial position within each frame), or presented intact. The second face was always presented in the normal format, so that the two conditions can be called parts-to-whole and whole-to-whole. Normal subjects performed better in the whole-to-whole than in the part-to-whole condition; they averaged 93 percent and 74 percent correct answers, respectively, thus providing further evidence that their perception of a whole face is not equivalent to the perception of its parts. LH showed abnormally little difference between the two conditions, scoring 74 percent and 73 percent correct answers, respectively. This finding is consistent with the hypothesis that he can no longer benefit from seeing faces as wholes.

3.7   General Conclusions

Starting with the clinical observations of people with brain damage, hypotheses about the functional architecture of visual object recognition were formulated and tested in controlled experiments with brain-damaged subjects. Questions about the nature of shape representation within this architecture were initially addressed using normal subjects, but as soon as some preliminary answers were obtained in this way, the experimental paradigms could be adapted for use with brain-damaged subjects and the linkage between types of representation and neural systems could be tested directly. Because our interest is in the functioning of the normal system, experiments with both brain-damaged and normal subjects are relevant to testing these hypotheses. Our decision to use a given population depends on theoretical and practical considerations.

Let us summarize what we have learned from the foregoing experiments. The selective impairment of face recognition in a prosopagnosic subject, LH, suggests that we are endowed with a specialized system for recognizing faces. This system is not necessary for (or is less important for) recognizing common objects, even when such objects form a large and visually homogeneous category. Furthermore, the system is anatomically distinct, in that it can be selectively damaged by head injury. Studies of subject CT suggest that this system can also be selectively disconnected from other brain areas. A pattern of impairment opposite to LH's was observed in an agnosic subject, CK, suggesting that the face recognition system does not merely elaborate the processing of the object system, but rather processes stimuli in parallel with it, and is at least partially functionally independent of the other system. How many specialized systems are there? Patterns of co-occurrence among disorders of face, object, and printed word recognition over many cases suggest that there are two underlying systems of representation. According to this interpretation, LH has moderate damage to one system—the one essential for face recognition, used for object recognition, and not needed for word recognition. CK has severe damage to the other system—the one essential for word recognition, used for object recognition, and not necessary for face recognition.

The two systems can be distinguished by the way they represent shape. Previous research has suggested that word recognition requires the ability to represent numerous shapes and that impaired visual word recognition results from a reduction in the number of shapes that can be represented within a short time. Research with normal subjects suggests that faces are recognized as single complex wholes that are not decomposed into separately represented parts. A final study with LH showed that the quality of his face perception was not dependent on the opportunity to perceive the face as a whole, which is consistent with the idea that he has an impairment in the holistic perception of faces. Referring back to the issue raised at the outset of this chapter, we can now offer a tentative answer: Face recognition and common object recognition depend on different systems that are anatomically separate, functionally independent, and differ according to the degree of part decomposition used in representing shape.

Suggestions for Further Reading

Farah, M. J. (1990). *Visual agnosia: Disorders of visual object recognition and what they tell us about normal vision.* Cambridge: MIT Press.

Farah, M. J., and G. Ratcliff (eds.). (1994). *The neuropsychology of high-level vision: Collected tutorial essays.* Hillsdale: Erlbaum Associates.

Heilman, K. M., and E. Valenstein (eds.). (1993). *Clinical neuropsychology* (3rd ed.). New York: Oxford University Press.

Shallice, T. (1988). *From neuropsychology to mental structure.* New York: Oxford University Press.

Problems

3.1 The neuropsychological studies described in this chapter all involved single cases, rather than groups of subjects. What are the advantages and disadvantages of this practice?

3.2 Why might the visual system have evolved different systems for recognizing different types of stimuli?

References

Biederman, I. (1987). Recognition-by-components: A theory of human image understanding. *Psychological Review* 94, 115–147.

Damasio, A. R., H. Damasio, and G. W. Van Hoesen (1982). Prosopagnosia: Anatomic basis and behavioral mechanisms. *Neurology* 32, 331–341.

Farah, M. J. (1990). *Visual agnosia: Disorders of object recognition and what they tell us about normal vision.* Cambridge, MA: MIT Press/Bradford Books.

Farah, M. J. (1991). Patterns of co-occurrence among the associative agnosias: Implications for visual object representation. *Cognitive Neuropsychology* 8, 1–19.

Farah, M. J., K. L. Klein, and K. L. Levinson (1995). Face perception and within-category discrimination in prosopagnosia. *Neuropsychologia* 33, 661–674.

Farah, M. J., and M. A. Wallace (1991). Pure alexia as a visual impairment: A reconsideration. *Cognitive Neuropsychology* 8, 313–334.

Farah, M. J., K. D. Wilson, H. M. Drain, and J. R. Tanaka (1995). The inverted inversion effect in prosopagnosia: Evidence for mandatory, face-specific processing mechanism. *Vision Research* 35, 2089–2093.

Johnston, J. C., and J. L. McClelland (1980). Experimental tests of a hierarchical model of word identification. *Journal of Verbal Learning and Verbal Behavior* 19, 503–524.

Morton J., and M. H. Johnson (1991). CONSPEC and CONLERN: A two-process theory of infant face recognition. *Psychological Review* 98, 164–181.

Pallis, C. A. (1955). Impaired identification of faces and places with agnosia for colors. *Journal of Neurology, Neurosurgery and Psychiatry* 18, 218–224.

Tanaka, J. W., and M. J. Farah (1993). Parts and wholes in face recognition. *Quarterly Journal of Experimental Psychology.*

Valentine, T. (1988). Upside-down faces: A review of the effect of inversion upon face recognition. *British Journal of Psychology* 79, 471–491.

Chapter 4
Visual Object Recognition
Irving Biederman

Try this experiment. Turn on your television with the sound off. Now change channels with your eyes closed. At each new channel, blink quickly. As the picture appears, you will typically experience little effort and delay (though there is some) in interpreting the image, even though it is one you did not expect and even though you have not previously seen its precise form. You will be able to identify not only the textures, colors, and contours of the scene but also the individual objects and the way in which the objects might be interacting to form a setting or scene or vignette. You will also know where the various entities are in the scene, so that you would be able to point or walk to any one of them if you were in the scene. Experimental observations confirm these subjective impressions (Intraub 1981; Biederman, Mezzanotte, and Rabinowitz 1982). People can usually interpret the meaning of a novel scene from a 100-millisecond (msec) exposure to it. However, they cannot attend to every detail; they attend to some aspects of the scene—objects, creatures, expressions, or actions—and not others. In this chapter, we focus primarily on our ability to recognize an object in a single glance on the basis of its shape.

Before we review the research and theory on object recognition, we will consider just what kinds of things a theory of object recognition might account for.

4.1 The Problem of Object Recognition

Object recognition is the activation in memory of a representation of a stimulus class—a chair, a giraffe, or a mushroom—from an image projected by an object to the retina. We would have very little to talk about in this chapter if every time we viewed an instance of a particular class it projected exactly the same image to the retina, as occurs, for example,

The writing of this chapter was supported by grants from the U.S. Air Force Office of Scientific Research (90-0274) and the McDonnell-Pew Foundation Program in Cognitive Neuroscience (T89-01245-029).

when the digits on a bank check are presented for reading by an optical scanner.

4.1.1   Pattern Variability

But there is a fundamental difference between reading digits on a check and recognizing objects in the real world: An object's orientation in depth can vary greatly, so that any one three-dimensional object can project an infinity of possible images onto the two-dimensional retina. We might see the object not only from a novel orientation but also when it is partially occluded by other surfaces—for example, behind foliage or draperies. Or the image of the object might fall on a different part of the retina or be of a different size. An object may be a novel instance of its class that does not exactly correspond to our previous experience, as, for example, when we view a new model of a chair or car. It is precisely this variation—and the apparent success of our visual system and brain at achieving recognition in the face of it—that makes the problem of pattern recognition so interesting.

4.1.2   Level of Classification

When we defined object recognition in the previous section as "the activation in memory of a stimulus class . . . ," we did not specify just what constitutes a class. If we look at an elephant, we can classify it at many levels of abstraction: as an entity, as a living thing, as an animal, as an elephant, as an Asian elephant, as Jumbo. You probably feel that elephant is the most natural class. But why?

Linguists have developed the concept of *basic level* to refer to the initial classification given to individual visual entities, for example a chair, a bird, or a mushroom. When shown a picture of a sparrow, most people answering quickly call it a bird not a sparrow or an animal. The basic level (bird) is a level of abstraction of visual concepts that maximizes between-category distinctiveness and within-category informativeness (Rosch et al. 1976). It can be distinguished from *subordinate* (sparrow) and *superordinate* (animal) levels of classification. Most of our knowledge of the visual world can be accessed through the basic level. Specifying the subordinate-level class—for example, that something is an African versus an Indian elephant or is a particular style of sofa—provides only a slight increase in informativeness at an enormous loss of distinctiveness. That is, the difference between an African and an Asian elephant is much smaller (and less significant) than the difference between an elephant and a sofa. (Face recognition, a special form of subordinate-level recognition, is discussed in Chapter 3.) The superordinate level, which classifies something as, for example, an animal or an article of furniture, sacrifices informativeness with only a slight gain in distinctiveness. The difference between (the classes) animals and furniture may be slightly greater than the differences, say, between an elephant and a rabbit or a sofa and a lamp, but that slight gain in distinctiveness comes at an enormous loss in the additional information we obtain from knowing that something is a lamp and not just an article of furniture, or an elephant and not just an animal. Basic-level terms are the first to enter a child's vocabulary, are used to a much greater extent than any other terms to describe objects, and are the highest level of abstraction whose objects share a characteristic shape (Rosch et al. 1976).

There are exceptions to the finding that people classify images more rapidly at the basic than at subordinate levels. Although a picture of a sparrow is classified as bird rather than a sparrow, a picture of a penguin is classified more quickly as a penguin than as a bird (Jolicoeur, Gluck, and Kosslyn 1984). The same holds true for ostrich, duck, and a number of other atypical instances of basic-level categories. Jolicoeur and his colleagues coined the term *entry level* to accommodate cases in which exemplars are initially classified at what would be, technically, a subordinate-level class. To a large extent, these exceptions have a different shape than the typical instances of the basic-level category. Some bird books display silhouettes of the entry-level subfamilies—ducks, songbirds (the prototypical basic-level class), hawks, and so on—to key the sections containing the subordinate-level information. In this chapter, we focus on the classification of an image into entry-level classes but also consider how subordinate-level classification might be accomplished.

4.1.3   How Many Entry-Level Objects Have to Be Modeled?

There are approximately three thousand entry-level terms for familiar concrete objects that can be identified on the basis of their shape rather than on surface properties of color or texture or on their position in a scene. These criteria, therefore, eliminate terms such as *fur* or *sand*. I arrived at this estimate by calculating the average number of entries per page meeting the criteria on a random sample of dictionary pages and multiplying it by the number of pages in the dictionary (Biederman 1987). This procedure yielded an estimate of approximately 1,600 terms, a result roughly consistent with linguists' estimates of the number of entry-level terms and naming words in the vocabulary of a six-year-old child. (A child of this age has a vocabulary of about ten thousand words, 10 percent of which are concrete nouns.) I then doubled this value to allow for idiosyncratic classes and objects not captured by the dictionary sample for a rough estimate of three thousand entry-level terms (or classes).

There may be an average of ten perceptual models for each of the three thousand entry-level, shape-based classes because (a) most objects require a few models for different orientations (such as the front and back of a house), and (b) some entry-level terms (such as lamp, house, or chair) have several readily distinguishable object models (Biederman 1988). Six-year-old children reveal full adult competence in naming the objects in their visual world; indeed, they often achieve naming competence by the age of three. As the six-year-old has been awake for about thirty thousand hours, my estimate indicates that the child learns a new object model at a rate of one per waking hour.

4.2   Representing the Image

The initial sensing of visual information is performed by the photosensitive cells (rods and cones) of the retina, which are activated by individual photons reflected by an object. Each receptor responds to photons from only a tiny portion (a few minutes of arc) of the visual field. The exact same pattern of receptor activation is never duplicated from one occasion of looking at an object to the next. Indeed, as noted in section 4.1.1, recognition can be quite tolerant of the considerable variability in an object's image caused by differences in viewpoint or occlusion. The object does not even have to be identical to one seen previously for us to achieve relatively effortless classification of its image. How is the activity of individual photoreceptors employed by the brain to create a representation of an object that allows it to be recognized under such highly varied conditions?

4.2.1   Representation of Shape Information in V1

Ganglion-cell neurons arising in the retina synapse with neurons in the lateral geniculate body that in turn send fibers to V1, the first visual cortical area to receive information employed for shape perception. Although the cells in the retina and lateral geniculate body have a center-surround organization—in that they respond best to a spot of light (or darkness) at an extremely small area (a few minutes of arc in the fovea) of the visual field—*simple cells* in V1, the first cortical visual area, respond to variation in luminance at a particular orientation (e.g., to a bar at a vertical orientation but not at a horizontal or oblique orientation). The tuning to orientation could arise from a mapping in which a V1 cell receives inputs from a collinear array of geniculate cells.

Simple cells respond to a restricted region (e.g., 0.5 to 2 degrees) of the visual field, for example a vertical dark "bar" with light-colored flanks 1 degree in length and 0.3 degrees in width that is centered 2 degrees left of fixation. (Cells that are tuned to larger-scale variations in luminance; e.g., a bar 1.5 degrees in length, would have larger receptive fields.) *End-stopped cells* in V1 respond maximally to an oriented stimulus (such as a bar) only if the stimulus terminates within the receptive field of the cell. End-stopped cells would presumably be maximally activated by contours that end at corners (or vertices).

Activation of simple and end-stopped cells are generally believed to provide the initial cortical activity of shape representation. Indeed, it would be possible to distinguish different shapes according to the differential activity of such cells. However, the identical shape presented at another position, size, or orientation also activates different cells; so we need some basis for representing shape that is not dependent on the particular V1 cells activated.

4.2.2   Invariant-Image Description

By a number of theoretical accounts, two related problems have to be addressed in order to form a representation allowing for invariant recognition. One problem is *grouping* or *binding*. When viewing an object like the one shown in Figure 4.1, we subjectively group contours *a* and *b* as part of one component, the brick, and *c* and *d* as part of another component, the cone, even though *a* and *c* are closer together and more similar in orientation than *a* and *b*. What principles allow such grouping?

A second problem is that of invariant description. It is particularly useful to have a representation that is the same whatever the viewpoint. We could then be fairly confident of what the object in Figure 4.1 looks like when it is rotated 30 degrees in depth. What information do we need to

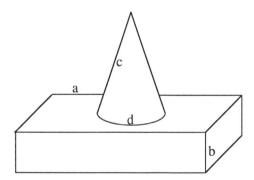

Figure 4.1
A vertical cone on a horizontal brick. This chapter concerns how we identify this image even though we probably have never seen it before. Why do we group segments *a* and *b* as part of one entity and *c* and *d* as part of another, despite the greater proximity of *a* and *d* (or *a* and *c*) and their greater similarity in orientation? (Adapted by permission of the publisher and authors from J. E. Hummel and I. Biederman, Dynamic binding in a neural network for shape recognition, 1987, *Psychological Review* 99, Figure 10, 489. Copyright 1987 by the American Psychological Association.)

do this? Objects in the real world have color, texture, and surface markings, but these sources of information are absent in the figure. There is some evidence that our capacity to recognize an object from different viewpoints is dependent on discontinuous edges of two types. Edges marking *orientation discontinuities* are formed by a sharp change in the orientation of abutting surfaces, such as occurs with the adjacent sides of a brick (segments *b* or *d* in Figure 4.1). Edges marking *depth discontinuities* are typically formed when one's line of sight grazes (that is, is tangent to) a curved surface so that there is a sharp jump in depth from surface to the background, as occurs with segment *c* in Figure 4.1. Sometimes the two types of edges coincide, as they do in segment *a*. A line drawing representing only these kinds of discontinuities (which can arise from differences in luminence, texture, color, etc.) can convey much of the three-dimensional shape of an object, as Figure 4.1 readily demonstrates. But how is it that a line drawing can convey the shape of an object? Or the fact that Figure 4.1 is a cone on top of a brick?

Subsections 4.2.3 and 4.2.4 describe the image information that might be employed to solve problems of viewpoint invariance and part structure. In section 4.3 we review theories about how neural computations might exploit this information.

4.2.3   Viewpoint-Invariant Properties

Viewpoint-invariant properties play a significant role in deriving a three-dimensional world from a two-dimensional image. Figure 4.2 illustrates several properties of image edges that are extremely unlikely to be a consequence of the particular alignment of eye and object. If the observer changes viewpoint or the edge or edges change orientation, assuming that the same region of the object is still in view, the image will still reflect that property. For example, a straight edge in the image is perceived as being a projection of a straight edge in the three-dimensional world. The visual system ignores the possibility that a (highly unlikely) accidental alignment of eye and a curved edge is projecting the image. Hence such properties have been termed *nonaccidental* (Lowe 1984). On those rare occasions when an accidental alignment of eye and edge does occur—for example, when a curved edge projects an image that is straight—a slight alteration of viewpoint or object orientation readily reveals that fact.

Figure 4.2 illustrates several nonaccidental properties. In the two-dimensional image, if an edge is straight (collinear) or curved, it is perceived as a straight or curved edge, respectively. If two or more two-dimensional image edges terminate at a common point, or are approximately parallel or symmetrical, then the edges projecting those images are similarly interpreted. For reasons that will be apparent when we consider some theories of object recognition, Figure 4.2 presents these viewpoint-

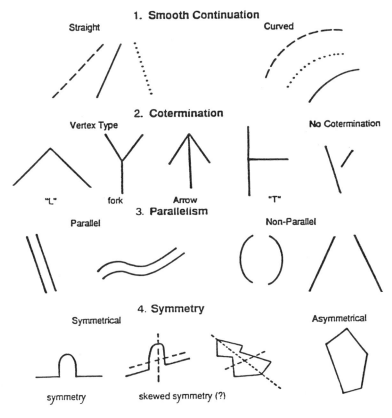

Figure 4.2
Contrasts in some viewpoint-invariant relations. In the case of parallelism, biases toward parallel and symmetrical percepts when images are not exactly parallel or symmetrical are evidenced. (Adapted by permission from D. Lower, Perceptual organization and visual recognition, Unpublished doctoral dissertation, Stanford University, 1984, Figure 4.2, 77.)

invariant properties as dichotomous *contrasts* (or differences). Any one edge can be characterized as straight or curved. We can describe the relation of two or more edges as coterminating or noncoterminating or parallel or nonparallel. The number of coterminating edges and whether they contain an obtuse angle also does not vary with viewpoint and can serve as a viewpoint-invariant classification of vertex type—L, Y (or fork), or arrow (or their curved counterparts) in Figure 4.2. In a strict sense, parallelism and symmetry varies with viewpoint and orientation, as occurs, for example, with perspective convergence. But there is a clear bias toward interpreting approximately parallel edges as parallel, especially when the surfaces are perceived as varying in depth (Ittelson 1952; King, Meyer, Tangney, and Biederman 1976). Within a tolerance range defined by the

128   Biederman

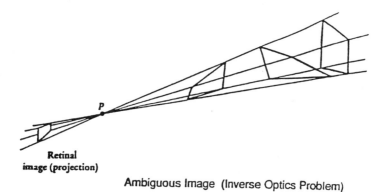

Retinal
image (projection)

Ambiguous Image (Inverse Optics Problem)

(b)

(c)

Figure 4.3
The Ames peephole perception demonstrations. (a) Illustration of the inverse-optics problem: A single image can be produced by an infinity of possible real-world objects. (b) Three
stimulus arrangements constructed by the Ames group. The upper left panel shows the
perspective lines from the peephole at the lower right. (c) The percepts from the stimuli in
the upper panels. The three stimulus arrangements produce identical percepts. (Adapted by
permission of publisher and author from R. N. Haber and M. Hershenson, *The psychology of
visual perception*, 1981, Figure 12.5, 284. Copyright 1981 by Holt, Rinehart, & Winston.)

cues for surface slant, pairs of image edges that could be parallel or symmetrical, given uncertainty as to the actual orientation of the edges to the eye, are interpreted as parallel or symmetrical (King et al. 1976), as suggested by Figure 4.3.

The psychological potency of these viewpoint-invariant properties was demonstrated when Ames and his associates constructed a set of peephole perception demonstrations in which subjects viewed three arrangements of wires through a peephole, as shown in Figure 4.3b (Ittleson 1952). Although all three stimulus arrangements shown projected the identical image of a chair, as shown in Figure 4.3c, in only one of them (the left-hand one) did the wires actually form a chair. In the middle arrangement the segments all had the same cotermination points as the chair, except that the surfaces were no longer parallel. In the right-hand arrangement the segments did not even coterminate, yet the perception of this stimulus was indistinguishable from the other two. (Peephole viewing eliminates cues for stereoscopic vision, motion parallax, and image variation that would have resolved the accidents of viewpoint.) These results provide strong evidence that the viewpoint-invariant properties shown in Figure 4.1 and the biases toward parallelism and symmetry are immediate and compelling and could thus serve as a basis for characterizing image edges for purposes of recognition.

4.2.4   Decomposing Complex Objects into Parts

Complex visual entities almost always invite a decomposition of their elements into simple parts. We readily distinguish the legs, tail, and trunk of an elephant or the shade from the base of a lamp. People's spontaneous descriptions of basic-level classes almost always include a specification of distinctive parts (Tversky and Hemenway 1984). The manner of the decomposition into parts does not depend on familiarity with the object in that different observers agree on the part decompositions of nonsense shapes (Biederman 1987; Connell 1985; Kimia, Tannenbaum, and Zucker 1992). Nor does the part decomposition depend on surface color or texture as the part structure is readily perceived in line drawings.

In general, whenever there is a pair of matched cusps (discontinuities at minima of negative curvature), people will express a strong intuition that the object should be segmented at that region (Connell 1985). This tendency of the visual system to segment complex objects at regions of matched concavities is not an arbitrary bias. Hoffman and Richards (1985) note a result from projective geometry—the *transversality principle*—that whenever two shapes are combined, their join is almost always marked by matched cusps, as illustrated in Figure 4.4a. (The cusp projects an L-vertex that will be largely viewpoint invariant.) Segmenting at such regions provides a basis for appreciating the part structure of objects, as shown for the

(a)

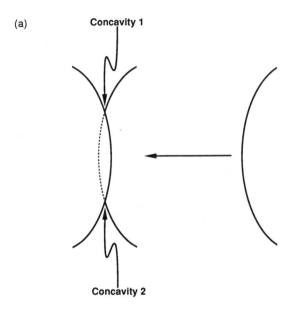

Concavity 1

Concavity 2

(b)

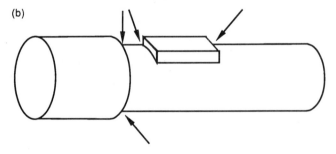

Figure 4.4
An illustration of the transversality regularity and how it can be applied to the segmentation of an object's parts.

flashlight in Figure 4.4b. Siddiqi, Tresness, and Kimia (1994) provide evidence that a narrowing of a shape without minima of negative curvature, which they call a *neck*, provides another basis for part decomposition. Indeed, an animal's neck provides a natural parsing region for separating the shoulders from the head. Matched cusps (or, more weakly, minima at negative curvature) and necks may provide much of the basis for the Gestalt principle of a good figure. If a shape is segmented at paired cusps or necks, the resulting parts will be convex or only singly concave. Such parts appear simple.

4.3   Theories of Object Recognition

Two major problems must be addressed by any complete theory of object recognition. The first is how to represent that information in the image so that it can activate a representation in memory under varied conditions. The second problem is how that stimulus representation is matched against—or indexes or activates—a representation of an object in memory.

With respect to the issue of representation of information in the image, different theories can be ordered along a continuum, according to the degree to which the image information is elaborated prior to matching (Dickinson, Pentland, and Rosenfeld 1992). Dickinson and colleagues refer to this continuum as *primitive complexity*. At one end of the continuum are simple points. Theories that posit the matching of such points exploit few of the principles of invariance or part decomposition described in section 4.2. Next on the continuum are schemes in which points are grouped into contours to provide a more complex primitive; even more complex primitives are groups of contours or surfaces; the most complex primitives of all are simple volumes. Dickinson et al. note a trade-off among various models between the ease of determining the indexing primitive and the ease of indexing an object model: the simpler the indexing primitive, the easier it is to determine that primitive but the more difficult it is to index an object from it. Thus the luminance of a small patch of points is easy to determine, but it is difficult to index an object from that patch. Once we know the convex volumes (parts and their relations) that might comprise an object, it is relatively easy to determine which object has those parts, but it can be very difficult for current vision systems to determine the convex volumes present in the image.

Theorists who have opted for simple primitives tend to focus on developing models that more readily allow activation of object representations from those primitives. Those who assume more complex primitives focus on schemes for more efficient and accurate extraction of the primitives from the image.

In this section we consider theories from three points along this continuum: (a) models that attempt recognition based on the outputs of simple cells (activated by small patches of pixels), either directly (Lades, Vorbrüggen, Buhmann, Lange von der Malsburg, Würtz, and Konen (1993) or with an intervening layer (Poggio and Edelman 1990); (b) a model by David Lowe (1987) of object recognition based on nonaccidental configurations of edges; and (c) a model by Irving Biederman and associates (Biederman 1987; Hummel and Biederman 1992) that assumes simple volumetric primitives roughly corresponding to an object's parts. The theories differ not only in the complexity of their matching primitives but in other characteristics as well. We also comment on these other characteristics when describing the theories in our overview in section 4.5.

One of the major advances in cognitive science over the past decade has been the development of theoretical formalisms, *neural networks*, that allow the expression of symbolic activity in terms of a pattern of activation over an aggregate of connected neuron-like elements. Several of the models considered in this section are of this type.

4.3.1   Matching of Simple Cell Outputs

The Lades et al. Face-Recognition System

Christoph von der Malsburg and his associates (Lades et al. 1993) initially developed their model as a face-recognition system, and it has enjoyed considerable success at that task. The model can be represented as a two-layer network, as illustrated in Figure 4.5. The input (or image) layer consists of an array of columns of individual units (or kernels), each roughly corresponding to a V1 simple cell. As described in section 4.2.1, a particular cell is tuned to variation in luminance at a particular orientation at a particular scale (i.e., spatial frequency) at a particular position of the visual field. The tuning of a simple cell can be approximated mathematically by a Gaussian-damped, sinusoidal filter termed a *Gabor filter*. A column of these filters, each tuned to different orientations and scales but with maximum responsiveness centered on the same region of the visual field (e.g., all those cells whose receptive fields are centered at 2 degrees left of fixation), is termed a *Gabor jet*. It roughly corresponds to the simple cells of a V1 hypercolumn.

In Figure 4.5 the Gabor jets are illustrated as a stack of disks centered at a single position in the visual field. The jets are arranged in a lattice, with each node of the lattice designating the center of the receptive field for a jet. In the implementation described here, each jet consists of filters at five scales and eight orientations (therefore, at 45-degree intervals) so that at each node forty filters comprise each jet. There are 5 × 9 nodes (jets) in the lattice. Other parameters could have been employed but these are

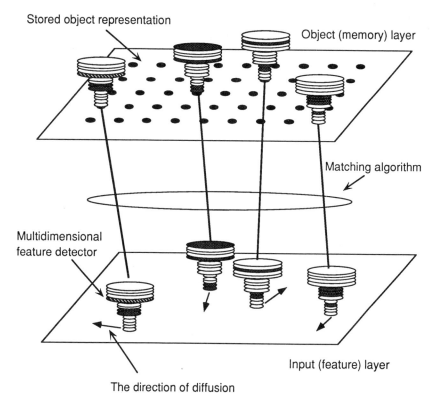

Stored object representation

Object (memory) layer

Matching algorithm

Multidimensional
feature detector

Input (feature) layer

The direction of diffusion

Figure 4.5
The architecture of the Lades et al. (1989) recognition system. Shown here are four of the
Gabor jets, each composed of a set of filters (each represented by a disk in the stack) from
a regular 5 × 9 matrix of jets. The filters differ in scale (spatial frequency) and orientation
tuning. Activation values of the original image are stored in the object layer. The figure
depicts the diffusion of the jets in the input layer, as indicated by the arrows, when that
layer is activated by a new image and the matching algorithm attempts to find the best
match against a previously stored image. (Reprinted with permission of the authors from
Fiser, Biederman, and Cooper [1994]. Copyright by József Fiser and Irving Biederman.)

sufficient for reasonably accurate face recognition, which was the original
goal of the system. The receptive fields of the largest filters are considerably larger than those indicated in Figure 4.5 in that they are affected by
luminance variation approximately two nodes away from the center of
their receptive fields.

A particular image results in activation of the different filters to various
extents. These values are stored along with the relative positions of the
adjacent jets. Figure 4.6a shows an image, a face, with the lattice superimposed over it. A new image is matched against the original by the

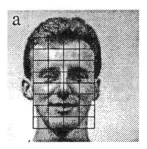
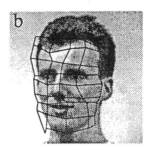

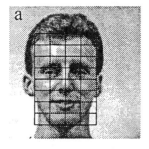
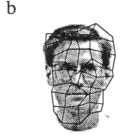

Figure 4.6
An example of the deformation of the Gabor-jet lattice when matching faces in the Lades et al. (1993) recognition system. (a) The target face stored in a gallery of images of the faces of 56 individuals, showing the original positioning and regularity of a lattice. (b) The upper face is an image of the same individual from a different orientation and slightly changed expression. The lower face is of a different individual at approximately the same orientation as the original. The superimposed meshes when the *b* images are matched against the *a* image show the degree of deformation for each of the matches. In this example, the system correctly matched the upper (b) face with the (a) face. (Reprinted with permission of the authors from Fisher, Biederman, & Cooper [1994]. Copyright by Józef Fiser and Irving Biederman.)

individual jets that, when activated by the new image, diffuse (gradually change their positions) to determine their own best fit, as illustrated by the arrows on the jets in the input layer in Figure 4.5. With faces, the same individual can be in a different orientation and expression (as shown in the upper panel of Figure 4.6b) or be the image could be of a different individual (as shown in the lower panel of Figure 4.6b). Although details of this matching algorithm are beyond the scope of this chapter, we note that similarity of a pair of images is (a) a positive function of the similarity of the activation values of the Gabor filters for corresponding jets (i.e., the jet in the third row, fourth column), and (b) a negative function of the degree to which a given jet has to be displaced, relative to its immediate neighbors, to find its best match in a new image. To the extent that the jets move independently, the resultant positions will depart from the original, regular positions, as suggested by the different directions of movement of the jets in Figure 4.5 and as shown in the deformed mesh in the upper and lower panels of Figure 4.6b (see also Figure 4.16). Typically, the greater the deformation of the lattice, the lower the similarity of the match. A test image is compared against a number of stored images. The most similar image is taken to be the recognition choice. There is no reduction in similarity if the face or object appears at a position in the visual field other than where it first appeared and or if it is of a different size. In that case the lattice just has to be repositioned or expanded or contracted with little or no distortion of the original positions of the jets. Similarly, variations in the overall illumination levels are factored out, although differences in the direction of illumination for two images of the same person reduce similarity.

In a test of the Lades et al. system, researchers prepared fifty-six pairs of images of the faces of fifty-six individuals. One image of each person was sorted in a gallery of faces and recognition was attempted with the other member of the pair (which could differ in expression and orientation). The average ranking of the correct face was 1.4 (chance would have been 27.5) (Fiser, Biederman, and Cooper 1994). In section 4.5 we will evaluate this system as an object recognizer.

Because activation values are dependent on the specific view or aspect of the object, the Lades et al. and the Poggio and Edelman (1990) models are said to be *view (or aspect) based*. As different views of an object are encountered, the system builds up different patterns of activation of the hidden units that represent the different poses. What happens if the object is seen from a slightly different view? To the extent that the new image is similar—in terms of the pattern of filter activation values—to a previously learned view, the model might exhibit graded generalization to the new view if the unit in the output layer corresponding to that object is more activated than units representing other objects. It is important to note that in the test of face recognition the rotation in depth was limited to approximately 30 degrees (as illustrated in Figure 4.6). With rotations beyond that value, the accuracy of face recognition declined significantly. The Poggio and Edelman (1990) model was designed, in part, to increase the capacity of a filter-matching model to handle greater variations of rotation in depth.

The Poggio and Edelman Radial Basis Function Model

The Lades et al. model attempts to match filter outputs directly to an object representation layer. Poggio and Edelman (1990) assume a first stage that is similar to that of the Lades et al. model (in that it does a simple filtering of the image); in addition, they assume a single hidden layer between that input stage and an output stage. Units in the hidden layer self-organize to take weighted activation values of the L1 filters to distinguish among a set of stimuli learned by the network. In the hidden layer of the network proposed by Poggio and Edelman (1990) these units are termed *radial basis functions* (RBFs); as they are designed to allow optimal classification of an image, a minimal number of these units allow classification of a large number of possible images. This model, then, provides a basis for determining when a new representation might be needed. In one exercise (Poggio and Edelman 1990), only two RBF units were sufficient to recognize ten to forty views of a bent paper clip over a 90-degree range of orientation. The object layer in the Lades et al. model is a representation of a particular view of an object, whereas the RBFs that emerge from experience with a series of views of an object need not (and typically do not) correspond to any particular view. The RBF thus constitutes a prototype for a modest range of views or deformations of an object.

The RBFs belonging to a single object are linked so that together they form a set of prototypes for an object. A significant challenge for the Poggio and Edelman model, however, is determining which object is projecting a new image so that an existing RBF can be modified, a new one created, or different RBFs linked. Currently, the model must be informed of this by some other system (or the programmer). In section 4.5 we consider several empirical tests of whether human object recognition can be predicted from filter outputs in the manner assumed by this class of models.

4.3.2 Model-Based Matching of Edges

The two models described in the previous section are pure "bottom-up" systems in that they assume a one-way flow of information from the initial image filtering to the representation of an object (or face). With an extremely large set of possible objects in a gallery or with variations in the shape and orientation of test objects to be matched against stored images, the speed and accuracy of correct recognition can decline greatly. A number of theorists have proposed schemes that reduce the degree of matching required by considering only those objects in memory that share certain features, which are initially extracted from the image, and only those poses of the object that are consistent with those features. Lowe (1987) offers a detailed proposal for how such a system might work. Whereas Lowe's proposal is limited to images with straight edges, Ullman's (1989) model, which has somewhat similar characteristics, has the potential for recognizing a broader class of objects, including those with curved surfaces. With both systems, a fully three-dimensional model of an object is stored rather than a large set of representations each based on a different view. Ullman's model employs an initial extraction of features to determine the precise orientation and scale of the object model to be matched against the image.

Lowe's SCERPO model is directed primarily toward determining the orientation and location of objects, even when they are partially occluded by other objects, under conditions in which exact three-dimensional object models are available. The SCERPO takes as input an image such as the one shown in Figure 4.7a, an image of a number of disposable razors in arbitrary orientations. The model detects edges by finding sharp changes in image intensity values across a number of scales (as discussed in Chapter 1). The results of this edge-detection stage are shown in Figure 4.7b. The edges are then grouped according to the viewpoint-invariant properties of collinearity, parallelism, and cotermination. A few of these image features are then tentatively matched against image features of the object model generated from a particular orientation of the object that would maximize the fit of those image features. From this initial hypothesis, the locations of additional image features (edges) are proposed and their presence in the image evaluated. Figure 4.7c shows the successful final matches for five orientations of the razors. These matches provide segments not detected initially by the edge finder (middle panel) and discard edges initially detected but not part of the object model (e.g., the glare edges on the handle of the razor extending horizontally in the lower part of the figure). SCERPO may provide a plausible scheme for characterizing human performance under conditions in which the initial extraction of image edges is uncertain, as in conditions of poor visibility or where the orientation of an object is unfamiliar.

Ullman's (1989) Alignment model first reorients all the object models that might be possible matches for the image and tests them for the fit of the image against the aligned models in memory. The alignment capitalizes on the formal result that three non-coplanar points are generally sufficient to determine the orientation of any object. In practice, the three points are typically viewpoint invariant in that they are selected at a point where there is a cotermination of edges. However, any salient points, or even general features, would be sufficient for alignment. Although it appears unlikely that people rotate (align) all possible candidate models in memory prior to matching, the alignment model offers a possible account of those cases in which recognition depends on reorienting a mental model. Ullman and Basri (1990) present a general theory of how a three-dimensional object can be represented as a combination of two-dimensional images so that it is recognized under such transformations as rotation in depth and non-rigid transformations.

(a)

(b)

Figure 4.7
Lowe's viewpoint consistency model can find objects at arbitrary orientations and occlusions. (a) The original image of a bin of disposable razors. (b) The straight line segments that SCERPO derived from the image. (c) Final set of successful matches between sets of image segments and five particular viewpoints of the model (shown as bright dotted lines). (Reprinted by permission of the publisher and author from D. Lowe, The viewpoint consistency constraint, 1987, *International Journal of Computer Vision* 1, 66, 70, Figures 4, 5, and 8. Copyright 1987 by Kluwer Academic Publishers.)

(c)

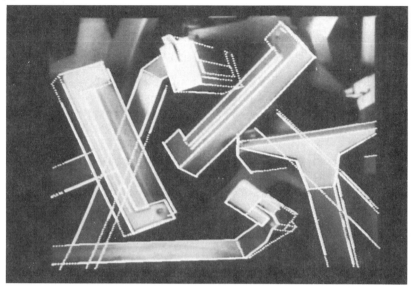

Figure 4.7 (cont.)

Matching Viewpoint-Invariant Parts

Biederman (1987; Hummel and Biederman 1992) proposed a theory of entry-level object recognition that assumes that a given view of an object is represented as an arrangement of simple, viewpoint-invariant, volumetric, primitives called *geons*. Five (of the twenty-four) geons are shown in the left panel of Figure 4.8. The relationships among the geons are specified, so that the same geons in different relations will represent different objects, as with the cup and pail in the right panel of Figure 4.8. The geons have two particularly desirable properties: they can be distinguished from each other from almost any viewpoint, and their identification is highly resistant to visual noise. We will consider in greater detail the segmenting of the image into regions to be matched with geons, the description of the image edges in terms of viewpoint-invariant properties, and the geon arrangement that emerges from the parsing and edge processing.

Geons from Viewpoint-Invariant Edge Descriptions

According to RBC, each segmented region is approximated by a geon. Geons are members of a particular set of convex or singly concave volumes that can be modeled as *generalized cones*, a general formalism for representing volumetric shapes (Binford 1971; Brooks 1981). A generalized cone is the volume swept out by a cross section moving along an axis.

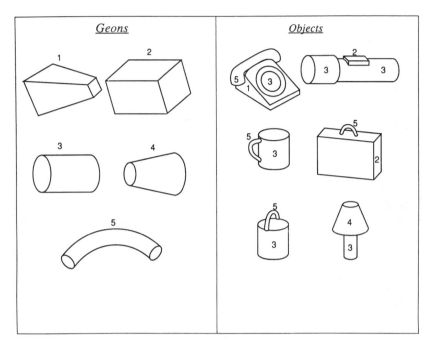

Figure 4.8
(Left) Five geons. (Right) Only two or three geons are required to uniquely specify an object. The relations among the geons matter, as illustrated by the pail and the cup.

The set of geons is defined so that they can be differentiated on the basis of dichotomous or trichotomous contrasts of viewpoint-invariant properties to produce twenty-four types of geons. The contrasts of the particular set of nonaccidental properties shown in Figure 4.2 were emphasized because they may constitute a basis for the generation of this set of perceptually plausible components. Figure 4.9 illustrates the generation of a subset of the twenty-four geons from contrasts in the nonaccidental relations of four attributes of generalized cones. Three of the attributes specify characteristics of the cross section: curvature (straight versus curved), size variation (constant [parallel sides], expanding [nonparallel sides], expanding and contracting [nonparallel sides with a point of maximum convexity]), and whether nonparallel-sided geons terminate in a point (in which case they have an L-vertex) or are truncated (in which case they will have arrow vertices). One attribute—straight versus curved—specifies the axis. These are modal types. It is possible that a given region of the image activates two or more geons, depending on the presence of particular image features.

When the contrasts in generating functions are translated into image features, it is apparent that the geons have a larger set of distinctive

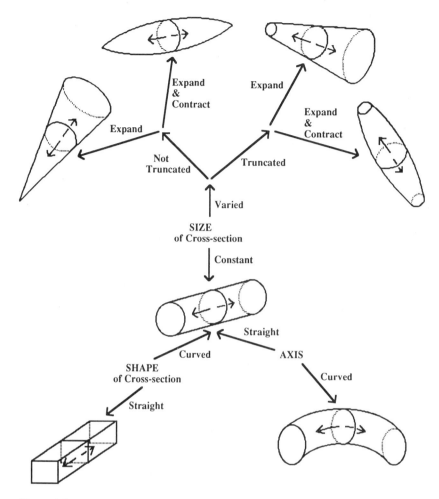

Figure 4.9
An illustration of how variations in three attributes of a cross section (curved versus straight edges; constant versus expanded versus expanded and contracted size; symmetrical versus asymmetrical) and one attribute of the shape of the axis (straight versus curved) can generate a set of generalized cones differing in nonaccidental relations. Constant-sized cross sections have parallel sides; expanded or expanded and contracted cross sections have sides that are not parallel. When the sides are not parallel they could be truncated (as with the cone) or end at a point (L-vertex). Curved versus straight cross sections and axes are detectable through collinearity or curvature. Shown here are the neighbors of a cylinder. The full family of geons has 24 members. (Adapted by permission of publisher and author from I. Biederman, Recognition-by-components: A theory of human image understanding, 1987, *Psychological Review* 94, 122, Figure 6. Copyright 1987 by the American Psychological Association.)

Brick   Cylinder

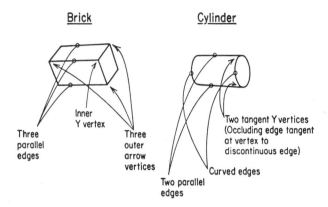

Figure 4.10
Some nonaccidental differences between a brick and a cylinder. (Reprinted by permission of the publisher and author from I. Biederman, Recognition-by-components: A theory of human image understanding, 1987, *Psychological Review* 94, 121, Figure 5. Copyright 1987 by the American Psychological Association.)

nonaccidental image features than the four that might be expected from a direct mapping of the contrasts in the generating function. Figure 4.10 shows some of the nonaccidental contrasts distinguishing a brick from a cylinder. The silhouette of a brick contains a series of six vertices, which alternate between Ls and arrows, and an internal Y-vertex. The vertices of the silhouette of the cylinder, by contrast, alternate between a pair of Ls and a pair of tangent Ys. The internal Y-vertex is not present in the cylinder (or in any geon with a curved cross section). These differences in image features would be available from a general viewpoint and, thus, could provide, along with other contrasting image features, a basis for discriminating a brick from a cylinder. Dickinson, Pentland, and Rosenfeld (1992) provide an extensive account of how nonaccidental configurations can be employed to determine a particular geon. Zerroug and Nevatia (1995) have derived a number of viewpoint-invariant properties of generalized cylinders that allow recovery of volumetric shape from a gray-level image. Interestingly, a number of these properties only hold if the volume can be described as a generalized cylinder with a cross section orthogonal to the axis.

Deriving the geons from contrasts in viewpoint-invariant properties renders the geons themselves largely invariant under changes in viewpoint. (Current theoretical work is exploring a redefinition of the geons in terms of those volumes that are maximally viewpoint invariant. Most likely the resultant volumes will largely correspond to the present set.) Because the geons are simple (viz., convex or only singly concave), lack sharp concavities, and have redundant image properties, they can be readily restored in the presence of visual noise. Therefore objects that are represented as an arrangement of geons will possess the same invariance to viewpoint and noise. Geon activation requires only categorical classification of edge characteristics for processing to be completed quickly and accurately. A representation that requires fine metric specification, such as the degree of curvature or length of a segment, cannot be performed with sufficient speed and accuracy by humans to be the controlling processing for object recognition.

Geon Relations and Geon Attributes

Much of the capacity to represent the tens of thousands of object images that people can rapidly classify from a small alphabet of geons derives from several viewpoint-invariant relations between pairs of geons and some coarse metric attributes of individual geons. Examples of relations that have been hypothesized are vertical position (above, below, beside), join type (end-to-end, end-to-middle centered, end-to-middle off-centered), relative size (larger, smaller, equal to), and relative orientation (parallel, orthogonal, oblique). These relations are defined for joined pairs of geons so that the same subset of geons represent different objects if they are in different relations to each other—like the cup and pail in Figure 4.8. Also specified are two coarsely coded metric aspects of the geons: (a) the relative-aspect ratio of the geon (five levels of the length of the axis compared to the diameter of the cross section) and (b) the orientation of the geon (e.g., vertical, horizontal, or oblique). There are eighty-one combinations of pairwise relations and fifteen attributes. A representation that specifies parts (geons), attributes, and relations independently and explicitly is termed a *structural description*.

Three-Geon Sufficiency

Object space and three-geon sufficiency. With twenty-four possible geons, eighty-one combinations of relations, and fifteen attributes, the variations in relations and aspect ratio can produce 10,497,600 possible two-geon objects ($24^2 \times 15^2 \times 81$). A third geon, with its possible attributes and its relations to one other geon, yields over 306 billion possible three-geon objects. This is two orders of magnitude greater than the number of seconds in a hundred-year lifetime.

If the 30,000 familiar object models estimated in section 4.1.3 are distributed homogeneously throughout the space of possible object models, then the extraordinary disparity between the number of possible two- or three-geon objects and the number of objects in an individual's object vocabulary—even if the estimate of 30,000 is short by an order of magnitude—means that an arrangement of two or three geons would almost always be sufficient to specify any object.

The theory thus implies a *principle of geon recovery*: if an arrangement of two or three geons can be recovered from the image, objects can be quickly recognized even when they are occluded, rotated in depth, novel, extensively degraded, or lacking customary detail, color, and texture. Experimental results support this expectation of geon theory (Biederman 1987): When only two or three geons of a complex object (such as an airplane or elephant) are visible, recognition can be fast and accurate (although, predictably, not as fast as with a complete image). You can try this for yourself by covering up parts of pictures of common objects. See if the object remains recognizable to a friend (who did not see the original) when only two or three parts are in view. The simple line drawings of objects shown in Figure 4.8 illustrate this expectation of three-geon sufficiency.

Just as three geons are usually sufficient for classification, objects composed of a single geon are often appropriate for several entry-level objects. In these cases, other information—such as color, texture, small details, or context—are required for classification (Biederman and Ju 1988). For example, distinguishing among a peach, a nectarine, and a plum requires that surface color and texture be specified. The expectation from geon theory would be that the identification of single-part objects would require more time than objects with distinctive geon configurations, as well as, to a much greater extent, specification of color and texture. Biederman, Hilton, and Hummel (1991) confirmed these expectations.

A Neural-Net Implementation of Geon Theory

Hummel and Biederman (1992) proposed a neural-net implementation of geon theory.

Problems and goals of the implementation. As discussed in section 4.2.1, the representation of the image at the first cortical stage, V1, probably consists of activation of a large number of cells, each tuned to variation of luminance at a particular orientation in a small (0.5 to 2.0-degree) region of the visual field. How could a structural description specifying parts (geons) and relations be derived from this activity? Moreover, given the

---

ent diagonals (*d*), and four different horizontals (*h*); horizontal position in the visual field (*Horiz. Pos.*), left (*l*) to right (*r*); vertical position in the visual field (*Vert. Pos.*), bottom (*b*) to top (*t*); and *size*, *small* (near 0 percent of the visual field) to *large* (near 100 percent of the visual field). Layers 4 and 5 represent the relative orientations, locations, and sizes of the geons in an image. Cells in layer 6 respond to specific conjunctions of cells activated in layers 3 and 5, and cells in layer 7 respond to complete objects, defined as conjunctions of cells in layer 6. (From J. E. Hummel, and I. Biederman, 1992, *Psychological Review*, 99, 486, Figure 7. Copyright 1992 by the American Psychological Association.)

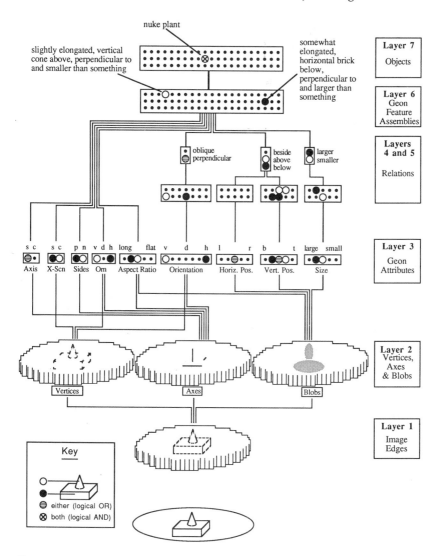

Figure 4.11
The architecture of the Hummel and Biederman (1992) neural-net implementation of geon theory, indicating the representation activated at each layer by the image in the key. In layers 3 and above, large circles indicate cells activated in response to the image and dots indicate inactive cells. Cells in layer 1 represent the edges (specifying discontinuities in surface orientation and depth) in an object's image. Layer 2 represents the vertices, axes, and blobs defined by conjunctions of edges in layer 1. Layer 3 represents the geons in an image in terms of their defining dimensions: Axis shape (*Axis*), straight (*s*) or curved (*c*); cross section shape (*X-Scn*), straight (*s*) or curved (*c*); whether the *Sides* are parallel (*p*) or nonparallel (*n*); coarse orientation (*Orn*), vertical (*v*), diagonal (*d*), or horizontal (*h*); aspect ratio, elongated (*long*) to flattened (*flat*); fine orientation (*Orientation*), vertical (*v*), two differ-

evidence for position, size, reflection, and depth invariance (described in section 4.4.3), how could the same description be derived when completely different cells are activated because of a change in viewpoint? The Hummel and Biederman network offered a possible answer to this question.

The network, whose overall architecture is shown in Figure 4.11, takes as input a line drawing representing the orientation and depth discontinuities of an object and activates units in the seventh layer (L7) that represent a viewpoint-invariant structural description of the object specifying its geons, geon attributes and the relations between geons. This description is activated regardless of whether the model has previously been exposed to the object. The model is meant to be a working hypothesis and is admittedly incomplete.

The binding problem. In the network, V1 is roughly modeled by a lattice of units in L1 that code whether a given contour is straight or curved and whether it passes through or terminates (is end-stopped) in the receptive field of a particular end-stopped cell. The end-stopped activity of two or more contours at a common point in the visual field activates L2 units representing such vertices as forks, arrows, and Ls.

The *binding problem*—determining what goes with what—is a major problem that needs to be solved to achieve invariant recognition. Consider again Figure 4.1, which depicts a vertical cone on top of a horizontal brick. We readily segment this object into two parts, the brick and the cone. In section 4.2.4 we considered what information the visual system might employ to segment an object into its parts. Here we focus on how grouping itself might be represented. If we have, in an oversimplified case, four neurons activated by *a, b, c,* and *d,* how does the neural activity code *a* and *b* to one group and *c* and *d* to another? Hummel and Biederman's solution (1992) was to induce the units activated by the contours of one geon to fire (approximately) synchronously and the units activated by another geon to fire synchronously but out of phase with each other. The signal that induces the synchronization is passed through links of nearby units that have collinear or parallel receptive fields or, for end-stopped cells, units that coterminate or have complementary orientations. In Figure 4.1, the synchronization would run around the vertices of each of the geons (through end-stopped cells) but would not pass from segment *a* of the brick to segment *d* of the cone because *a* and *d* do not coterminate. Thus the activity does not pass from the stem to the top of a T-junction. Instead, end-stopped cells activated by *a* and its extension on the other side of the cone would be synchronized because they have complementary orientations. Details of the algorithms that induce the synchrony are discussed in Hummel and Biederman (1992).

Using temporal asynchrony for representing geon attributes and relations. The units activated in L1 activate units in L2 representing vertices, axes, and blobs, providing information about the approximate size of the part and its center of mass. The L2 units are enumerated throughout the visual field so that a given vertex detector—for example a Y-vertex at a given orientation—will be available for each hypercolumn. All the L2 units, in turn, activate a single set of geon attribute units in L3. For example, all Y-vertex units send activation to the unit representing a straight cross section. The temporal correlation in the firing of the units activated by one object part induced in L1 and L2 is maintained through the first six layers of the model. For the example shown in Figure 4.11, all the units marked by filled circles represent information about the part that is a brick. All these fire together and out of phase with the units marked by the open circles representing the cone. Only thirty-six cells in the L3 layer are required to specify information for each part such as the geon type (one of eight possibilities) its orientation, and aspect ratio. The binding is thus achieved without positing additional units for "anding" that would, for example, posit a small vertical cone detector for each position of the visual field. Because the binding is temporary, these same cells can be used to code other parts of the object as well as the parts of other objects, no matter where they are in the visual feld.

L4 and L5 derive invariant relations of vertical position, size, and relative orientation. The temporal correlation is maintained so that the "above" cell fires in phase with the units representing the cone. The same pattern results if the object is presented at another region in the visual field or at another site. The outputs of L3 that represent the distributed values of a geon, its orientation and aspect ratio, and the outputs of L5 representing their interrelations, provide a vector that self-organizes a unit in L6 termed a *geon feature assembly.* Units in L7 are object cells that self-organize to an integration over successive outputs from L6. These operations produce a parts-based structural description that is subsequently used directly as a basis for viewpoint-invariant recognition. The model's recognition performance conforms well to the results from the shape-priming experiments described in section 4.4.3 in that it manifests invariance to translation, rotation, and orientation in depth.

Binding through temporal correlation may provide some insight into the underlying neural basis of the attentional bottlenecks, discussed in Chapter 2. As the number of objects or object parts increases, insufficient temporal resolution to keep them out of phase with each other may be available and accidental simultaneous firing of the units representing two or more parts could occur. At this point, attention would be required to inhibit some of the activity producing the accidental synchrony.

4.4   Empirical Tests of Geon Theory

In this section, we describe experimental tests of two assumptions that distinguish geon theory from other theories of object representation. These assumptions are (1) that objects are represented in terms of their simple parts, and (2) that the parts are characterized by differences in viewpoint-invariant properties. In most of these experiments, the subjects named briefly presented (e.g., 100-msec.) object pictures. The flash of the picture was followed by a mask, an array of meaningless straight and curved line segments, to reduce persistence of the image. Naming reaction times and errors were the primary dependent variables.

4.4.1   Is the Representation Part-Based?

According to RBC, an object is representated in terms of its geons, which are activated by such local image features as vertices and edges. But, if the geons are activated by image features, why not just represent an object in terms of those features? In the real world, objects are often partially occluded by other surfaces, as when we view a car behind some light foliage. The pattern of occlusion of small regions of the parts can vary dramatically when we change viewpoint or when the wind shakes the leaves, but the various parts of the car likely would remain identifiable. To represent objects in terms of local image features, we would need a different representation for each arrangement of occluding contour or for each slightly altered orientation of the object.

Subjective impression is consistent with a parts-based representation. To see this, identify the two contour-deleted images in the second column of Figure 4.12 while covering up the images in the third column. Now look at the images in the third column, covering up those in the second column. Do these images look the same as those you just viewed? Now compare them. You will note that they are actually different images, each member of a pair having different vertices and edges. Despite the differences at the vertex and contour level, however, you probably saw the members of each pair as identical.

To test this issue of features versus parts experimentally, Biederman and Cooper (1991a) created pairs of *complementary images* of object pictures by deleting every other edge and vertex from each geon to create the two images of each object shown in Figure 4.12. The two images, when superimposed, form the intact picture shown in the far left column with no overlap in contour. The complementary images were created in such a way that each part (or geon) of the object can be recovered (or fail to be recovered) from each of the images. Although the complementary images share no edges and vertices, they presumably activate the same compo-

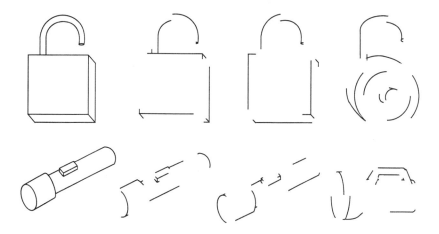

Figure 4.12
Complementary-feature images. From an original intact image (left column), two complementary-feature images (middle two columns) were formed by deleting every other vertex
and edge from each part so that each image had 50 percent of its contour. When superimposed, the two complements comprise the original intact image with no overlap in contour.
(Right column) Same-name, different-exemplar images from another complementary pair.
Subjects never viewed the intact image. Assuming that an image in the second column was
shown on the first (priming) block, the other member of the complementary pair (in the
third column) would be an instance of complementary-feature priming and the right figure
would be a different-exemplar control. (Adapted by permission of the author and publisher
from I. Biederman and E. E. Cooper, Priming contour-deleted images: Evidence for intermediate representations in visual object recognition, 1991, *Cognitive Psychology* 23, 397,
Figure 1. Copyright 1991 by Academic Press.)

nents. Because the amount of contour deleted from each image is substantial and includes vertices, it is unlikely that a local process of good continuation could not complete the contour of these images (see Biederman and
Cooper 1991a, for a more complete discussion of this issue).

On a first block of trials, subjects viewed a number of brief (200-msec)
presentations of one member from each complementary pair, naming it as
quickly and as accurately as possible. On the second block, they saw either
the identical image, its complement, or a same-name, different-exemplar
image (also contour-deleted) from a category with the same name and
basic-level concept but a different shape as shown in the far right column
of Figure 4.12). The mean correct naming reaction times and error rates
graphed in Figure 4.13 were markedly lower for the identical image than
for the different exemplars, indicating that a portion of the priming was
indeed visual and not just conceptual or lexical (i.e., the result of faster
accessing of the name). The critical comparison, however, concerned the
relative performance of the complementary condition. If priming was a

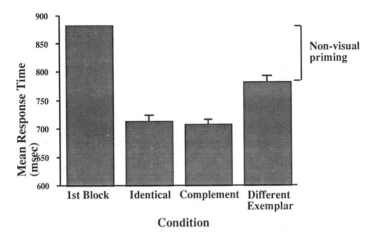

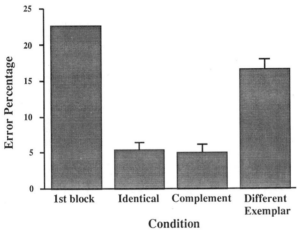

Figure 4.13

Mean correct reaction times and error rates of the complementary-feature priming experiment. The advantage in naming speed and accuracy of the identical image (when the second image of Figure 4.12 was also shown on the second block) compared to the different-exemplar condition provided a measure of visual priming. Any advantage of the identical over the complementary condition would be evidence of feature priming. There was none. Therefore, there was no contribution to the magnitude of priming from the representation of the lines and vertices. (Adapted by permission of the author and publisher from I. Biederman and E. E. Cooper, Priming contour-deleted images: Evidence for intermediate representations in visual object recognition, 1991, *Cognitive Psychology* 23, 399, Figure 2. Copyright 1991 by Academic Press.)

function of repetition of the specific vertices and edges in the image, then the complementary condition would have been equivalent to the different-exemplar condition, as neither shared any features with the original image. Remarkably, there was no difference in performance in naming complementary and identical images, indicating that none of the priming could be attributed to the specific vertices and lines actually present in the image.

What then caused the priming? Before we can attribute the priming to activation of the parts (in relations to other parts), which were common in the two conditions, we have to evaluate whether the priming could have been a consequence of activation of a semantic model of a subordinate category. Although the different-exemplar condition had the same basic-level category as the same-exemplar conditions, the identical and complementary conditions differed from the different-exemplar condition at the entry or subordinate level. If, for example, priming was due to activation of the concept of a grand piano or of a square lock rather than just piano or lock, then the advantage of the same-exemplar conditions could have been obtained without a contribution from activation of the parts. To test this possibility, we ran an experiment in which complementary images were created by deleting half the parts of the objects, as shown in Figure 4.14. With these stimuli, presumably, the same subordinate category would be activated from either member of a complementary pair, but through different parts. (This experiment required the use of objects that had at least four parts to look complete.) The design was otherwise identical to that of the previous study. As with the first experiment, performance with identical images was better than with different exemplars. Now, however, performance with the complements was equivalent to that with the different exemplars, indicating that none of the priming could be attributed to a subordinate semantic model. By eliminating subordinate-level concept priming as a factor in the first experiment, we obtained results suggesting that all the priming can be attributed to a representation of the parts of the object (and their interrelations) and none to the activation of the image features or subordinate-level concepts.

When does an object become unrecognizable? If object recognition is mediated by a representation of an object's parts, recognition should be particularly difficult if contour is deleted in locations that reduce the recoverability of the parts from the image. The images in the right-hand column of Figure 4.15 provide confirmation of this prediction. One method of reducing the recoverability of the parts is by deleting the cusps (the parsing regions discussed in section 4.2.4) to the point where the remaining contours bridge the cusp through smooth continuation. An example of how this method can interfere with the recovery of the parts is shown with the cup in Figure 4.15c. In this figure, the cusp between the top of the handle and the back of the lip of the bowl (shown in Figure 4.15a) has been

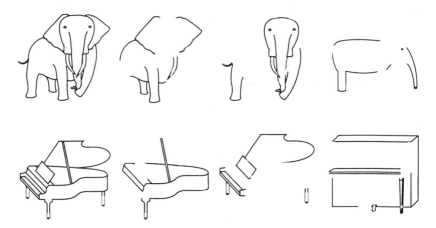

Figure 4.14
Complementary-part images. From an original intact image (left column), two complementary images (middle two columns), each composed of half the parts, were formed by deleting parts so that each image contained approximately 50 percent of the contour. If superimposed, the two complements would comprise the original intact image with no overlap in contour. (Right column) A same-name, different-exemplar image from another complementary pair. Subjects never viewed the intact image. Assuming that an image in the second column was shown on the first (priming) block, the other member of the complementary pair (in the third column) would be an instance of complementary-parts priming and the right figure would be a different-exemplar control. Unlike the results shown in Figure 4.13, the complementary and different-exemplar conditions were equivalent, both having reaction times and error rates that were considerably higher than the identical condition. This result indicates that none of the complementary-feature priming (Figures 4.12 and 4.13) could be attributed to the activation of a subordinate concept. (Adapted by permission of the author and publisher from I. Biederman and E. E. Cooper, Priming contour-deleted images: Evidence for intermediate representations in visual object recognition, 1991, *Cognitive Psychology* 23, 403, Figure 4. Copyright 1991 by Academic Press.)

deleted; the remaining contours would be continuous if joined through smooth continuation. Another technique is to delete a segment of a vertex so that a three-segment vertex, such as a fork or a tangent Y, or a T-junction, becomes a two-segment L-vertex. Examples of such operations are provided by the goblet in Figure 4.15c, in which the tangent Y-vertices at the junction of the sides and lip of the bowl have been converted to L-vertices by the deletion of the front of the lip of the goblet. Similarly, the T-vertices formed at the junction of the scissors blades or where the legs of the stool occlude the cross brace have been converted to L-vertices by deletion of one of the parts of the top of the T. Alternatively, the parsing regions can be deleted from several parts so that the remaining contours form an inappropriate vertex with more segments than

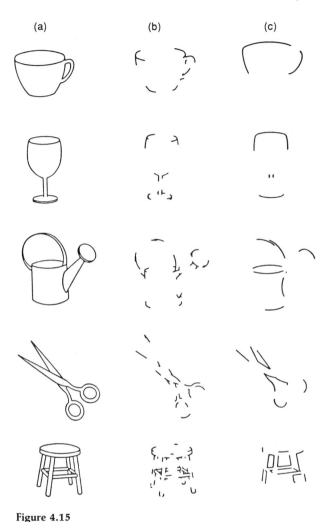

Figure 4.15
Example of five stimulus objects in the experiment on the perception of degraded objects. Column (a) shows the original intact versions. Column (b) shows the recoverable versions. The contours have been deleted in regions where they can be replaced through collinearity or smooth curvature. Column (c) shows the nonrecoverable versions. The contours have been deleted at regions of concavity so that collinearity or smooth curvature of the segments bridges the concavity. In addition, vertices have been altered (e.g., from Ys to Ls. (Modified by permission of the author and publisher from I. Biederman, Recognition-by-components: A theory of human image understanding, 1987, *Psychological Review* 94, 135, Figure 16. Copyright 1987 by the American Psychological Association.)

any of the original vertices, as with the watering can, where contours from the handle, opening, and base of the spout all appear to meet at a common vertex. If the same amount of contour is deleted, but in regions where the parts can still be activated, as in Figure 4.15b, objects remain identifiable.

The median accuracy of recognition of the nonrecoverable images, even after five seconds of viewing, was 0 percent (Biederman 1987). Given sufficient time (a few hundred msec), the recoverable images could be recognized perfectly. Actually, even when more contour is removed from the recoverable images than from the nonrecoverable images the former remain recognizable. You can test this by covering up parts of the objects in the middle column, say the right or left half, and determining whether you, or a person who has not seen the original versions, can still identify the objects. Recognition should be possible as long as enough contour remains to recover two or three parts of the object.

4.4.2   Viewpoint-Invariant versus Metric Properties

The complementary image and recoverable-nonrecoverable experiments provide evidence that the priming effects can be explained by a representation that specifies the parts of the object. But are these parts geons? A fundamental assumption of geon theory is that viewpoint-invariant differences, such as straight versus curved or parallel versus nonparallel, are given more weight than an equivalent amount of metric variation, such as aspect ratio. But how can metric and viewpoint-invariant contour variation be equated so that such an assumption can be investigated? Cooper and Biederman (1993) reasoned that the greater salience of viewpoint-invariant differences would be produced not in V1 but by the neural tissue more exclusively devoted to object recognition in extrastriate cortex, say in inferotemporal cortex (IT) (see Chapter 5). Consequently, they scaled the similarity of their stimuli according to the Lades et al. (1993) model of V1 simple-cell hypercolumns.

The positioning of the lattice over an original image is shown in the images in the left-hand column of Figure 4.16. Distorted lattices are shown in the middle and right columns of the figure, reflecting the reduced similarity of the images due to changes from the original in a viewpoint-invariant property (middle column) or a metric property (right column). In general, the two sources of error (filter similarity and lattice distortion) are correlated so that the more distorted the lattice, the less the similarity of the image to the original.

In the Cooper and Biederman (1993) experiment, subjects judged whether a pair of sequentially presented images of simple objects (containing only two or three parts), as illustrated in Figure 4.16, had the same name (and basic-level category). The images were each shown for 100

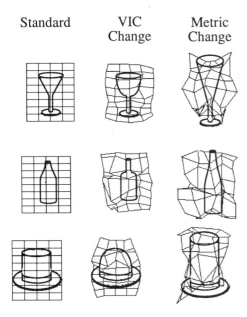

Standard   VIC   Metric
                  Change   Change

Figure 4.16
Examples of the Cooper and Biederman (1993) stimuli testing sensitivity in object recognition to viewpoint-invariant contrasts (VIC changes) compared to aspect-ratio differences
(metric changes) of a single part. The superimposed grids over the VIC and metric-changed
stimuli show the deformation of the Gabor jet lattices from the Lades et al. system
when these images were matched against the standard. The grid over the standard shows
the original positions of the Gabor jets. The greater the deformation, the lower the similarity between the standard and a changed image. Overall, the metric-changed stimuli were
slightly less similar to the standard than the VIC-changed stimuli. (Adapted by permission
from E. E. Cooper, and I. Biederman (1993). Geon differences during recognition are more
salient than metric differences. Poster presented at the Meetings of the Psychonomic
Society, Washington, D. C. Copyright 1993 by Eric E. Cooper.)

msec, with an intervening mask. When the images had the same name,
which occurred on half the trials, one of parts differed in either a nonaccidental property (as illustrated in the middle column of Figure 4.16) or
in aspect ratio (right column). Dissimilarity, as specified by the Lades et
al. model, between the original images and those that differed in a nonaccidental property was slightly less than the dissimilarity between the
original and metric changed images. If a difference in a viewpoint-invariant
property results in greater dissimilarity for object classification than a
difference in a metric property (when equated according to V1 similarity),
then it should be more difficult (i.e., slower and less accurate) to judge
whether two images belong to the same category when they differ in a
viewpoint-invariant property, compared to two images that differ in aspect
ratio. The results clearly supported this implication of geon theory.

4.4.3   Viewpoint Invariance and Its Neural Underpinnings

One of the most striking characteristics of human object recognition is its apparent invariance over changes in viewpoint. When we see an object at one position on the retina, at a given size and at a given orientation in depth, we can typically recognize it as the same object when, on a subsequent occasion, it is in another position, size, and orientation. Experimental results from name-priming studies confirm these impressions. The design of these studies are similar to that for the complementary-image experiment in that objects are presented in two blocks of trials, with the subject naming the objects as fast as possible. Biederman and Cooper (1991b; 1992) found that naming reaction times and error rates on the second block of trials were unaffected by a change in the position of the object—say from 2 degrees to the left of fixation to 2 degrees to the right of fixation—a 65 percent change in size, or a mirror reflection. As in the complementary-images experiment, slower naming times and higher error rates for images with the same name but a different shape documentated the fact that the priming in these experiments was largely visual, rather than conceptual or lexical. Similarly, Biederman and Gerhardstein (1993) found that rotation in depth had no effect on name priming as long as the object could be readily described as an arrangement of distinctive geons and the original geons remained in view. The rotation, of course, altered the aspect ratio and degree of curvature of the objects.

M. Goodale (Chapter 5) provides an overview of the research that, over the past two decades, has established that there are at least two major cortical visual systems for the processing of shape. Both cortical pathways start in V1, but one extends dorsally to the posterior parietal region (PP), while the other extends ventrally, through V2, V4, and IT. Why would there be two systems for representing shape? It has been argued, by Goodale (Chapter 5) and Biederman and Cooper (1992), among others, that the dorsal pathway supports shape representations for motor interactions, where position, size, and orientation in depth must be precisely specified. To sit in a chair, we must specify where the chair is, its size, and its orientation in depth. In contrast, it would seem to be advantageous for a recognition system to be invariant over position, size, and orientation. Cells in both PP and IT have large receptive fields, but the shape of the receptive fields in PP—peaked, with just a partial overlap—may be most efficient for providing coarse coding of position, whereas the relatively flat, almost completely overlapping, receptive fields in IT may be designed to produce positional invariance (O'Reilly, Kosslyn, Marsolek, and Chabris 1990).

What are cells in IT tuned to? Tanaka and his associates (Tanaka 1993; Kobatake and Tanaka 1994) have recently shown that a number of cells in area TE, a part of IT in the macaque brain presumed to mediate object recognition, respond to complex object features—such as two black horizontal bars superimposed over two adjacent corners of a square—but not to simple features—such as the horizontal bars or oriented lines—that form parts of complex features. (Cells earlier in the ventral pathway respond best to the simple features.) The set of complex features that Tanaka discovered appear to be largely viewpoint invariant in that it would be quite easy to distinguish most of them from almost any viewpoint. For example, other complex features are an upside-down T and a large circle with a small square attached to its bottom—both of which could be readily distinguished from the square with two horizontal bars. For the most part, these complex features appear to be the kinds of representations that would be created by the geon feature assembly layer (L6) in the Hummel and Biederman network.

4.4.4   Are the Experimental Results Predictable from Theories That Perform Matching Based on Filter Representations?

Fiser, Biederman, and Cooper (1994) assessed whether the Lades et al. (1989) recognition system could account for the empirical results presented in this section. It could not. Whereas the system's recognition performance was equivalent (and reasonably accurate) for recoverable and nonrecoverable stimuli, human subjects do not recognize nonrecoverable images. The system showed a marked decrement in recognition of complementary-feature images compared to identical images, but people show equivalent degrees of priming for both types of images. As discussed in section 4.4.2, the system also did not manifest the greater sensitivity that humans reveal to viewpoint-invariant differences relative to metric differences. It is likely that the Poggio and Edelman system would perform in an identical manner to the Lades et al. system. Because a rotation in depth always produces a change in initial filter values, the Lades et al. and Poggio and Edelman systems could not manifest the invariance documented by Biederman and Gerhardstein (1993) in their experiment with depth-rotated images.

4.5   An Extension of RBC to Scene Perception

The mystery about the perception of scenes is that the exposure duration required to have an accurate perception of an integrated real-world scene is not much longer than what is typically required to perceive individual objects. The recognition of a visual array as a scene requires not only identification of the various entities but also a semantic specification of the interactions among the objects as well as an overall semantic characterization of the arrangement. Perception of a scene is not, however, necessarily derived from an initial identification of the individual objects comprising that scene (Biederman 1988). That is, in general we do not first identify a stove, refrigerator, and coffee cup, in specified physical relations and then come to the conclusion we are looking at a kitchen.

Some demonstrations and experiments suggest that geon theory may provide a basis for explaining rapid scene recognition. Robert Mezzanotte (described in Biederman 1988) showed that a readily interpretable scene could be constructed from arrangements of single geons that only pre-served the approximate overall aspect ratio of the object. In these kinds of scenes, some examples of which are shown in the upper portion of Figure 4.17, none of the entities were identified as anything other than a simple volumetric body (e.g., a brick) when shown in isolation. Most important, Mezzanotte found that such settings can be recognized sufficiently quickly to interfere with the identification of intact objects inappropriate to the setting.

It is possible that quick understanding of a scene is mediated by the perception of *geon clusters*, an arrangement of geons from different objects

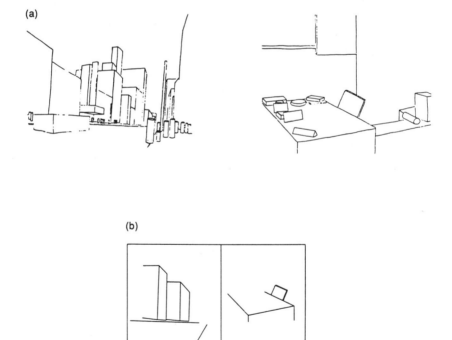

Figure 4.17
(a) Two of Mezzanotte's scenes, "City Street" and "Office." (b) Possible geon clusters for the scenes in (a).

that preserve the relative size and aspect ratio and relations of the largest visible geon of each object. In such cases, the individual geon will be insufficient to allow identification of the object. However, just as an arrangement of two or three geons almost always allows identification of an object, an arrangement of two or more geons from different objects may produce a recognizable cluster. The cluster acts very much as a large object does. The lower section of Figure 4.17 shows possible geon clusters for the scenes in the upper section of that figure. If this hypothesis is true, fast scene perception should be possible only in scenes where such familiar object clusters are present. Although this account of scene recognition awaits rigorous experimental test, you may be able to gauge it for yourself with the television experiment described in the opening paragraph of this chapter. Are there some scenes that you cannot identify from a single glance? My own experience is that such scenes are those which do not contain a familiar geon cluster.

4.5   Overview of Theories of Object Recognition

Although we used the complexity of the primitive as an organizing theme to present the three proposals for object recognition in section 4.4, several other distinguishing characteristics of theories can help provide a general framework for understanding theorizing in this area. Some of these characteristics will be described in this section. We should also note that the three kinds of theories of object recognition we have presented in this chapter are not necessarily mutually exclusive. (Nor are they frozen entities; theorists are actively involved in advancing their development.) It is possible that they address different aspects of object recognition or object recognition under different conditions. Because there are many ways in which an image can be classified on the basis of its shape, there is probably no single route to classification. It might be more profitable to consider the conditions under which one or the other forms of processing might be involved. We do this while discussing the distinguishing characteristics of the theories.

4.5.1   Viewpoint Dependence and the Development of Invariant Representations

The representations of Lades et al. (1993) and Poggio and Edelman (1990) theories are viewpoint dependent in that recognition is achieved by template deformation or interpolation or extrapolation during matching. The template is created directly from a single image in the Lades et al. system and from a set of similar images in the Poggio and Edelman system. Both models are self-organizing systems in which connection weights are automatically determined by the stimuli presented.

The representation activated in L7 in the Hummel and Biederman (1992) geon model is viewpoint invariant. One cannot play the model backwards and determine the precise orientation of the image that resulted in the activation of layers 3 through 7. Whereas it is clear how the RBF units in the Poggio and Edelman model develop as a consequence of experience with a set of images, the development of the intermediate layers in the Hummel and Biederman model are largely unspecified. It is possible, of course, that the units allowing for structural description of an object are genetically determined; such an assumption, however, flies in the face of an enormous amount of recent evidence indicating that the development of neural connectivity is activity dependent. Genetics provides only a rough scaffolding within which certain kinds of very common stimulation, such as long edges, serve to determine the course of neuronal interactions.

The self-organizing systems of the kind proposed by Lades et al. and Poggio and Edelman could thus describe the development of a network that maps filter values onto hidden units that, in turn, allow the activation of a structural description of an object. Over the first months or year of life, these units would self-organize in response to recurrent activity driven by the statistical regularities of experience—for example, that co-terminating edges terminate at a common point in depth. Later units would respond to the part structure and viewpoint-invariant properties of an object, such as those expressed in the Hummel and Biederman system. This visual experience would necessarily be one of specific images, which is all anyone experiences, and the neural activity derived from those images would sculpt the connections to intermediate units. Because geons are largely viewpoint invariant, they will be better able to activate a unit tuned to them for a greater proportion of the viewing experience than a unit tuned to a shape that is highly irregular. Consequently, units tuned to geons might have a better chance of achieving a stable self-organization. Once developed, intermediate (geon) units might be employed for any new object in much the same way, perhaps, that units for phoneme (or syllable) recognition developed in infancy can be employed for representing novel words and names.

4.5.2   Bottom-Up versus Bottom-Up and Top-Down Systems

The Lades et al., Poggio and Edelman, and Biederman models assume that one-way, bottom-up processing proceeds from image to activation of the representation of the object. The Lowe and Ullman models assume model-based matching in which some initial processing constrains further testing. Does object recognition always proceed as a bottom-up, one-way street? Probably not. For example, there are times when we are stumped in our first pass at looking at an object. At such times, we might perform some mental operation (such as the mental rotation discussed by Kosslyn in

Chapter 7) to determine whether some possible object corresponds to the image. In general, however, the extraordinary speed of object recognition in the absence of any context argues against a central role for top-down information.

4.5.3   Metric versus Viewpoint Invariant Differences

Whereas geon theory assigns great weight to viewpoint-invariant properties, the filter-matching models do not. Lowe's SCERPO model employs viewpoint-invariant properties in its initial hypothesis selection but thereafter the matching is performed on representations that are metrically specified. Viewpoint-invariant differences appear to be of little use in distinguishing faces, whereas the fine metric detail captured by filter-matching models enjoy considerable success in this regard. Interestingly (as discussed by Farah in Chapter 3), object and face recognition may involve different cortical loci.

What about subordinate differences among objects, such as those that distinguish a Mazda 626 from a Honda Accord? Some investigators (e.g., Bülthoff and Edelman 1992), have argued that an application of RBF theory can provide an account of how human recognition performance degrades with the introduction of differences in orientation among highly similar objects, such as a set of paper clips that differ only in the angles between their segments. Although there may be occasions when we have to rely on fine metric detail, Biederman and Shiffrar (1988) argue that when faced with a problem of discriminating among highly similar objects, people most often search for some viewpoint-invariant difference, albeit at a small scale. In the case of cars, it may be the logo or nameplate. Imagine that you have to distinguish among a set of chairs from a dining room set that are all the identical model. How would you do it? Most likely you would search for a scratch or stain or irregularity that is viewpoint invariant in that you could employ it as long as a particular surface is in view.

4.5.4   Concluding Remarks and Future Directions

Shortly after I assumed a new position at the University of Minnesota in September 1987, I was interviewed by a reporter for the University newspaper. A student photographer listened quietly to the interview until I described a possible application of my research in object recognition to development of a robot vision system that could be employed for inventory control. She then blurted out, "I would *love* to have such a robot. It could pick up after me!" Despite intense research activity during the intervening seven years, we still do not have an object recognition system that can come anywhere close to matching the capabilities of the human.

But the photographer's remark underscores one reason why there is little doubt that this intense activity will continue. The potential payoffs of an artificial object-recognition system are enormous.

Closer to the hearts of most vision scientists, however, is the appreciation that visual recognition is simply too central and extraordinary an activity, as described in the first paragraph of this chapter, to remain a mystery. About 50 to 65 percent of the primate cortex is devoted to vision. A considerable fraction of this is in the extrastriate cortex, in regions presumed to be involved in higher-level vision. The absolute amount of cortex is, of course, much greater in humans than it is in the monkey, and the proportion of cortex devoted to higher level vision appears to be greater as well. By most accounts, demands on vision for information about the physical world, such as the detection of motion or visual representations for motor interactions, do not noticeably differ between human and monkey. What then is the function of the larger expanse of human extratriate cortex? Nobel Prize winner Francis Crick has recently argued (1994) that vision can provide the basis for an attack on that most ineffable quality of mind: consciousness. Mysteries like these just won't go away.

The vision scientists involved in this mission are psychophysicists, computer scientists, neuroscientists, and cognitive neuroscientists. Discussions among investigators flow freely over issues of computational and neural-net modeling, perceptual data, and neural coding and structuring. Part of the joy in working on visual recognition is the great diversity of talent and knowledge that one is exposed to everyday. What are the likely directions of this activity? In theorizing, there is little doubt that symbolic psychological theories have given way to theories of subsymbolic operations, as expressed in neural-net modeling. This is a natural course of progression for an activity that is assumed to operate neurally. In my opinion this trend will continue, if not accelerate.

Suggestions for Further Reading

An excellent treatment of many of topics discussed in this chapter, and of presumed neural mechanisms, can be found in Kosslyn's recent book (1994). A somewhat more popular treatment of these issues is in Kosslyn and Koenig 1992. A delightful treatment of vision that, nonetheless, comes to full grips with the deep problems of neural representation and consciousness is in Crick 1994. A description of a fully self-organizing, neural-net system for recognition is presented in Waxman, Seibert, Bernardon, and Fay 1993.

Problems

4.1 Consider the features that might be used to distinguish between English capital letters say, between an A and an H, a C and a V, and a C and an H. Characterize their differences in terms of viewpoint-invariant properties. Which pair would you expect to be least confusable? Why? Because print is displayed on flat surfaces there is little or no requirement for depth invariance. Why are viewpoint-invariant properties relevant?

4.2   After reading Chapter 3 on face recognition, discuss how the Lades et al. (1993) face-recognition system might provide a basis for creating an integrated (or holistic) representation of a face.

4.3   Which layer(s) of the Hummel and Biederman (1992) network (Figure 4.11) are likely to be the locus (or loci) of priming, as revealed in the complementary-feature and complementary-parts experiments described in section 4.4.1?

References

Biederman, I. (1987). Recognition-by-components: A theory of human image interpretation. *Psychological Review* 94, 115–147.

Biederman, I. (1988). Aspects and extensions of a theory of human image understanding. In Z. Pylyshyn, ed., *Computational processes in human vision: An interdisciplinary perspective.* New York: Ablex.

Biederman, I., and E. E. Cooper (1991a). Priming contour-deleted images: Evidence for intermediate representations in visual object recognition. *Cognitive Psychology* 23, 393–419.

Biederman, I., and E. E. Cooper (1991b). Evidence for complete translational and reflectional invariance in visual object priming. *Perception* 20, 585–593.

Biederman, I., and E. E. Cooper (1992). Size invariance in visual object priming. *Journal of Experimental Psychology: Human Perception and Performance* 18, 121–133.

Biederman, I., and P. C. Gerhardstein (1993). Recognizing depth-rotated objects: Evidence and conditions for three-dimensional viewpoint invariance. *Journal of Experimental Psychology: Human Perception and Performance* 19, 1162–1182.

Biederman, I., H. J. Hilton, and J. E. Hummel (1991). Pattern goodness and pattern recognition. In J. R. Pomerantz, and G. R. Lockhead, eds., *The perception of structure,* 73–95. Washington, D.C.: American Psychological Association.

Biederman, I., and G. Ju (1987). Surface versus edge-based determinants of visual recognition. *Cognitive Psychology* 20, 38–64.

Biederman, I., R. J. Mezzanotte, and J. C. Rabinowitz (1982). Scene perception: Detecting and judging objects undergoing relational violations. *Cognitive Psychology* 14, 143–177.

Biederman, I., and M. M. Shiffrar (1987). Sexing day-old chicks: A case study and expert systems analysis of a difficult perceptual-learning task. *Journal of Experimental Psychology: Learning, Memory, and Cognition* 13, 640–645.

Brooks, R. A. (1981). Symbolic reasoning among 3-D models and 2-D images. *Artificial Intelligence* 17, 205–244.

Binford, T. O. (1971). Visual perception by computer. *IEEE Systems Science and Cybernetics Conference.* Miami, December 1971.

Bülthoff, H. H., and S. Edelman (1992). Psychophysical support for a 2-D view interpolation theory of object recognition. *Proceedings of the National Academy of Sciences* 89, 60–64.

Connell, J. H. (1985). Learning shape descriptions: Generating and generalizing models of visual objects. Unpublished master's thesis, Cambridge: Massachusetts Institute of Technology.

Cooper, E. E., and I. Biederman (1993). Geon differences during recognition are more salient than metric differences. Poster presented at the Meeting of the Psychonomics Society, Washington, D.C., November.

164   Biederman

Crick, F. (1994). *The astonishing hypothesis: The scientific search for the soul.* New York: Scribner's.

Dickinson, S. J., A. P. Pentland, and A. Rosenfeld (1992). From volumes to views: An approach to 3-D object recognition. *Computer Vision, Graphics, and Image Processing: Image Understanding* 55, 130–154.

Edelman, S. (1993). Representation, similarity, and the chorus of prototypes. Weizmann Institute (Rehovat, Israel) Technical Report CW93-10.

Fiser, J., I. Biederman, and E. E. Cooper (1994). Are the direct outputs of Gabor filters sufficient for human object recognition or are they only the prior stage for intermediate representations? Poster presented at the Annual Meeting of the Association for Research in Vision and Ophthalmology. Sarasota, FL., May.

Gibson, E. J. (1969). *Principles of perceptual learning and development.* New York: Appleton-Century-Crofts.

Haber, R. N., and M. Hershonson (1981). *The psychology of visual perception.* New York: Holt, Rinehart, and Winston.

Hoffman, D. D. and W. Richards (1985). Parts of recognition. *Cognition* 18, 65–96.

Huttenlocher, D. P., and S. Ullman (1987). Object recognition using alignment. In *Proceedings of the First International Conference on Computer Vision*, IEEE Computer Society, 102–111. London, June.

Hummel, J. E., and I. Biederman (1992). Dynamic binding in a neural network for shape recognition. *Psychological Review* 99, 480–517.

Intraub, H. (1981). Identification and naming of briefly glimpsed visual scenes. In D. F. Fisher, R. A. Monty, and J. W. Senders, eds. *Eyemovements: Cognition and Visual Perception.* Hillsdale, NJ: L. Erlbaum.

Ittelson, W. H. (1952). *The Ames demonstrations in perception.* New York: Hafner.

Jolicoeur, P., M. A. Gluck, and S. M. Kosslyn (1984). Picture and names: Making the connection. *Cognitive Psychology* 16, 243–275.

Kimia, B. B., A. R. Tannenbaum, and S. W. Zucker (1995). Shapes, shocks, and deformations, I: The components of shape and the reaction-diffusion space. *International Journal of Computer Vision.*

King, M., G. E. Meyer, J. Tangney, and I. Biederman (1976). Shape constancy and a perceptual bias toward symmetry. *Perception & Psychophysics* 19, 129–136.

Kobatake, E., and K. Tanaka (1994). Neuronal selectivity's to complex object features in the ventral visual pathway of the macaque cerebral cortex. *Journal of Neurophysiology* 71, 856–867.

Kosslyn, S. M. (1994). *Image and brain: The resolution of the imagery debate.* Cambridge, MA: MIT Press.

Kosslyn, S. M., and O. Koenig (1992). *Wet mind: The new cognitive neuroscience.* New York: Free Press.

Lades, M., J. C. Vorbrüggen, J. Buhmann, J. Lange, C. von der Malsburg, R. P. Würtz, and W. Konen (1993). Distortion invariant object recognition in the dynamic link architecture. *IEEE Transactions on Computers* 42, 300–311.

Lowe, D. (1984). Perceptual organization and visual recognition. Unpublished doctoral dissertation, Stanford University.

Lowe, D. G. (1987). The viewpoint consistency constraint. *International Journal of Computer Vision* 1, 57–72.

Marr, D. (1982). *Vision: A computational investigation into the human representation and processing of visual information.* New York: W. H. Freeman.

O'Reilly, R. C., S. M. Kosslyn, C. J. Marsolek, and C. F. Chabris (1990). Receptive field characteristics that allow parietal lobe neurons to encode spatial properties of visual input: A computational analysis. *Journal of Cognitive Neuroscience* 2, 141–155.

Poggio, T., and S. Edelman (1990). A network that learns to recognize three-dimensional objects. *Nature* 343, 263–266.

Rosch, E., C. B. Mervis, W. D. Gray, and P. Boyes-Braem (1976). Basic objects in natural categories. *Cognitive Psychology* 8, 382–439.

Siddiqi, K., K. Tresness, and B. B. Kimia (1994). Parts of visual form: Ecological and psychophysical aspects. Laboratory for Engineering, TR LEMS-104, Brown University.

Tanaka, K. (1993). Neuronal mechanism for object recognition (1993). *Science* 262, 685–688.

Tversky, B., and K. Hemenway (1984). Objects, parts, and categories. *Journal of Experimental Psychology: General* 113, 169–193.

Ullman, S. (1989). Aligning pictorial descriptions: An approach to object recognition. *Cognition* 32, 193–254.

Ullman, S., and R. Basri (1990). Recognition by a linear combination of models. A. I. Memo No. 1152, Artificial Intelligence laboratory, MIT.

Waxman, A. M., M. Seibert, A. M. Bernardon, and D. A. Fay (1993). Neural systems for automatic target learning and recognition. *Lincoln Laboratory Journal* 6, 77–116.

Zerroug, M., and R. Nevatia (1995). Volumetric descriptions from a single intensity image. *International Journal of Computer Vision*, in press.

Chapter 5

The Cortical Organization of Visual Perception and Visuomotor Control

Melvyn Alan Goodale

In February 1988, a young woman suffered irreversible brain damage as a result of near-asphyxiation by carbon monoxide. She was only thirty-four years old, held a B.A. degree in business management and a private pilot's licence, and was working as a freelance commercial translator. Although she was British, she had recently moved to northern Italy. Unfortunately, the bathroom in her new home was equipped with a propane water heater that was improperly vented. One day, as she was taking a shower, so much carbon monoxide built up in the room that she became anoxic, collapsed, and lost consciousness. By the time she was discovered, she was in a deep coma and was rushed to hospital. As she slowly regained consciousness, it became clear that her visual system had been badly damaged by the anoxia. Not only was she unable to recognize the faces of her relatives and friends or the visual appearance of common objects, but she was also unable to discriminate between such simple geometric forms as triangles and circles. At the same time, she had no difficulty identifying people from their voices or objects placed in her hands. Her perceptual problems appeared to be exclusively visual. Formal testing carried out some six months after the accident revealed that DF, as the patient has become known, was unable to describe the orientation and form of any visual contour, no matter how that contour was defined. She could not identify shapes whose contours were defined by differences in luminance or color or by differences in the direction of motion or the plane of depth. Not surprisingly, DF was also unable to recognize shapes that were defined by the similarity or proximity of individual elements of the visual array. Sadly, her ability to deal with shape and pattern has improved very little since the time of the original testing.

5.1 Acting without Perceiving

DF's difficulties with "form vision" cannot be explained by deficits in fundamental sensory processing. Even when she was tested in the first year after the accident, she could detect targets presented as far out as 30 degrees into her peripheral visual fields. Remarkably, her ability to identify colors was also essentially unimpaired. Her spatial contrast sensitivity was also normal above five cycles per degree and was only moderately impaired at lower spatial frequencies.[1] (Of course, even though she could detect the presence of the spatial frequency gratings in these tests, she could not report their orientation.) It is interesting that many individuals with peripheral visual problems, whose visual fields are much more restricted than DF's and who have much poorer resolution acuity, have no difficulty recognizing the shapes and patterns that DF finds impossible to identify. DF's deficit, then, is not primarily "sensory" in nature but instead seems to be much more of a "perceptual" problem. She simply cannot perceive shapes and forms, even though her visual system appears to have the requisite low-level sensory information at its disposal (Milner, et al. 1991).

The profound nature of DF's deficit is revealed in her drawings of common objects (illustrated in Figure 5.1). Not only was she unable to identify the simple line drawings in the left-hand column of this figure, but she was also unable to copy them, at least in a recognizable way. Notice, however, that she could depict some aspects of the drawings, such as the dots indicating the print in the line drawing of the open book, even though she could not duplicate the overall shape or arrangement of the elements of the drawing. DF's inability to copy the drawings is not due to a problem in controlling the movements of the pen or pencil; when she was asked on another occasion to draw an object from memory, she was able to do so reasonably well, as the drawings on the right-hand side of Figure 5.1 illustrate. Needless to say, when shown any of the drawings she had done herself, whether the ones retrieved from memory or those copied from another drawing, she had no idea what they were and commented that they all looked like "squiggles." This failure to recognize even what she had drawn herself means that she sometimes makes errors in drawing from memory—particularly when she lifts her pen from the page. On these occasions, she does not always return the pen to the right place to continue the drawing; as a consequence she produces drawings in which the components of the object are displaced from one another. An example of this kind of error is illustrated in Figure 5.2.

What is remarkable about DF is that, despite her profound visual deficits, she has amazingly intact visuomotor abilities. Thus, even though she

1. Spatial contrast sensitivity was measured by asking DF to discriminate between a field of alternating dark and light bars (a spatial frequency grating) and a field of homogeneous gray of the same overall brightness. The spatial frequency of the grating (the number of light-dark cycles per degree of visual angle) and the relative contrast of the light and dark bars was varied systematically over a broad range of frequencies and contrasts.

Model   Copy   Memory

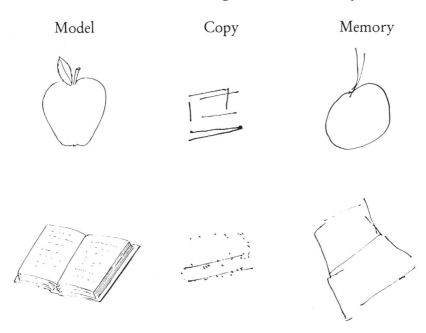

Figure 5.1
The patient DF's attempts to draw from models and from memory. DF was unable to identify the line drawings of an apple and an open book shown on the left. In addition, her copies were very poor. Note that she did incorporate some elements of the line drawing (e.g., the dots indicating the text in the book) into her copy. When asked on another occasion to draw an apple or an open book from memory, she produced a respectable representation of both items (right-hand column). When she was later shown her own drawings, she had no idea what they were. (Adapted by permission from P. Servos, M. A. Goodale, and G. K. Humphrey. The drawing of objects by a visual form agnosic: Contribution of surface properties and memorial representations, 1993, *Neuropsychologia* 31, 251–259.)

cannot recognize a very familiar object on the basis of its visual form, she can grasp that object under visual control as accurately and as proficiently as people with normal vision. She can open doors, shake hands, pick up a pencil, and grasp a preferred item of food, all under visual control. In all these activities, her visually guided behavior appears essentially normal.

Formal testing has confirmed this striking dissociation. For example, when DF was shown the slot illustrated in Figure 5.3, which could be placed in any one of a number of different orientations around the clock, she showed great difficulty in recognizing its orientation. In fact, she could not even indicate the orientation by rotating her hand or a hand-held card to match that of the slot. For those of us with normal vision, of course, this is a trivial task.

Figure 5.2
A misaligned drawing by the patient DF. Partway through an effort to draw a fork from memory, DF took her pen from the page. When she placed her pen back on the page and continued her drawing, she did not align the second part of the drawing with the first.

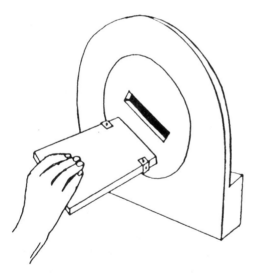

Figure 5.3
Apparatus used to test sensitivity to orientation in the patient DF. The slot could be placed in any one of a number of orientations around the clock. Subjects were required either to rotate a hand-held card to match the orientation of the slot or to "post" the card into the slot, as shown in this figure.

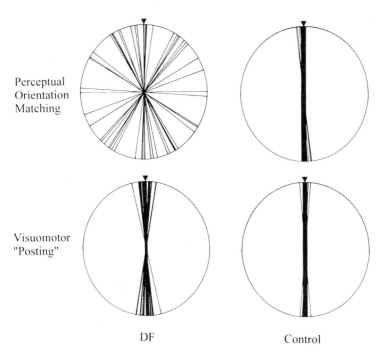

Figure 5.4
Polar plots of the orientation of the hand-held card on the perceptual matching task and the visuomotor "posting" task, for DF and an age-matched control subject. The correct orientation on each trial has been rotated to vertical. Note that although DF was unable to match the orientation of the card to that of the slot in the perceptual matching card, she did rotate the card to the correct orientation as she attempted to insert it into the slot on the "posting" task.

The performance of DF and a control subject on this perceptual matching task can be directly compared by examining the polar plots shown in Figure 5.4, top, in which the orientation of the hand-held card on each trial is plotted with respect to the actual orientation of the slot. As the figure shows, DF's responses bear very little relation to the slot's orientation. A slight change in the task requirements, however, produced a dramatic change in her performance. When DF was asked not to match the hand-held card to the orientation of the slot, but instead to reach out and "post" the card in the slot, she was able to do so accurately with little difficulty (see Figure 5.4, bottom). Moreover, she began to rotate the card in the correct direction as soon as she began her movement. Thus, in one task, in which she was asked to match the orientation of the card to that of the slot, DF was profoundly impaired; in another task, in which she was asked to insert the card in the slot, she was virtually normal.

What is the essential difference between these tasks that led to such remarkable differences in performance? Why could DF use visual information in one situation but not the other? What does the pattern of spared and damaged visual abilities in patients like DF tell us about the organization of the human visual system? The answers to these questions form the focus of the present chapter. We begin by asking yet another question: What is vision used for anyway?

5.2   The Function(s) of Vision

Most organisms are sensitive to electromagnetic radiation and human beings are no exception. We are particularly well equipped to deal with reflected and emitted radiation in that portion of the electromagnetic spectrum from 400 to 750 nanometers (nm), the so-called visible spectrum. So sensitive are the individual elements of the human retina that they can respond to one quantum of light—the minimum amount of radiant energy that can exist (Wald 1950). But sensing light is not enough. The ability to direct one's behavior with respect to the pattern of light striking the peripheral receptors requires a considerable amount of processing on the part of the central nervous system, from the retina to the cerebral cortices. So vital is this processing in the higher primates that an estimated 60 percent of neocortex in the monkey brain can be described as visual cortex (Maunsell and Newsome 1987).

It is commonly assumed that the function of all this visual machinery is to provide some sort of internal representation of the external world to serve as the perceptual foundation for thought and action. Indeed, the idea that the ultimate function of vision is to deliver our perception of the world has been an underlying assumption, most often implicit, for much of the theorizing about vision and about the organization of the visual system. Thus, while much attention has been paid to the modular and parallel organization of the visual-input pathways, it has been assumed, for the most part, that all these different processes are part of a single monolithic system dedicated to transforming the raw visual image into a unified percept of the visual world. In this chapter, I will challenge this assumption and present evidence suggesting that there is no single multipurpose representation and that visual input is transformed in different ways for different purposes (see also Chapter 3, this volume). In particular, I propose that what we ordinarily think of as our perceptual experience of the visual world depends on processes and pathways that are largely independent of those mediating the visual control of skilled actions within that world.

5.3   Motor Control: The Origins of Visual Systems

Vision did not evolve to enable organisms to "see"; it evolved to control their movements with respect to distal stimuli.[2] After all, natural selection has little to do with how well an animal sees the world; it operates at the level of the overt behavior that enables the organism to avoid predators, find mates, forage for food, and move from one part of the environment to another. We can get some idea about what the early visual systems of our vertebrate ancestors might have been like by looking at the organization of the visual pathways in such simpler living vertebrates as frogs and toads. The first thing we note is that many of the visual-control systems for the different patterns of behavior these animals exhibit have quite independent neural substrates. In the frog, *Rana pipiens*, for example, visually guided prey-catching and visually guided locomotion around barriers are separately mediated by different pathways—from the retina right through to the effector systems controlling the behavior (Ingle 1973; 1982; 1983). The visual control of prey-catching depends on circuitry involving retinal projections to the optic tectum, while the visual control of locomotion around barriers depends on circuitry involving retinal projections to particular regions of the pretectal nuclei. Each of these retinal targets projects, in turn, to different premotor nuclei in the brain stem and spinal cord. In fact, accumulating evidence from studies in both frog and toad suggest that there are at least five separate visual-motor *modules*, each responsible for a different kind of visually guided behavior and each having distinct input and output pathways (Ingle 1983). While the outputs of these different modules have to be coordinated, in no sense are all these actions guided by a single visual representation of the world residing somewhere in the animal's brain. There is no monolithic visual system that provides a unified representation for the motor system; there is, instead, a set of relatively independent (and parallel) visuomotor channels. (For an extended discussion of these and related issues, see Goodale 1983a and 1988.)

The same kind of visuomotor modularity exists in mammals. The first hint of this modularity comes from the fact that the axons of the retinal ganglion cells project to a number of different target nuclei in the brain, each of which has a distinctive set of outputs (see Figure 5.5). There is good reason, therefore, to suppose that these different retinal pathways are responsible for rather different patterns or classes of behavior. In rodents,

2. When I use the phrase "vision evolved to . . .", I do not mean to invoke a teleological account of evolution. It is simply a shorthand way of saying that the heritable differences in the visual control of motor output might enable some members of a population to reproduce more efficiently than others and that, therefore, visual abilities might change (or even "improve") over generations.

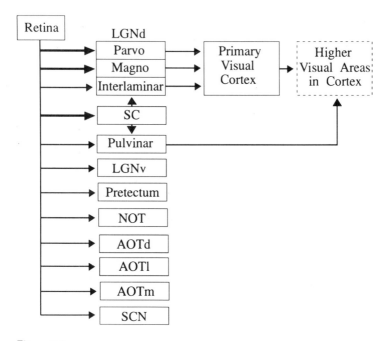

Figure 5.5
Schematic diagram of the projections from the retina to the brain in the typical mammal.
LGNd: lateral geniculate nucleus, pars dorsalis; LGNv: lateral geniculate nucleus, pars
ventralis; SCN: suprachiasmatic nucleus; NOT: nucleus of the optic tract.

for example, the visual control of orientation to food and novel stimuli has
been shown to be mediated by subcortical pathways that are quite independent from those controlling avoidance of obstacles during locomotion
(Goodale 1983b; Goodale and Milner 1982; Goodale 1988). Moreover,
the organization of these visuomotor pathways is remarkably similar to
the circuitry controlling analogous behaviors in the frog. Thus, different
patterns of visually guided behavior exhibited by vertebrates are mediated
by separate pathways from visual receptors through to motor nuclei, each
pathway processing a particular constellation of inputs and each evoking a
particular combination of effector outputs.

5.4 Perception and Action in the Primate Brain

The simple input-output modules described above, no matter how subtle
their interactions, can only go so far in controlling visual behavior. The
visual world of human beings (and many other vertebrates for that matter)
is much more complex than that of the frog, and requires more flexible

information processing. After all, we see the world; we do not simply react to visual stimuli within it. In other words, representational systems have evolved in our brains—systems that permit us to identify objects and events, attach meaning and significance to them, and establish their causal relations. Moreover, vision provides important inputs into these representational or *perceptual* systems.[3]

Perceptual systems are clearly very different from the simple visuomotor modules of amphibia. They are not linked to specific motor outputs but to cognitive systems involving memory, semantics, planning, and communication. As we discuss in greater detail in a later section, the emergence of large areas of cerebral cortex in the primate brain devoted to visual processing reflects, in part, the development of these perceptual systems. But, while perception permits the formation of goals and the decision to engage in a goal-directed action without reference to particular motor outputs, the actual *execution* of an action may nevertheless be mediated by dedicated visuomotor modules that are not dissimilar in principle to those found in frogs and toads.

5.4.1   The Emergence of Cortical Visual Systems

For example, in primates, the control of saccadic eye movements, like the control of the orientation movements directed at food objects in frogs and rodents, has been shown to depend on sensorimotor transformations carried out in the optic tectum (or superior colliculus, as it is known in mammals), which receives direct input from the retina and projects to premotor nuclei in the brain stem (for review, see Sparks and Mays 1990). Other visually modulated behaviors, such as modifications of the vestibulo-ocular reflex, the elicitation of optokinetic nystagmus, and the visual control of posture, also depend on separate and relatively independent visuomotor pathways that are largely subcortical. Nevertheless, there is evidence that in the primate brain the outputs from these basic visuomotor modules and the transformations they perform on incoming sensory information can be modified by higher-order input from the cerebral cortex. The basic tectal circuitry controlling saccadic eye movements, for example, is modulated

---

3. The term *perceptual* (or more properly, its root word, *perception*) is used in many different ways in both visual science and everyday language. Sometimes perception refers to any processing derived from sensory input; at other times, it is used in a more restricted sense. In this latter usage, the term is often identified with our phenomenological experience of the world, allowing us to assign meaning and significance to external objects and events. In this chapter, use the word in this more restricted sense. Perception then is seen as subserving the recognition and identification of objects and events and their spatial and temporal relations. In this sense, perception provides the foundation for the cognitive life of the organism, allowing it to construct long-term memories and models of the environment.

by a number of cortical areas that project not only to the superior colliculus itself but also directly to premotor and motor nuclei in the brain stem (see, e.g., Bruce 1990). Thus, although the superior colliculus still plays a central role in the control of saccadic eye movements in the primate, this phylogenetically ancient structure has become a part of a more complex control system that involves circuitry in the neocortex but is still dedicated to the control of this particular motor output.

In fact, a number of different visuomotor control systems have emerged in the primate cerebral cortex to support a range of skilled behaviors, such as visually guided reaching and grasping—behaviors in which a high degree of coordination is required between movements of the fingers, hands, upper limbs, head, and eyes. As we have already seen, some of these new control systems work by modulating phylogenetically older visuomotor modules in the subcortical brain; others appear to be quite independent of this ancient circuitry. In any case, the primitive input-output visuomotor pathways of the early vertebrates, which were little more than simple servomechanisms, have been upgraded to much more subtle control mechanisms capable of mediating the visual guidance of the complex motor outputs that primates are capable of producing. Nevertheless, these systems, as complex as they are, do not require that the controlling stimuli be perceived, any more than the operation of the simple input-output modules of the frog require perception of the relevant stimuli.

Thus, the massive expansion of the areas devoted to visual processing in the primate cerebral cortex can be viewed as reflecting two closely related developments: (1) the emergence of *perceptual systems* for identifying objects in the visual world and attaching meaning and significance to them; and (2) the emergence of complex visuomotor control systems or *action systems* that permit the execution of skilled actions directed at those objects. As we shall see in the next section, it has recently been proposed that this functional division of labor is paralleled by an anatomical segregation in the cortical visual pathways subserving perception and action in the primate brain (Goodale and Milner 1992).

5.4.2   Two Streams of Processing in Primate Cerebral Cortex

Ungerleider and Mishkin (1982) were the first to suggest that the projections from striate cortex to extrastriate visual areas in the monkey cerebral cortex could be subdivided into two functional "streams of processing." They identified a so-called ventral stream, which leaves visual cortex and projects via a series of cortico-cortical projections to the infero-temporal cortex, and a dorsal stream, which projects from striate cortex to regions of the posterior parietal cortex. Their original conception of these pathways is illustrated in Figure 5.6. On the basis of this anatomy, together

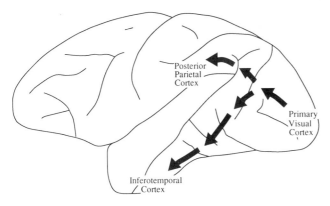

Figure 5.6
Ungerleider and Mishkin's original conception of the two streams of visual processing in the primate cerebral cortex.

with evidence from electrophysiological and behavioral studies in the monkey and neuropsychological studies in humans, Ungerleider and Mishkin proposed that the ventral stream of visual processing plays a special role in the identification of objects, while the dorsal stream is responsible for localizing objects in visual space.

More recently, however, David Milner and I (Goodale and Milner 1992) reinterpreted the division of labor between the ventral and dorsal streams in terms of the distinction outlined earlier between perception and action. We believe the critical difference between the two streams lies not so much in the nature of the visual information they receive but in the transformations they perform upon that information. As a consequence, our account places less emphasis on input distinctions (object vision versus spatial vision or "what" versus "where") and more on the requirements of the output systems that each stream of processing serves. In our view then, the ventral stream plays the major role in the perceptual identification of objects, while the dorsal stream mediates the required sensorimotor transformations for visually guided actions directed at those objects. Processing within the ventral stream enables us to identify an object, such as a ripe pear in a basket of fruit; but processing within the dorsal stream provides the critical information about the location, size, and shape of that pear so that we can accurately reach out and grasp it with our hand. Notice that in this account information about object attributes, such as size, shape, orientation, and spatial location, are processed by both streams but that the nature of that processing is very different. A schematic diagram outlining the projection pathways and interconnections of the ventral (perceptual) stream and the dorsal (action) stream is set out as Figure 5.7.

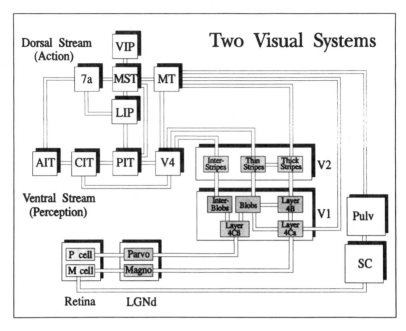

Figure 5.7
Schematic diagram of Goodale and Milner's conception of the two streams of visual processing in the primate cerebral cortex. LGNd: lateral geniculate nucleus, pars dorsalis; SC: superior colliculus; Pulv: pulvinar; PIT: posterior inferotemporal cortex; CIT: central inferotemporal cortex; AIT: anterior inferotemporal cortex; MT: middle temporal area; MST: medial superior temporal area; LIP: lateral intraparietal sulcus; VIP: ventral intraparietal sulcus. For a discussion of the functions of these areas, see Milner and Goodale (1995).

But why should perception and action require different kinds of visual processing? Why couldn't all processing be handled by a single set of visual pathways? To answer these questions, it is necessary to examine the different output requirements for visual perception, as opposed to the visual control of action. Consider first the task of the perceptual systems. As was suggested earlier, the fundamental task of these systems is to parse the visual array into different objects and events, classify those objects and their interrelations, and attach meaning and significance to them. Such operations are essential for accumulating a knowledge base about the world, exchanging information with conspecifics, and choosing among different courses of action. In short, without perception, cognitive operations would be impossible. As a consequence, perception tends to be concerned with the enduring characteristics of objects (and their interrelations), so that they can be recognized when they are encountered again in different visual contexts or from different vantage points. To generate these long-term representations, perceptual mechanisms must be object-based; in other words, the visual coding for the purposes of perception must deliver the identity of the object independent of any particular viewpoint or viewing condition. As a consequence, human perception is characterized by *constancies* of shape, size, colour, lightness, and location. The generation of these object-based representations might depend on networks of multiple views, so that object identity can be accessed by transforming or interpolating any particular view of an object with respect to a particular network (see, e.g., Tarr and Pinker 1989; Bülthoff and Edelman 1992). Alternatively, a particular view of an object could be transformed to some sort of canonical or prototypical view (Palmer et al. 1981). Whatever the particular coding mechanisms might be (and such mechanisms could vary across different classes of objects), the essential task for the perceptual system is coding (and later recovering) object identity. It is objects, not object views, that the perceptual system is ultimately designed to deliver.

This is not the case for the visuomotor mechanisms that support actions directed at objects. Here, the underlying visuomotor transformations have to be viewer-centered; in other words, both the location of the object and its orientation and motion must be encoded relative to the observer. (One constancy that must operate, however, is object size; in order to scale the grip during grasping, the underlying visuomotor mechanisms must be able to compute the real size of the object independent of its distance from the observer.) Finally, because the position and orientation of a goal object in the action space of an observer is rarely constant, such computations must be calculated *de novo* every time an action occurs. To use a computer-based metaphor, action systems must do most of their work online in real time, while perceptual systems do most of their work offline.

To summarize then, while similar (but, as we shall see, not identical) visual information about object shape, size, local orientation, and location is available to both systems, the transformational algorithms applied to these inputs are uniquely tailored to the function of each system. As a consequence of these differences in processing requirements, two cortical visual systems have evolved: a ventral stream for visual perception and a dorsal stream for the visual control of skilled actions.

Electrophysiological studies of the dorsal and ventral streams in the monkey tell a story that is remarkably consistent with this dualistic account of cortical visual function. What follows is a brief review of some of the more instructive studies of single-unit activity in these two pathways. (For a more detailed accounts of this literature, see Goodale 1993; Milner and Goodale 1993, 1995.)

The Dorsal Action System

Ever since Hyvärinen and Mountcastle carried out their pioneering work on posterior parietal cortex—a major terminus for projections in the dorsal stream—it has been clear that the responses of neurons in this region, in contrast to those in the ventral stream, are greatly dependent on the concurrent behavior of the animal with respect to the visual stimulus (Hyvärinen and Poranen 1974; Mountcastle, Lynch, Georgopoulos, Sakata, and Acuña 1975). Their early work showed that separate subsets of cells in posterior parietal cortex are implicated in visual fixation, pursuit and saccadic eye movements, visually guided reaching, and the manipulation of objects. Indeed, progress in demonstrating the response selectivity of cells in this region had to await the development of reliable single-cell recording techniques for use with awake monkeys. Recordings from anesthetized monkeys showed almost no consistent activity in the posterior parietal regions—even though it was clear that this area receives visual inputs. In other words, unless the monkeys were able to move their eyes and/or hands in relation to the visual stimulus, nothing much would happen. In fact, in a review of work in this field, Andersen (1987) pointed out that most neurons in these areas "exhibit both sensory-related and movement-related activity."

The behavioral modulation of the visual activity of cells in this region can be quite subtle. For example, recent work by Duhamel, Colby, and Goldberg (1992) has shown that some cells in area LIP (lateral intraparietal sulcus), which lies in the posterior parietal region, show transient shifts in their receptive field just before the animal makes a saccadic eye movement, so that stimuli that fall within the receptive field after the eye movement is completed begin to modulate the cell's activity before the eye movement occurs. In addition, many cells respond when an eye movement brings the site of a previously flashed stimulus into the cell's receptive field. These results suggest that networks of cells in the posterior parietal cortex anticipate the retinal consequences of saccadic eye movements and update the cortical representation of visual space to provide a continuously accurate representation of the location of objects in the world with respect to the observer. The egocentric coding provided by these and other movement-modulated visual cells in posterior parietal cortex is extremely short term and is updated every time the animal moves. Such coding would be of little use in establishing the position of objects in world coordinates relative to other objects in the world. But short-term coding of the egocentric position of objects is just the sort of information that is critical for the programming and online control of such skilled movements as goal-directed grasping. Converging evidence for this conclusion comes from the long history of work showing that deficits in visually guided reaching movements follow damage to these regions of parietal cortex in the monkey. (For review, see Milner and Goodale 1993, 1995.)

Some of the cells in the posterior parietal region that have been shown to fire when the monkey manipulates an object also appear to be visually sensitive to such structural features of the object as size and orientation—features that determine the posture of the hand and fingers during the grasping movement (Taira, Mine, Georgopoulos, Murata, and Sakata 1990; Sakata, Taira, Mine, and Murata 1992). These neurons, which are located in a region in posterior parietal cortex close to and overlapping area LIP (see Figure 5.7), are tied both to object properties and to the movements of the hands and fingers that are appropriate for those properties. At present, it is not known where the source of the shape information for this visual coding is located. It is unlikely, however, that the shape coding in these cells is dependent on input from the high-level modules in the ventral stream that support perceptual report, for monkeys with massive deficits in object recognition following damage to infero-temporal cortex seem as capable as normal animals of reaching out and grasping objects (Klüver and Bucy 1939; Pribram 1967).

There are many motion-sensitive cells in the dorsal pathway, particularly in areas MT and MST (see Figure 5.7). Many of these cells, by virtue of their sensitivity to changes in optic flow produced by object movement, could theoretically provide the necessary inputs for continually updating information about the orientation and structural features of objects in the three-dimensional egocentric space of the observer (see Newsome, Wurtz, and Komatsu 1988). Some of these cells seem capable of monitoring limb position during manual prehension (Mountcastle et al. 1975), while motion-sensitive cells in the temporal lobe have been reported not to respond to such self-produced visual motion (Hietanen and Perrett 1993).

The posterior parietal region is strongly linked to those premotor regions of the frontal cortex directly implicated in eye-movement control, reaching movements of the limbs, and grasping actions of the hands and fingers, and with areas that have been implicated in short-term coding of spatial location (for review, see Milner and Goodale 1993, 1995; Cavada and Goldman-Rakic 1993). Moreover, the different networks in the posterior parietal area that are modulated by different motor outputs (saccades, pursuit eye movements, fixation, reaching, and manipulation) appear to be linked to corresponding functional areas in the premotor region. It may be that different combinations of these networks are recruited for the production of different visuomotor acts (Milner and Goodale 1993; Stein 1992). The posterior parietal region also sends prominent projections to a number of pontine motor nuclei, which are linked, in turn, with the cerebellum (Stein and Glickstein 1992).

Finally, it should be noted that the dorsal stream receives visual inputs from both primary visual cortex and the superior colliculus (through nuclei in the thalamus, such as the pulvinar) (Gross 1991). Most of these inputs originate from the so-called magnocellular projections to LGNd or the superior colliculus.[4] The fact that there is a large input from the superior colliculus, a structure known to be intimately involved in the control of saccadic eye movements (Sparks and Mays 1990), again speaks to the important role of the dorsal stream in the visual control of motor output.

The Ventral Perceptual System

Unlike the dorsal stream, the ventral stream receives most of its visual inputs from primary visual cortex (Gross 1991). These projections are about equally divided between the magnocellular pathway and the parvocellular pathway (Ferrera, Nealey, and Maunsell 1992). In contrast to the cells in posterior parietal cortex described earlier, visually driven cells in inferotemporal cortex are unaffected by anesthesia and the concurrent behavior of the animal.

Many of the cells in inferotemporal cortex and in neighbouring areas show remarkable categorical specificity for visual stimuli (Gross 1973; Tanaka 1992; Perrett, Hietanen, Oram, and Benson 1992), and some of them maintain their selectivity irrespective of viewpoint, retinal image size, and even color (Hasselmo, Rolls, Baylis, and Nalwa 1989; Perrett, Oram, Harries, Bevan, Benson, and Thomas 1991; Hietanen, Perrett, Oram, Benson, and Dittrich 1992). In some respects then, these cells demonstrate the so-called object-centered coding first postulated by Marr (1982) in his classic account of vision. The majority of these object-centered cells are sensitive to faces or other biologically important stimuli. Most of the cells in inferotemporal cortex are not object-centered, however, and are sensitive instead to particular views of objects or particular object features. Nevertheless, cells in this region that are responsive to similar object features are clustered together in a columnar arrangement, much like the familiar columnar structure in primary visual cortex (Fujita, Tanaka, Ito, and Cheng 1992). These and other cells in inferotemporal cortex typically have exceptionally large receptive fields, usually including the fovea and often extending across the vertical meridian. Such ensembles of cells may participate in the generation of the object-based representations described earlier.

4. The retinal projections to the dorsal lateral geniculate nucleus (LGNd), the superior colliculus, and other central structures are divided into two main types: magnocellular projections, which arise from large retinal ganglion cells and project in turn to layers of large cells in the LGNd as well as to the superior colliculus; and parvocellular projections, which arise from smaller ganglion cells and project to layers of the LGNd containing small cells. The magnocellular or broadband channel, as it is sometimes called, is broadly tuned for wavelength and shows high temporal but low spatial resolution. The parvocellular, or color-opponent, channel is selectively tuned to different wavelengths and shows low temporal but high spatial resolution. Another pathway has recently been discovered from the interlaminar regions of LGNd to primary visual cortex (Hendry and Yoshioka 1994). This system appears to have heterogeneous functional properties, large receptive fields, and rather long response latencies.

It is well known that bilateral lesions of inferotemporal cortex typically produce severe deficits in visual recognition and discrimination learning (Dean 1982; Ungerleider and Mishkin 1982). More recent studies have shown that lesions confined to visual areas that represent earlier stages of visual processing in the ventral stream, such as area V4 (see Figure 5.7), also result in marked deficits in the learning of discriminations based on the orientation and form of visual stimuli (Heywood, Gadotti, and Cowey 1992; Walsh, Butler, Carden, and Kulikowski 1992). There is also evidence that some aspects of color-constancy mechanisms may depend on the integrity of V4 (Walsh, Carden, Butler, and Kulikowksi 1993). But, as was mentioned earlier, there is no evidence that monkeys with ventral-stream lesions have any problems controlling the actual visually guided movements they direct at goal objects. Indeed, Pribram (1967) once remarked that monkeys with bilateral inferotemporal lesions were still able to pluck gnats out of the air with great facility.

These electrophysiological and behavioral observations are entirely consistent with the idea that networks of cells in the ventral stream, in sharp contrast to the action systems of the dorsal stream, are more concerned with the enduring characteristics of objects than they are with the moment-to-moment changes in the visual array. As we saw earlier, such object-based descriptions form the basic raw material for recognition memory and other long-term representations of the visual world. There is extensive evidence for the neural encoding of such visual memories in the neighbouring regions of the medial temporal lobe and related limbic areas (see Fahy, Riches, and Brown 1993; Nishijo, Ono, Tamura, and Nakamura 1993). In addition, within inferotemporal cortex itself, there is evidence to suggest that the responsivity of cells can be modulated by the reinforcement history of the stimuli employed to study them (see Richmond and Sato 1987; Sakai and Miyashita 1992). There is also recent evidence that cells in this region may play a role in comparing current visual inputs with internal representations of recalled images, which are themselves presumably stored in other regions (Eskandar, Richmond, and Optican 1992; Eskandar, Optican, and Richmond 1992).

A New Division of Labor

The electrophysiological evidence reviewed above suggests that the ventral and dorsal streams of processing carry out very different transformations on their respective visual inputs. Transformations in the ventral stream appear to abstract the enduring features of objects so that long-term and object-based representations of objects (and their relations) can be eventually constructed, presumably with the help of semantic processes located in related networks elsewhere in the brain. Although some ventral systems appear to be dedicated to a limited class of biologically relevant objects, such as faces, the different kinds of representations constructed and delivered by the ventral stream and its associated structures are essentially object based and are not constrained by the particular disposition of the object with respect to the observer or the particular actions in which the observer is engaged. In contrast, the transformations carried out in the dorsal stream are very much concerned with computing the location and disposition of objects and object features within specific egocentric frames of reference. Each of these egocentric frames of reference is linked to a particular effector system; some are related to saccadic eye movements, others to fixation and pursuit, and still others to reaching and grasping movements. These different visuomotor systems are recruited in various combinations for the performance of particular goal-directed actions. In summary, the evidence supports the recent proposal of Goodale and Milner (1992) that the ventral stream mediates the visual perception of objects, while the dorsal stream mediates the visual control of skilled actions directed at those objects.

We can now begin to understand the reasons for the remarkable dissociation in the visual abilities of DF, the young woman we met at the beginning of the chapter—that is, why she is unable to identify an object, such as a pencil, or describe its size, shape, and orientation, even though she can reach out and pick up the pencil quite accurately, rotating her hand and opening her fingers the right amount well before she makes contact with it. It would appear that DF's perceptual systems have been compromised in some way by the anoxia she experienced; they no longer have access to information about the shape and form of objects, and thus she has no perceptual experience of these object qualities. At the same time, visually driven action systems in DF's brain, particularly those controlling reaching and grasping, appear to be working normally and can still utilize information about object shape and orientation to control the posture of the hand and fingers during execution of a grasping movement. In the next section, we return to the discussion of DF (and other patients) to see how well the pattern of deficits and spared visual abilities in these individuals conforms to the distinction between vision for perception and vision for action that has been developed here.

5.5   The Neuropsychology of Perception and Action

Individuals who have sustained damage to visual areas of the cerebral cortex sometimes have great difficulty identifying objects or drawings, even though their low-level visual abilities are largely intact. Such individuals are said to have *visual agnosia* (for a review, see Farah 1990). In the late nineteenth century, the neurologist Lissauer (1890) made a distinction between two kinds of visual agnosia. Patients suffering from what he termed *apperceptive agnosia* were thought to be unable to achieve a coherent percept of the structure of an object. In contrast, patients with what he termed *associative agnosia* were thought to be able to achieve such percepts but unable to recognize the object. Thus, according to Lissauer, an associative agnosic would be able to copy a drawing even though he or she might not be able to identify it, while an apperceptive agnosic would be unable to copy or identify the drawing. Lissauer argued that the apperceptive agnosic had a disruption at a relatively early stage of perceptual processing, while the disruption in the associative agnosic was at a higher, cognitive level of processing at which percepts are normally associated with stored semantic information. Although more recent writers have questioned whether or not the different disorders of object recognition can be explained by such a simple serial scheme (see Humphreys and Riddoch 1987a and Chapter 3, this volume), Lissauer's distinction is still used as a convenient, if crude, initial means of classifying individual patients.

In the context of the present arguments, apperceptive visual agnosia would appear to reflect damage to the human equivalent of the ventral stream, insofar as it is this stream that mediates our perception of objects and events in the world. Associative agnosia, on the other hand, might be characterized as a disconnection between those structures coding the semantic features of objects and the visual processing networks in the ventral stream itself—even though those networks might be quite intact. Thus, Lissauer's original conception of agnosia can be mapped onto the successive stages of processing in the perceptual systems of the ventral stream and beyond.

5.5.1   The Patient DF: Damaged Perception but Spared Visuomotor Abilities

DF, of course, would be classified as an apperceptive agnosic; her perception of objects is profoundly disturbed and she is incapable of recognizing or copying even the simplest of geometric forms. Some writers have referred to this kind of visual deficit as *visual form agnosia* (Benson and Greenberg 1969), which emphasizes the fact that the patient appears to have no access to visual information about the form and shape of objects. But, as we saw at the beginning of this chapter, other aspects of DF's behavior reveal a remarkable sensitivity to object shape.

For example, even though DF could not rotate a hand-held card to match the orientation of a slot placed in different positions around the clock, she had no difficulty posting the card in the slot (Figure 5.4). The matching task and the posting task might appear superficially similar. After all, in both cases, DF had to rotate the card to match the orientation of the target slot. But there is a crucial difference. In the matching task, what DF was doing by rotating the card was reporting her perception of the slot; the particular form in which the reporting was done was essentially arbitrary. We might have asked her, for example, to draw the slope of the slot on a piece of paper. In the posting task, however, she was performing an action and was thereby engaging a visuomotor system that initiates an obligatory rotation of the hand in order to perform the action successfully. Acts such as this are nonarbitrary and, as I argued earlier, require no explicit perception of the orientation of the hand or the slot. Presumably, our primate ancestors had to deal with this problem throughout their evolutionary history and, as a consequence, evolved dedicated visuomotor systems for its solution. DF's visuomotor systems work just fine; it is her perception of the slot's orientation that is the problem.

A similar dissociation between perception and action was evident in DF's visual processing of object size. Thus, when she was presented with a pair of rectangular blocks of the same or different dimensions (from two identical sets of five different blocks varying from a square to an elongated rectangle but with the same overall area), she was unable to distinguish between them. Even when she was asked to indicate the width of a single block by means of her index finger and thumb, her matches bore no relationship to the dimensions of the object and showed considerable trial-to-trial variability (see Figure 5.8, top). In contrast, when she was asked simply to reach out and pick up the block, the aperture between her index finger and thumb changed systematically with the width of the object, just as it does in normal subjects (see Figure 5.8, bottom).[5] In other words, DF scaled her grip to the dimensions of the object she was about to pick up, even though she appeared to be unable to perceive those object dimensions. DF's visual form agnosia may make it difficult for her to perceive the size and orientation of objects, but she is still able to access information about these object features to control her grasping movements.

But object size and orientation are not the only object features that control the parameters of grasping movements. Even casual observations of grasping reveal a remarkable sensitivity to the shape of an object as well; when we reach out to pick up an object, even one we have not seen before, we place our fingers on the surface of the object in such a way that it will not slip from our fingers when we exert lifting and gripping forces. Some locations on the surface of an object will provide more stable "grasp points" than others; that is, the locations where the friction required to prevent slip will be minimal. To choose these points, some sort of analysis of the overall shape of the object must be carried out. But does the visual

5. To make these measurements, changing positions of DF's hand and fingers were monitored by optoelectronic recording (WATSMART, Northern Digital Inc., Waterloo), in which the position of infrared light-emitting diodes (IREDs) attached to the thumb, index finger, and wrist were recorded by two infrared-sensitive cameras at 100 Hz. The three-dimensional trajectories of each IRED and the change in the size of the grip aperture (the distance between the IREDs on the index finger and thumb) were later reconstructed offline.

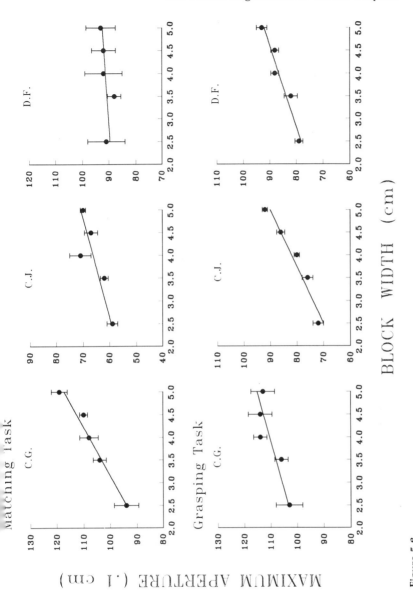

Figure 5.8
Aperture between the index finger and thumb in the perceptual matching task and the grasping task for DF and two control subjects presented with rectangular blocks of different dimensions. See text for description of tasks. In interpreting these graphs, it is the slope of the function that is important, rather than the absolute values plotted, since the placement of the infrared-light-emitting diodes on the fingers and the size of the hand varied somewhat from subject to subject.

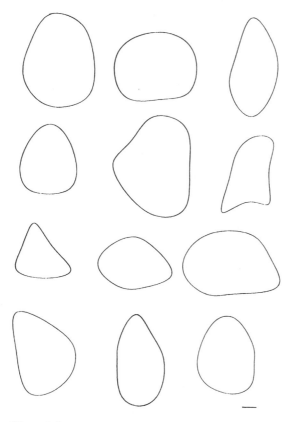

Figure 5.9
The twelve different shapes used in both the same-different visual discrimination task and the grasping task. The line at the bottom right-hand corner indicates 1 cm.

analysis of object shape for the purposes of controlling a grasping movement depend, like the related analysis of object size and orientation, on visual mechanisms that are relatively independent of those underlying the perceptual identification of objects?

To answer this question, we compared DF's ability to discriminate between objects of different shape with her ability to position her fingers correctly on the boundaries of those same objects when she was required to pick them up. The shapes (illustrated in Figure 5.9) were based on templates used to develop algorithms for the control of grasping in two-fingered robots working in novel environments (Blake 1992). Such shapes were chosen because they have smoothly bounded contours and an absence of clear symmetry. Thus, the determination of stable grasp points requires an analysis of the entire contour envelope of the shape. When DF was presented with pairs of these shapes, she was unable to determine

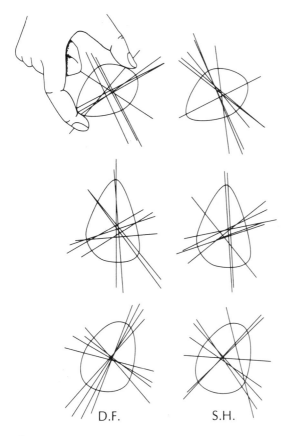

Figure 5.10
Diagrams showing the grasp lines (joining points where the thumb and index finger first made contact with the shape) selected by DF and a control subject (SH) when picking up three of the twelve shapes in the grasping task. The four different orientations at which each shape was presented have been rotated so that they are aligned. Notice the similarity between DF's grasp lines and those of the control subject. (Adapted by permission from M. A. Goodale. Visual pathways supporting perception and action in the primate cerebral cortex (1993). *Current Opinion in Neurobiology* 3, 578–585.)

whether the two shapes were the same or different; this failure was evident whether or not the two shapes on "same" trials had the same relative orientation or different orientations. But when DF was asked to pick up these objects using her index finger and thumb in a "precision grip," she had no difficulty placing her finger and thumb on stable grasp points on the circumference of the object.

As Figure 5.10 shows, the grasp points she selected were remarkably similar to those chosen by a neurologically intact control subject. In addition, DF showed the same systematic shift in the selection of grasp points as the control subject when the orientation of the object was changed. Moreover, in both DF and the control subject, the line joining the two grasp points passed through the center of mass of the object along an axis that defined the maximum or minimum diameter of the shape; in addition, the grasp points were often located on regions of the object boundary with the greatest convexity or (concavity)—regions that would be expected to yield the most stable grasp points. Thus, despite her inability to perceive the shape of objects, DF remained remarkably sensitive to object shape during the execution of a grasping movement.

In summary then, the brain damage that DF suffered as a consequence of anoxia appears to have interrupted the normal flow of shape and contour information into her perceptual systems without affecting the processing of shape and contour information by her visuomotor control systems. But where is the damage in DF's brain? If, as David Milner and I have proposed, the perception of object and events is mediated by the ventral stream of visual projections to inferotemporal cortex, then DF should show evidence of damage relatively early in this pathway. In fact, magnetic resonance imaging of DF's brain carried out just over a year after her accident revealed that most of the damage in the posterior portion of her cerebral cortex is located in the ventrolateral regions of the occipital cortex, a region containing visual areas that have been implicated in the human homologue of the ventral stream of processing. At the same time, her primary visual cortex, which provides input for both the dorsal and ventral streams, appears to be largely intact. Presumably, while input from primary visual cortex to the ventral stream has been compromised in DF, input from this structure to the dorsal stream is essentially intact. Of course, as was mentioned earlier, the dorsal stream, unlike the ventral stream, also receives input from the superior colliculus through the pulvinar (see Figure 5.7). Thus, input from both the superior colliculus and the lateral geniculate nucleus (via primary visual cortex), which converge on areas of the dorsal stream, could continue to mediate well-formed visuomotor reponses in DF.

There is some additional evidence, albeit rather indirect, to suggest that primary visual cortex in DF is still capable of processing orientation information and could therefore provide the required inputs to the action systems in the dorsal stream. In a recent study, Keith Humphrey, Rick Gurnsey, and I tested DF for the so-called McCollough effect (McCollough 1965). In this task, the subject is first exposed to alternating horizontal and vertical colored grating patterns for fifteen minutes or so, each grating being present for about ten seconds. One of the patterns, say the horizontal, is always red; the other is always green. Following this adaptation period, the subject is presented with test patterns containing either horizontal or vertical black-and-white gratings or both. Subjects typically experience strong color aftereffects when viewing these patterns such that they see colors in each black and white grating complementary to the colors viewed in the adaptation period. Thus, in the present example, they would report that the horizontal black-and-white grating appear "greenish" and the vertical one "pinkish." In other words, they show an orientation-contingent color aftereffect. Despite the fact that DF in earlier testing could not discriminate between the two black-and-white gratings used in the test phase, after adaptation with the colored gratings, she too showed a normal and vivid orientation-contingent color aftereffect (Humphrey, Goodale, and Gurnsey 1992). This striking observation must mean that, at some stage in her visual system, color and orientation processing are interacting. Since there is no evidence of color processing in the superior colliculus, the presence of a McCollough effect in DF argues that this orientation processing is preserved somewhere within the geniculostriate pathway, presumably in association with the same neural systems that ultimately yield her color phenomenology. One might speculate that while this orientation information is unable to reach the ventral perceptual systems in DF's damaged brain, it may still be conveyed to processing networks in her dorsal action stream.

One must be very cautious, however, about drawing strong conclusions about anatomy and pathways from patients like DF. Her deficits arose, not from a discrete lesion, but from anoxia. As a consequence, the brain damage in DF, while localized to some extent, is much more diffuse than it would be in a patient with a stroke or tumor. Yet, for reasons that we do not fully understand, anoxia tends to affect some visual pathways more than others. Thus, in some patients like DF, who have developed apperceptive or visual form agnosia following anoxia, the pathways mediating the perception of color and fine visual texture have been spared, even though the pathways mediating the perception of object shape and form are severely compromised.[6] At the same time, the anoxia also spared those pathways supplying the visuomotor systems with information about the shape, size, and orientation of goal objects. Yet, while the striking dissociation between perceptual and visuomotor abilities in DF can be mapped onto the distinction between the ventral and dorsal streams of visual processing that David Milner and I propose, that mapping can only be tentative. Our proposal is strengthened, however, by observations in other patients whose pattern of deficits is complementary to DF's and whose brain damage can be confidently localized to the dorsal stream.

6. It should be noted that although DF's ability to recognize line drawings of objects is extremely poor, she does much better with real objects. Systematic testing with real objects, however, reveals that for the most part she is basing her identification on the surface properties of the object, such as color, visual texture, and specularities. In this respect, she is similar to a number of other patients with apperceptive or visual form agnosia (Farah 1990).

5.5.2   Patients VK and RV: Visuomotor Deficits but Spared Perceptual Abilities

Patients with damage to the superior portions of the posterior parietal cortex often exhibit *optic ataxia*; that is, they are unable to use visual information to reach out and grasp objects in the hemifield contralateral to the lesion. At the same time, they often have no difficulty recognizing or describing objects that are presented in that part of the visual field. On the face of it, these clinical observations fit well with the "what" versus "where" dichotomy originally proposed by Ungerleider and Mishkin (1982), in which the posterior parietal cortex was thought to play the major role in spatial vision, localizing stimuli in space. But a closer examination of the behavior of patients with posterior parietal lesions reveals a constellation of visuomotor deficits that is more consistent with the idea that this structure is part of a cortical system mediating the visual control of skilled action (Goodale and Milner 1992).

Some patients who have been diagnosed as suffering from optic ataxia not only have difficulty reaching in the right direction but also show deficits in their ability to position the fingers or adjust the orientation of their hand when reaching toward an object in their contralesional field (Perenin and Vighetto 1988). At the same time, such patients are quite able to describe the orientation of objects presented in the field contralateral to their lesion (Jeannerod 1988). Patients with posterior parietal lesions can also show problems in scaling their grip to the sizes of objects they intend to pick up. We recently examined a seventy-six-year-old woman (VK) who had bilateral lesions in the occipitoparietal region of both hemispheres as the result of separate strokes (Jakobson, Archibald, Carey, and Goodale 1991). Even though VK was able to identify line drawings of common objects with little difficulty, the size of her grasp was only weakly related to the size of the objects she was asked to pick up; she often opened her hand as wide for small objects as she did for large ones (Figure 5.11).

Moreover, compared to age-matched control subjects, VK took much longer to initiate and execute the movement and also made a large number of adjustments in grip aperture as she closed in on the target (Figure 5.12). Her difficulty was not due to a motor deficit per se; she showed no evidence of motor weakness in the hand we tested, and her ability to tap her finger quickly (a measure of motor control) was in the normal range for her age. VK's deficit was truly visuomotor in nature.

In addition to problems dealing with object size and orientation, we might expect that some patients with damage to the posterior parietal region would show deficits in the visuomotor processing of object shape. This possibility was recently tested on a fifty-five-year-old woman (RV), who like VK, had sustained bilateral lesions to the occipitoparietal region

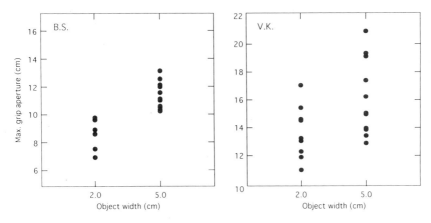

Figure 5.11
Graphs illustrating maximum grip aperture achieved by the patient VK and an age-matched control subject (BS) when picking up blocks of two different sizes. Note that VK often opens her hand as wide for the smaller object as she does for the larger one.

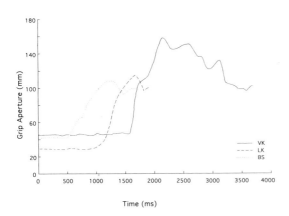

Figure 5.12
Representative grasp profiles for the patient VK and two age-matched control subjects (BS and LK). Note that VK shows a number of in-flight adjustments to the aperture of her grip—adjustments rarely seen in the grasp profiles of normal subjects.

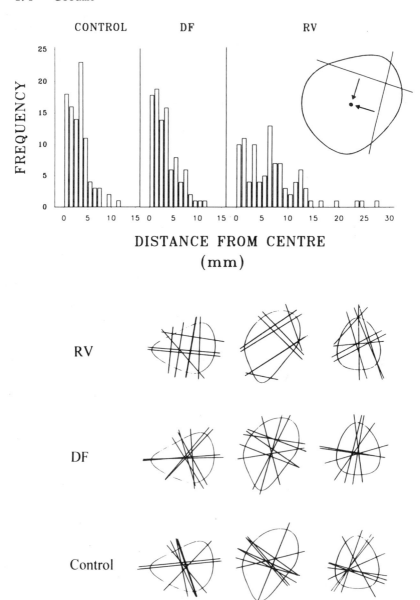

Figure 5.13
The diagram at the bottom of this figure shows the grasp lines (joining points where the thumb and index finger first made contact with the shape) selected by the visual agnosic patient (DF), the optic ataxia patient (RV), and the control subject (SH) when picking up three of the twelve shapes. The four different orientations in which each shape was presented have, again, been rotated so that they are aligned. Again, notice the similarity between DF's grasp lines and those of the control subject, and how their grasp lines differ

as a result of two separate strokes she experienced in one week. RV had no difficulty recognizing line drawings of common objects, achieving a perfect score on a twenty-item test. She had good motor strength and her finger-tapping scores were within the normal range. When presented with pairs of the smoothly contoured shapes we had used with DF (Figure 5.9), she made only a few errors on a same-different discrimination test, achieving a mean score of 85 percent. (DF, you will remember, was at chance level on this test.) Yet when asked to reach out and pick up each of the shapes, RV placed her index finger and thumb on very unstable grasp points on the shape boundary. Figure 5.13 illustrates her grasps, alongside those of DF and a normal control subject; note that the lines joining RV's opposed grasp points do not consistently pass through the center of mass of the shape as they do for the two other subjects; nor are the grasp points themselves located at such obviously stable locations on the shape boundary as regions of high convexity. In short, RV appears to be grossly deficient in her use of object-shape information for guiding her grip.

Observations such as these suggest that it is not only the spatial location of the object that is inaccessible for controlling movement in patients with dorsal stream lesions but the intrinsic characteristics of the object as well. It is not enough to say that these patients have deficits in spatial vision, as Ungerleider and Mishkin (1982) might claim. In fact, in one clear sense they do not; not only can they identify objects in the field contralateral to the lesion, they can also describe the relative location of those objects, even though they cannot pick them up (Jeannerod 1988). Of course, this pattern of deficits is quite consistent with our proposal that the posterior parietal cortex plays a critical role in the visuomotor transformations required for skilled actions such as visually guided prehension (Goodale and Milner 1992). As was emphasized earlier, the transformations underlying actions such as manual prehension require more than spatial location information; they also require information about such structural features of the goal object as size, shape, and local orientation. In this connection, it should be pointed out that not all patients with damage to the posterior parietal region have difficulty shaping their hand to correspond to the structural features and local orientation of the target object. Some have difficulty with hand postures, some with controlling the direction of their grasp, and some with "foveating" the target. Indeed, depending upon the size and locus of the lesion, a patient can demonstrate any

---

from those of RV, who often chooses very unstable grasp points. The frequency distributions above illustrate the distance between the grasp lines and the centre of mass of the shape for DF, RV, and the control subject (SH) for all twelve shapes. The inset shows how these distances were computed. Notice again that whereas DF and SH's grasp lines tend to pass through or close to the center of mass of the shape, this is not the case for RV.

combination of these visuomotor deficits. (For a review, see Jeannerod 1988 and Milner and Goodale, 1993, 1995.) Different subregions of the posterior parietal cortex, it appears, support different visuomotor components of a skilled act, a result that would be predicted from the electrophysiological evidence reviewed earlier.

The neuropsychological studies reviewed in this section show that the pattern of deficits and spared abilities in patients with lesions of the dorsal stream of visual processing are essentially complementary to those observed in DF, who is thought to have damage confined largely to the ventral stream. This double dissociation lends considerable support to the proposed division of labor outlined earlier, in which it was argued that the ventral stream mediates the perception of the visual world while the dorsal stream mediates the visual control of skilled actions in that world (Goodale and Milner 1992).

5.6   Limits on Visuomotor Processing

As I discussed in section 5.4, a direct implication of this proposal is that rather different transformations of visual information are carried out by each stream. These differences, it was argued, arise from differences in the output requirements of each stream. In this section, I examine how these differences might be reflected in the way DF and other subjects deal with particular kinds of visually guided acts. The performance of DF (and others) on these tasks provide us with even more insight into the functional architecture of the systems controlling perception and action.

5.6.1   Limits Arising from the Nature of the Response

DF, who is very poor at discriminating the local orientation, shape, and dimensions of objects, can still use this information to control grasping movements directed at those objects. Thus, as we have already seen , she can use outline shape to select the optimal grasp points on a smoothly contoured object when asked to pick it up. She can also compute the principal axis of a simple shape, such as a slot when required to rotate her hand to insert a hand-held card into that slot. But can she use outline shape to control the rotation of her hand when the shape of the card and its corresponding slot is more complex, and involves matching more than one principal axis? In a recent experiment in our laboratory (Goodale, Jakobson, Milner, Perrett, Benson, and Hietanen 1994), we attempted to answer this question by asking DF to insert a T-shaped object into T-shaped aperture; successful hand rotation on this task required her to combine the two linear components of the aperture into a composite visual pattern. (In perceptual tests, of course, DF was quite unable to identify

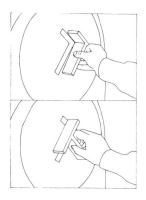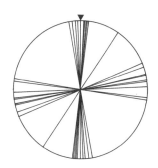

Figure 5.14
The diagram on left the illustrates DF's correct and incorrect postings of the T-shaped object into the T-shaped slot. The polar plot on the right illustrates the position of the T-shaped object as DF attempts to insert it into a T-shaped slot at different orientations. The correct orientation on each trial has been rotated to vertical. Note that on about half the trials DF rotates the hand-held T to the correct orientation and on the remaining trials she rotates it to a position 90 degrees from the correct orientation.

reliably the letter *T* or any other letter or similar pattern.) On approximately half the trials, she was quite accurate at orienting the hand-held T to match the orientation of the T-shaped slot. On the remaining trials, however, she attempted to insert the stem of the T-shaped form into the top of the slot; that is, she made an error of approximately 90 degrees (see Figure 5.14). It was as if the visuomotor transformations guiding the rotation of DF's hand were able to compute only a single orientation at a time and that, as a consequence, they were unable to combine the orientations of the stem and the top of the T appropriately.

Yet, as we saw earlier, when asked to pick up an object using a precision grip, she was able to use information about the entire outline of the shape to select the optimal grasp points. In other words, one visuomotor system, that controlling the posture and placement of the fingers during grasping, appears to be sensitive to relatively complex shape information while another, the system controlling the rotation of the wrist (when the hand is holding an object), is not. In fact, the transformations underlying the rotation of the wrist appear to be sensitive to only one principal axis of the goal object at a time. From an evolutionary perspective, this makes good sense. Rotating the wrist so that the extended hand can fit into an aperture or grab a branch quickly would have been a common event in the lives of our primate ancestors and, therefore, might be expected to utilize relatively hard-wired, dedicated systems (presumably located in the dorsal stream) that can compute a single optimal orientation quickly and with some precision. Such a dedicated system would be of little use when an observer is required to rotate the hand to match the axes of a complex-hand-held object with those of an equivalent pattern. In that case, it might make more sense to coopt more flexible perceptual systems (perhaps in the ventral stream) that have evolved to analyze such objects for other reasons. DF, of course, does not possess such systems, and as a consequence she is unable to match the T-shaped card with the T-shaped aperture.

5.6.2   Limits in the Processing of Edges and Contours

In all the visuomotor tasks DF performed well, the controlling visual stimuli were luminance defined. Other information about object boundaries might have been available in any one task, but a luminance-difference cue was common to all tasks. In fact, when the visual stimuli defining the slot in the "posting" task were systematically manipulated in a recent experiment (which used a kind of "virtual" slot), it was found that a difference in luminance was the only cue that DF could use reliably to control the rotation of her hand as she aligned the hand-held card with the slot (Goodale et al. 1994). For example, contours defined by the so-called Gestalt principles of organization—similarity, proximity, and good continuity—while they are clearly evident to the normal observer, provided no useful information for DF's residual visuomotor system (Figure 5.15).

The overriding tendency of DF's visuomotor system to rely on luminance-defined edges may reflect the fact that the dorsal stream is not capable of utilizing more subtle cues to demarcate the boundaries of goal objects. These sorts of visual stimuli may be available only to perceptual systems, which are concerned with parsing the world into discrete objects. As was mentioned in section 5.4, the efficient online control of action requires that visual information be reliably and rapidly processed. If the shape of a goal object can be determined only by reference to more computationally expensive algorithms, then visuomotor systems in the dorsal stream may have to rely on input from a more circuitous route involving perceptual machinery in the ventral stream.

5.6.3   Temporal and Spatial Limits in Visuomotor Control

Because observers (and sometimes objects) are often moving, the egocentric coordinates of a goal object can change considerably from moment to moment. As a consequence, it would be efficient to compute the required coordinates for action immediately before movements are initiated and quite inefficient to store such coordinates (or the resulting motor programs) for more than a few milliseconds before executing the action (as the actual coordinates of the goal object could change dramatically in that time). In short, to work properly, visuomotor systems in the dorsal stream

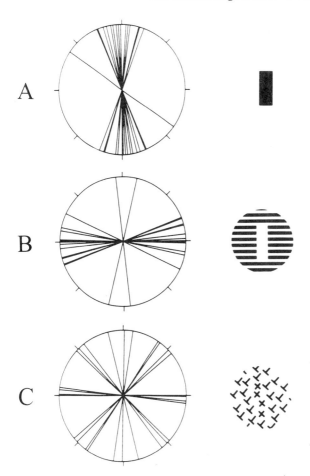

Figure 5.15
Polar plots illustrating the orientation of the hand-held card as DF attempts to place it on a "virtual" slot defined by different visual features. The correct orientation on each trial has been rotated to vertical. Panel A shows her responses to a luminance-defined slot; Panel B shows her responses to a slot defined by a missing rectangle on an orthogonal background of high-contrast stripes; and Panel C shows her responses to a slot defined by the similarity of patterned items. Note that the only slot that DF can deal with in this situation is the one well-defined by luminance (Panel A). When faced with a slot defined by a missing rectangle within a spatial frequency grating (Panel B), her visuomotor response is captured by the orientation of the background grating.

have to work almost entirely online. Thus, movements directed to remembered objects (objects that were present, but are no longer) might be expected to look rather different from movements directed to objects in real time.

In a recent experiment (Goodale, Jakobson, and Keillor 1994), we examined the kinematics (i.e., the spatiotemporal organization) of grasping movements made by normal subjects to a "remembered" object. The experiment was run as follows: subjects were first shown a rectangular block, the size and distance of which varied from trial to trial. Automated shutters located in front of their eyes were then closed for two seconds and the block was removed. The shutters then opened and the subjects were required to reach out and pretend to pick up the block as if it were still there. In other words, subjects were being asked to pantomime a grasping movement two seconds after last seeing the intended goal object. Subjects performed these pantomimed actions in a manner that was quite different from the way in which they executed natural, goal-directed grasping movements. Their mimed actions, compared to normal reaches, consistently reached lower peak velocities, tended to last longer, followed more curvilinear trajectories, and undershot target location. Moreover, subjects consistently opened their hand less when miming than when reaching for objects that were physically present. Nevertheless, their grip aperture was still highly correlated with the size of the object they had viewed just seconds before.

The programming of pantomimed movements such as these must rely, not on current visual information, but rather on a stored representation of the previously seen object and its spatial location. But what is the nature of this representation? If, as I argued above, visuomotor systems operate only in real time, the stored information driving pantomimed actions must depend on another system, one designed specifically for representing objects in their spatial locations over longer periods of time—in short, the perceptual system presumed to mediate object recognition. If this is the case, then DF, who appears to have no perception of object size, shape, and orientation, should have real trouble pantomiming grasping movements to objects seen only two seconds earlier. This is indeed the case. As Figure 5.16 illustrates, after a delay of two seconds, DF appeared to have lost all information about object size needed to preshape her hand in flight. Of course, this was to be expected as DF had no percept of the object in the first place. Thus, when no object was present to drive her real-time visuomotor control systems, she could not fall back on the stored information about object size available to normal subjects.

For the perceptual systems (i.e., those involved in visual learning and recognition), a retention interval of two seconds is trivial. Clearly we are capable of remembering the characteristics of objects we have seen only

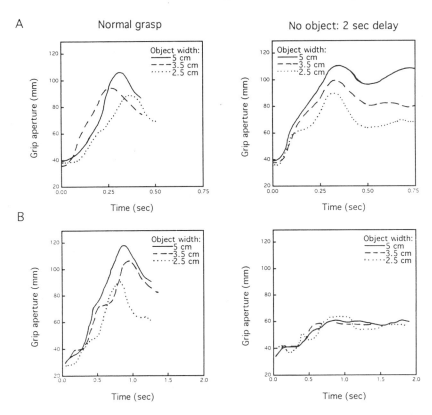

Figure 5.16
Representative profiles for grasps made by a control subject (A) and DF (B) to blocks of three different sizes when reaching normally in real time or after a two-second delay. Note that although the control subject continues to scale in the delay condition, DF shows no evidence of grip-scaling in this condition.

once for extremely long periods of time. The visuomotor coordinates needed to program a given movement, however, may have to be updated even over intervals as short as two seconds as the relative positions of the observer and the object change. Thus it would be counterproductive to store these coordinates for any significant period of time; far better that they be calculated immediately before each action occurs. It is perhaps not surprising, then, that increasing the delay from two to thirty seconds had no appreciable effect on the performance of either the normal subjects or DF.

The control of manual prehension depends on visuomotor systems that not only operate in real time but work with coordinate systems that locate the object in egocentric frames of reference. Thus, we would expect them to be ill equipped to deal not only with a temporal delay between seeing the object and directing an action toward it but also with a significant spatial displacement of the required output coordinates. We were not surprised to find therefore that requiring normal subjects to pantomime a grasping movement beside an object, as opposed to grasping it directly, resulted in the same kind of change in the kinematics of the movements as was seen in the temporally delayed responses. Moreover, since these pantomimed movements were presumably also driven by perceptual representations of the target object, rather than by online visuomotor control systems, DF was unable to perform them convincingly (Goodale, Jakobson, and Keillor 1994).

Results such as these provide convincing evidence that the visuomotor systems underlying goal-directed actions are different from those underlying pantomimed actions, which appear to depend more on the perceptual systems mediating object recognition. Nevertheless, even though the execution of a goal-directed action may depend on dedicated, online, control systems in the dorsal stream, the selection of appropriate goal objects and the action to be performed must depend in part on the perceptual machinery of the ventral stream. In other words, the two systems must work together in the production of normal goal-directed behavior.

5.7   Some Remaining Issues and Puzzles

5.7.1   Space and the Parietal Lobe

In their original conception of the division of labor between the two streams, Ungerleider and Mishkin (1982) argued that "spatial vision" was mediated largely by the dorsal stream of visual processing. One of the important pieces of behavioral evidence for this claim was the observation that monkeys with posterior parietal lesions were impaired on the so-called landmark task, in which an animal is required to choose one of two covered foodwells on the basis of the proximity of a landmark object placed somewhere between the two (Pohl 1973; Ungerleider and Brody 1977). Although it is commonly assumed that animals with inferotemporal lesions are unimpaired on the landmark task, even the early studies (Pohl 1973; Ungerleider and Brody 1977) showed that such animals are nonetheless impaired relative to control animals, although not so severely as monkeys with posterior parietal lesions. Monkeys with parietal damage are particularly impaired on a version of the landmark task in which the task is made more difficult over successive training days by moving the landmark closer to the midpoint between the two foodwells. But then again, even normal monkeys have difficulty when the landmark is moved farther and farther away from the correct response site. Part of the problem seems to be that if the animal fails to look at the landmark, its behavior falls to chance (Sayner and Davis 1972). In fact, looking at and touching the landmark before making their foodwell choice is a strategy that many normal monkeys adopt to solve the problem. Since monkeys, like humans, often show deficits in the visual control of their saccadic eye movements (e.g., Lynch and McLaren 1989) and/or limb movements (for review, see Ettlinger 1990) following posterior parietal lesions, such animals would be less likely to engage in this strategy and, as a consequence, might fail to choose the correct foodwell. This explanation for the poor performance of monkeys with parietal lesions is supported by the observation that such animals are also impaired on tasks in which the cue is separated from the foodwell but on which not its location but one of its object features (such as its color) determines the correct foodwell choice (Bates and Ettlinger 1960; Lawler and Cowey 1987; Mendoza and Thomas 1975). In summary, the impairment on landmark tasks following dorsal-stream lesions is most likely due to disruption in the circuitry controlling particular visuomotor outputs, such as shifts in gaze and goal-directed reaching, rather than a general disturbance in spatial vision. In fact, there is little other evidence to suggest that monkeys with posterior parietal lesions show deficits in spatial perception. (For a detailed discussion of this issue, see Milner and Goodale 1995)

Humans, like monkeys, also show disturbances in the visual control of skilled movements (optic ataxia) following damage to the posterior parietal cortex (see section 5.5.2). But damage to the posterior parietal region has been also been associated with other kinds of visual deficits, some of which have a strong spatioperceptual component. For example, lesions to this region, particularly in the right hemisphere, often result in a phenomenon known as visual neglect, in which patients ignore or fail to attend to visual stimuli presented to the side of space contralateral to the lesion. What is seldom appreciated, however, is the fact that the critical region for neglect is located much more ventrally in the parietal cortex than the critical region for optic ataxia. Indeed, the area most commonly implicated in visual neglect following damage to the posterior portion of the brain is a region at the junction of the occipital, temporal, and parietal lobes (Heilman, Watson, Valenstein, and Damasio 1983; Vallar and Perani 1986) —an area that is several centimeters below the area typically associated with optic ataxia (Perenin and Vighetto 1988).

The phenomenon of neglect, then, may reflect the emergence of spatial processing in the human brain that is rather different from that carried out in the dorsal stream, which appears to be largely concerned with visuomotor control. In fact, the human circuitry that corresponds to the dorsal stream in monkey may be confined largely to the more superior region of the posterior parietal cortex. Yet there are large regions of the posterior parietal cortex in humans that cannot be subsumed within the old primate dorsal stream. New systems in these and associated regions appear to have emerged for mediating many of the complex visuocognitive abilities (and more representational motor acts) that characterize our species. There is considerable evidence, for example, that high-level spatial skills such as map reading, maze learning, and mental rotation are particular sensitive to parietal damage (for review, see Milner and Goodale 1995).[7] The circuitry mediating these high-level skills may have coopted (in the evolutionary sense) some of the transformational algorithms that originally evolved for the control of movement. Thus, some forms of mental rotation and map reading (in which one moves from one set of coordinates to another) may make use of the viewer-centered representations that those dorsal-stream algorithms deliver. But, unlike the ancient visuomotor networks from which they may have evolved, these new networks can generate experential representations that can be manipulated for more cognitive purposes.

In summary, the posterior parietal cortex in humans is a complex structure containing not only a variety of different visuomotor networks (the ancient dorsal stream), each with rather different egocentric coordinate systems, but also a number of other spatial mechanisms with more cognitive functions (which may have evolved from the older dorsal-stream mechanisms). Moreover, spatial processing is not unique to the dorsal stream or to the posterior parietal cortex but is an ubiquitous feature of visual processing throughout cortex. After all, the ventral stream too must carry out spatial processing; in many cases, the identification of an object can only occur if the relative spatial location of object features (and other objects in the scene) has been determined. Indeed, it makes no sense to speak of spatial vision when space and time are the only dimensions the visual system can use to code real-world events and control real-world behavior.

5.7.2   Cooperation between the Two Streams

Although visuomotor systems in the dorsal stream are critical to the execution of efficient goal-directed grasping movements, it is unlikely that these pathways are involved in the initial selection of the goal object. After all, selecting the goal requires object recognition and access to semantic information about the significance of objects in the visual array—a process in which visual mechanisms in the ventral, not the dorsal, stream

7. Route finding using landmarks, however, is not typically associated with damage to the posterior parietal cortex. Patients with *topographical agnosia*, as this route-finding deficit is sometimes called, typically have lesions in areas corresponding to the ventral stream (Milner and Goodale 1995). Their deficit is more allocentric in nature and is often complicated by a failure to recognize familiar landmarks.

play a leading role. The problem of addressing the correct goal object for action is one that was identified over twenty years ago by P. M. Milner (1974), well before any distinction was made between the two streams of visual processing. Nevertheless, a solution to the problem would appear to require that the two streams cooperate in the planning and execution of a goal-directed action. Perhaps, the goal object is "flagged" in some way following identification by the ventral stream and this indexing is transmitted to the appropriate networks in the dorsal stream. Just how the appropriate networks are themselves identified is another problem. In other words, it is not clear how the particular action that one wishes to perform is translated into activation of some visuomotor networks in the dorsal stream and not others.

The problems of indexing a goal object and activating an appropriate action are clearly issues related to *attention* (see Chapter 2, this volume). Eventually, any comprehensive account of the functions of the two streams must deal with attention and the possibility that different attentional mechanisms might be at work in the dorsal and ventral pathways. Electrophysiological studies in the monkey have shown that cells in both streams can be affected by attention. Thus, cells in the dorsal stream (in posterior parietal cortex, for example) have been shown to be modulated by switches of attention to different parts of the visual field (Bushnell, Goldberg, and Robinson 1981). Cells in the ventral stream (in area V4, for example) also appear to have attentional properties, albeit of a somewhat different kind (Moran and Desimone 1985). In both cases, however, the modulation appears to reflect a switch in the spatial location of attention. In the ventral stream, the cell is enhanced for a *stimulus* in the attended place; in the dorsal stream, the cell is selective for a planned *response* to the attended place. Unfortunately, however, the interactions between these putative attentional systems in the two streams (as well as their relationship with prefrontal and premotor attentional mechanisms) is poorly understood. (For a discussion of these and related issues, see Milner and Goodale 1995.)

There is also the problem of how semantic information about the goal object is used to modulate the action itself. The kinematics of a particular action often reflect the function of the goal object to which that action is directed as well as its size, shape, orientation, and location. For example, when we pick up a hammer our grip reflects not only sensory information about the specific disposition of the hammer in our egocentric space but also the fact that we know its a hammer. We pick it up in such a way that we are ready to hammer nails with it. Presumably then, high-level systems controlling the production of skilled movements need to have access to the products of the ventral stream's processing so that they can modulate the relevant visuomotor networks. In the monkey there are anatomical connections from the inferotemporal cortex that could transmit highly processed visual information to inform parieto-frontal circuits that are involved in praxis (Seltzer and Pandya 1980; Cavada and Goldman-Rakic 1989; Andersen et al. 1990; Harries and Perrett 1991). Moreover, the ventral and dorsal streams are themselves interconnected (Felleman and Van Essen 1991). Whether or not such pathways in the human brain are involved in the functional modulation of action is unknown. Recently, however, a patient has been described who behaves as if her praxic system has a selective loss of these object-identity inputs (A. Sirigu et al., personal communication). The patient's two cortical visual systems seem to be essentially intact. She is able to pick up objects quite efficiently and has no difficulty recognizing familiar objects objects. In other words, she suffers from neither optic ataxia nor visual agnosia. But when the patient is shown a familiar object she will often attempt to pick it up using a grasp that, while efficient, is inappropriate for the use of the object. That is, although the two systems appear to be functioning adequately in isolation, they appear to be disconnected from each other.

The patient DF, of course, cannot generate functionally appropriate grasps toward objects she does not recognize, even though the grasping movements she produces are efficient with respect to the size, shape, orientation, and location of the objects. Nevertheless, because her perception of color and fine texture is intact, she is sometimes able to identify familiar objects. On these occasions, she can grasp the objects in a manner appropriate to their use. Thus, unlike Sirigu's patient, the connections between DF's ventral stream and her praxic and visuomotor systems appear to be intact. Her failure to generate grasps appropriate to the function of a goal object appear to be due to a disruption in form perception much earlier in the visual processing carried out within ventral stream.

5.7.3   The Two Cortical Streams and Consciousness

Even though DF can pick up an object under visual control with considerable facility, she cannot "see" the very object features that are driving her grasping movements. In other words, she has no conscious experience of the shape, size, and orientation of objects. Her inability to give a perceptual report of these object features is not due to some disconnection between visual perception and speech. As Figures 5.4, top and 5.8, top illustrate, even when she uses a manual response, she is unable to indicate anything at all about these characteristics of the target object. In fact, she is unable to demonstrate any recognition of objects on the basis of their shape, size, or orientation, no matter what form of perceptual report including forced-choice responding, is required.

As we saw earlier, DF's problem is probably due to damage early in the ventral stream of visual processing. It could be argued, therefore, that visual processing in this stream, which appears to provide the perceptual foundation for cognitive operations in the primate brain, is closely linked with visual phenomenology and awareness. In contrast, the visual processing that is carried out in the dorsal stream, despite the complex computations demanded by its role in the control of action, is not normally available to awareness. Indeed, there is considerable evidence from studies in normal subjects that manipulations of visual stimuli that produce large changes in motor output are often totally unavailable to perceptual report. (For a review of this literature, see Goodale 1988b.)

All of this discussion, of course, is highly speculative. Exactly what is meant by consciousness and how it relates to visual perception is not immediately clear. Nor is it clear what the function of consciousness might be.[8] What is apparent, however, is that patients like DF have maintained the ability to control skilled actions on the basis of visual stimuli they can no longer report. Despite this caveat, however, there are reasons to believe that perceptual processing within the ventral stream is linked with awareness. DF, for example, makes a strong distinction between what she can "see" and what she cannot "see." She can "see" color and fine visual texture; she cannot "see" shape. When she tries to guess about shape, even using forced choice, her response are always at chance level. In other words, DF has both recognition and awareness in relation to color, but she has neither of these in relation to shape.

One might speculate about why processing in the ventral stream is directly linked to awareness whereas processing in the dorsal stream is not. Since the ventral stream, according to Goodale and Milner, is strongly linked with semantics, perhaps the association with conscious experience is a requirement of the cognitive processes served by these visual pathways. In other words, the perceptual representations generated by the ventral stream are the visual component of conscious experience. As emphasized earlier, these representations utilize the enduring characteristics of objects and their relations. In contrast, the action systems of the dorsal stream depend on visuomotor transformations within egocentric or viewer-centered frames of reference—transformations that are not ordinarily available to consciousness. Indeed, intrusions of such viewer-centered information into visual phenomenology could disrupt the continuity of object identities across changing viewing conditions.

Even though the ventral stream appears to be more closely linked with conscious experience than is the dorsal stream, it does not follow that this stream is somehow the seat of consciousness. It is likely that processing in the dorsal stream and other visual pathways modulates the contents of our

8. These important issues are beyond the scope of the present chapter. The reader is referred to a recent discussion of these problems in which neuropsychological findings are examined in the context of consciousness and awareness (Milner and Rugg 1992).

experience and that other networks also contribute to the unity of our consciousness of the world. As we have already seen, there is evidence that the posterior parietal and inferotemporal regions are heavily interconnected. In addition, both streams project to areas in the superior temporal sulcus that receive input from a number of other modalities as well (e.g., Boussaoud, Ungerleider, and Desimone 1990; Morel and Bullier 1990; Baizer, Ungerleider, and Desimone 1991). Such interconnectivity may provide some of the integration necessary for essential unity and cohesion of our perceptual experience. Yet there may be other regions, such as parts of the frontal lobe, that play a superordinate role in the control of awareness. At the same time, the neuropsychological evidence reviewed in this chapter would suggest that activation of the ventral pathway may be a necessary condition for conscious visual experience.

Suggestions for Further Reading

For a more detailed discussion of all the issues raised in this chapter, readers are directed to a recent book by Milner and Goodale (1995). In this book, we discuss the distinction between perception and action and the mapping of this distinction onto the ventral and dorsal streams of cortical processing within a comparative and evolutionary account of visual functions. Several reviews articles—e.g., Goodale (1993), Goodale and Milner (1992), Milner and Goodale (1993)—also deal with these issues.

Readers interested in Ungerleider and Mishkin's (1982) original account are directed to their chapter in the published proceedings of a conference on the analysis of visual behavior held at Brandeis University in 1978. For more recent accounts of the patterns of connectivity within the cortical visual areas, see Boussaoud, Ungerleider, and Desimone (1990) and Felleman and Van Essen (1991).

A recent monograph by Farah (1990) provides a good survey of the classical and recent literature on visual form agnosia and related disorders. Although Farah tends to adopt a determinedly cognitive approach to visual agnosia with little discussion of the biology of vision, she does offer an excellent review of the different varieties of this fascinating disorder. Humphreys and Riddoch (1987b) have written a wonderful account of a patient with visual associative agnosia whose behavior can be usefully contrasted with that of patient DF.

Finally, readers interested in learning more about optic ataxia and other deficits associated with posterior parietal damage might wish to consult Jeannerod (1988). This book also offers an excellent account of the functional and neural organization of the systems underlying manual prehension.

Problems

5.1   We have seen that patient DF, who suffers from visual form agnosia, is unable to distinguish between two different objects, such as a square and a rectangle, in a same-different discrimination. Suppose she is presented with a square and a rectangle and, instead of being asked whether the two shapes were the same or different, she was asked simply to "pick up the square." Would her discrimination performance improve? In other words, would she tend to pick up the square more often than the rectangle, even when the left-right position of the two shapes is varied from trial to trial? Outline the reasons why her performance might be expected to improve (or why it would not)?

5.2   Visual illusions provide important insights into how the brain's perceptual machinery organizes the raw visual image falling on the retina. Is there any reason to suppose that the so-called action system described by Goodale and Milner (1992) would be subject to the same visual illusions as the perceptual system? Should some visual illusions affect the computations of both systems, and others only one? Why (or why not)?

5.3   Studies of resolution acuity and other psychophysical functions typically require subjects to indicate whether or not they "see" a difference between two visual targets. Do the results of such studies establish the psychophysical limits of the visual processes supporting different visuomotor systems, such as visually guided prehension and visually controlled postural adjustments? How might you design an experiment to determine whether or not this is the case?

5.4   How might you design an experiment to demonstrate that people are not always able to perceive visual stimuli that are nevertheless controlling their motor output?

References

Andersen, R. A. (1987). Inferior parietal lobule function in spatial perception and visuo-motor integration. In V. B. Mountcastle, F. Plum and S. R. Geiger, eds., *Handbook of physiology*, Section 1: *The nervous system*, vol. 5: *Higher functions of the brain*, Part 2, 483–518. Bethesda, MD: American Physiological Association.

Andersen, R. A., C. Asanuma, G., Essick, and R. M. Siegel (1990). Corticocortical connections of anatomically and physiologically defined subdivisions within the inferior parietal lobule. *Journal of Comparative Neurology* 296, 65–113.

Baizer, J. S., L. G. Ungerleider, and R. Desimone (1991). Organization of visual input to the inferior temporal and posterior parietal cortex in macaques. *Journal of Neuroscience* 11, 168–190.

Bates, J. A. V., and G. Ettlinger (1960). Posterior biparietal ablations in the monkey. *Archives of Neurology* 3, 177–192.

Benson, D. F., and J. P. Greenberg (1969). Visual form agnosia. *Archives of Neurology* 20, 82–89.

Blake, A. (1992). Computational modelling of hand-eye coordination. *Philosophical Transactions of the Royal Society* 337, 351–360.

Boussaoud, D., L. G. Ungerleider, and R. Desimone (1990). Pathways for motion analysis: Cortical connections of the medial superior temporal and fundus of the superior temporal visual areas in the macaque. *Journal of Comparative Neurology* 296, 462–495.

Bruce, C. J. (1990). Integration of sensory and motor signals in primate frontal eye fields. In G. M. Edelman, W. E. Gall and W. M. Cowan, eds., *Signal and sense: Local and global order in perceptual maps*, 261–314. New York: Wiley-Liss.

Bushnell, M. C., M. E. Goldberg, and D. L. Robinson (1981). Behavioral enhancement of visual responses in monkey cerebral cortex. I. Modulation in posterior parietal cortex related to selective attention. *Journal of Neurophysiology* 46, 755–772.

Bülthoff, H. H., and S. Edelman (1992). Psychophysical support for a two-dimensional view interpolation theory of object recognition. *Proceedings of the National Academy of Science* 89, 60–64.

Cavada, C., and P. S. Goldman-Rakic (1989). Posterior parietal cortex in rhesus monkey: II. Evidence for segregated corticocortical networks linking sensory and limbic areas with the frontal lobe. *Journal of Comparative Neurology* 287, 422–445.

Cavada, C., and P. S. Goldman-Rakic (1993). In T. P. Hicks, S. Molotchnikoff, and T. Ono, eds., *The visually responsive neuron: From basic neurophysiology to behavior. Progress in Brain Research*, vol. 95. Amsterdam: Elsevier.

Dean, P. (1982). Visual behavior in monkeys with inferotemporal lesions. In D. J. Ingle, M. A. Goodale, and R. J. W. Mansfield, eds., *Analysis of visual behavior*. Cambridge, MA: MIT Press.

Duhamel, J.-R., C. L. Colby, and M. E. Goldberg (1992). The updating of the representation of visual space in parietal cortex by intended eye movements. *Science* 255, 90–92.

Eskandar, E. M., L. M. Optican, and B. J. Richmond (1992). Role of inferior temporal neurons in visual memory II. Multiplying temporal waveforms related to vision and memory. *Journal of Neurophysiology* 68, 1296–1306.

Eskandar, E. M., B. J. Richmond, and L. M. Optican (1992). Role of inferior temporal neurons in visual memory I. Temporal encoding of information about visual images, recalled images, and behavioral context. *Journal of Neurophysiology* 68, 1277–1295.

Ettlinger, G. (1990). "Object vision" and "spatial vision": The neuropsychological evidence for the distinction. *Cortex* 26, 319–341.

Farah, M. J. (1990). Visual agnosia: Disorders of object vision and what they tell us about normal vision. Cambridge, MA: MIT Press.

Felleman, D. J., and D. C. Van Essen (1991). Distributed hierarchical processing in the primate cerebral cortex. *Cerebral Cortex* 1, 1–47.

Fahy, F. L., I. P. Riches, and M. W. Brown (1993). Neuronal signals of importance to the performance of visual recognition memory tasks: Evidence from recordings of single neurones in the medial thalamus of primates, 401–416. In T. P. Hicks, S. Molotchnikoff, and T. Ono, eds., *The Visually Responsive Neuron: From Basic Neurophysiology to Behavior. Progress in Brain Research* vol. 95. Amsterdam. Elsevier.

Ferrera, V. P., T. A. Nealey, and J. H. R. Maunsell (1992). Mixed parvocellular and magnocellular geniculate signals in visual area V4. *Nature* 358, 756–758.

Fujita, I., K. Tanaka, M. Ito, and K. Cheng (1992). Columns for visual features of objects in monkey inferotemporal cortex. *Nature* 360, 343–346.

Goodale, M. A. (1983a). Vision as a sensorimotor system. In T. E. Robinson, ed., *Behavioral approaches to brain research*, 41–61. New York: Oxford University Press.

Goodale, M. A. (1983b). Neural mechanisms of visual orientation in rodents: Targets versus places. In A. Hein and M. Jeannerod, eds., *Spatially Oriented Behavior*, 35–61. Berlin: Springer-Verlag: Berlin.

Goodale, M. A. (1988). Modularity in visuomotor control: From input to output. In Z. Pylyshyn, ed., *Computational processes in human vision: An interdisciplinary perspective*, 262–285. Norwood, NJ: Ablex.

Goodale, M. A., L. S. Jakobson, and J. M. Keillor (1994). Differences in the visual control of pantomimed and natural grasping movements. *Neuropsychologia* 36, 1159–1178.

Goodale, M. A., L. S. Jakobson, A. D. Milner, D. I. Perrett, P. J. Benson, and J. K. Hietanen (1994). The nature and limitations of orientation and pattern processing supporting visuomotor control in a visual form agnosic. *Journal of Cognitive Neuroscience* 6, 578–585.

Goodale, M. A., and A. D. Milner (1982). Fractionating orientation behavior in rodents. In D. J. Ingle, M. A. Goodale, and R. J. W. Mansfield, eds., *Analysis of visual behavior*, 549–586. Cambridge, MA: MIT Press.

Goodale, M. A., and A. D. Milner (1992). Separate visual pathways for perception and action. *Trends in Neuroscience* 15, 20–25.

Goodale, M. A., A. D. Milner, L. S. Jakobson, and D. P. Carey (1991). A neurological dissociation between perceiving objects and grasping them. *Nature* 349, 154–156.

Gross, C. G. (1973). Visual functions of inferotemporal cortex. In R. Jung, ed., *Handbook of sensory physiology*, vol. 7, part 3B, 451–482. Berlin: Springer-Verlag.

Gross, C. G. (1991). Contribution of striate cortex and the superior colliculus to visual function in area MT, the superior temporal polysensory area and inferrior temporal cortex. *Neuropsychologia* 29, 497–515.

Hasselmo, M. E., E. T. Rolls, G. C. Baylis, and V. Nalwa (1989). Object-centered encoding by face-selective neurons in the cortex in the superior temporal sulcus of the monkey. *Experimental Brain Research* 75, 417–429.

Heilman, K. M., R. T. Watson, E. Valenstein, and A. T. Damasio (1983). Localization of lesions in neglect. In A. Kertesz, ed., *Localization in neuropsychology*, 471–492. New York: Academic Press.

Hendry, S. H. C., and T. Yoshioka (1994). A neurochemically distinct third channel in the macaque dorsal lateral geniculate nucleus. *Science* 264, 575–577.

Heywood, C. A., A. Gadotti, and A. Cowey (1992). Cortical area V4 and its role in the perception of color. *Journal of Neuroscience* 12, 4056–4065.

Hietanen, J. K., and D. I. Perrett (1993). Motion-sensitive cells in the macaque superior temporal polysensory area. I. Lack of response to the sight of the monkey's own limb movement. *Experimental Brain Research* 93, 117–128.

Hietanen, J. K., D. I. Perrett, M. W. Oram, P. J. Benson, and W. H. Dittrich (1992). The effects of lighting conditions on responses of cells selective for face views in the macaque temporal cortex. *Experimental Brain Research* 89, 157–171.

Humphrey, G. K., M. A. Goodale, and R. Gurnsey (1991). Orientation discrimination in a visual form agnosic: Evidence from the McCollough effect. *Psychological Science* 2, 331–335.

Humphreys, G. W., and M. J. Riddoch (1987a). The fractionation of visual agnosia. In G. W. Humphreys and M. J. Riddoch, eds., *Visual object processing: A cognitive neuropsychological approach*. London: L. Erlbaum.

Humphreys, G. W., and M. J. Riddoch (1987b). *To see but not to see: A case study of visual agnosia*. Hillsdale, NJ: L. Erlbaum.

Hyvärinen, J., and A. Poranen (1974). Function of the parietal associative area 7 as revealed from cellular discharges in alert monkeys. *Brain* 97, 673–692.

Ingle, D. J. (1973). Two visual systems in the frog. *Science* 181, 1053–1055.

Ingle, D. J. (1982). Organization of visuomotor behaviors in vertebrates. In D. J. Ingle, M. A. Goodale and R. J. W. Mansfield, eds., *Analysis of visual behavior*, 67–109. Cambridge, MA.: MIT Press.

Ingle, D. J. (1983). Brain mechanisms of localization in frogs and toads. In J. P. Ewert, R. R. Capranica, and D. J. Ingle, eds., *Advances in vertebrate neuroethology*, 177–226. New York: Plenum Press.

Jakobson, L. S., Y. M. Archibald, D. P. Carey, and M. A. Goodale (1991). A kinematic analysis of reaching and grasping movements in a patient recovering from optic ataxia. *Neuropsychologia* 29, 803–809.

Jeannerod, M. (1988). *The neural and behavioral organization of goal-directed movements*. Oxford: Oxford University Press.

Klüver H., and P. C. Bucy (1939). Preliminary analysis of functions of the temporal lobes of monkeys. *Archives of Neurological Psychiatry* 42, 979–1000.

Lawler, K. A., and A. Cowey (1987). On the role of posterior parietal and prefrontal cortex in visuo-spatial perception and attention. *Experimental Brain Research* 65, 695–698.

Lissauer, H. (1890). Ein Fall von Seelenblindheit nebst einem Beitrag zur Theorie derselben. *Archiv für Psychiatrie und Nervenkrankheiten* 21, 222–270.

Lynch, J. C., and J. W. McLaren (1989). Deficits of visual attention and saccadic eye movements after lesions of parieto-occipital cortex in monkeys. *Journal of Neurophysiology* 69, 460–468.

Marr, D. (1982). *Vision*. San Francisco: Freeman.

Maunsell, J. H. R., and W. T. Newsome (1987). Visual processing in monkey extrastriate cortex. *Annual Review of Neuroscience* 10, 363–401.

McCollough, C. (1965). Color adaptation of edge-detectors in the human visual system. *Science* 149, 1115–1116.

Mendoza, J. E., and R. K. Thomas (1975). Effects of posterior parietal and frontal neocortical lesions in squirrel monkeys. *Journal of Comparative and Physiological Psychology* 89, 170–182.

Milner, A. D., and M. A. Goodale (1993). Visual pathways to perception and action. In T. P. Hicks, S. Molotchnikoff, and T. Ono, eds., *The visually responsive neuron: From basic neurophysiology to behavior. Progress in Brain Research*, vol. 95, 317–338. Amsterdam: Elsevier.

Milner, A. D., and M. A. Goodale (1995). *The visual brain in action.* Oxford: Oxford University Press.

Milner, A. D., D. I. Perrett, R. S. Johnston, P. J. Benson, T. R. Jordan, D. W. Heeley, D. Bettucci, F. Mortara, R. Mutani, E. Terazzi, and D. L. W. Davidson. (1991). Perception and action in "visual form agnosia." *Brain* 114, 405–428.

Milner, A. D., and M. D. Rugg, eds. (1992). *The neuropsychology of consciousness.* London: Academic Press.

Milner, P. M. (1974). A model for visual shape recognition. *Psychological Review* 81, 521–535.

Moran, J., and R. Desimone (1985). Selective attention gates visual processing in the extrastriate cortex. *Science* 229, 782–784.

Morel, A., and J. Bullier (1990). Anatomical segregation of two cortical visual pathways in the macaque monkey. *Visual Neuroscience* 4, 555–578.

Mountcastle, V. B., J. C. Lynch, A. Georgopoulos, H. Sakata, and C. Acuña. (1975). Posterior parietal association cortex of the monkey: Command functions for operations within extrapersonal space. *Journal of Neurophysiology* 38, 871–908.

Newsome, W. T., R. H. Wurtz, and H. Komatsu, H. (1988). Relation of cortical areas MT and MST to pursuit eye movements. II. Differentiation of retinal from extraretinal inputs. *Journal of Neurophysiology* 60, 604–620.

Nishijo, H., T. Ono, R. Tamura, and K. Nakamura (1993). Amygdalar and hippocampal neuron responses related to recognition and memory in monkey. In T. P. Hicks, S. Molotchnikoff, and T. Ono, eds., *The visually responsive neuron: From basic neurophysiology to behavior. Progress in Brain Research*, vol. 95, 339–358. Amsterdam: Elsevier.

Palmer, S., E. Rosch, and P. Chase (1981). Canonical perspective and the perception of objects. In J. Long and A. Baddeley, eds., *Attention and Performance IX*, 135–151. Hillsdale NJ: L. Earlbaum.

Perenin, M.-T. and A. Vighetto (1988). Optic ataxia: A specific disruption in visuomotor mechanisms. I. Different aspects of the deficit in reaching for objects. *Brain* 111, 643–674.

Perrett, D. I., J. K. Hietanen, M. W. Oram, and P. J. Benson (1992). Organization and functions of cells reponsive to faces in the temporal cortex. *Philosophical Transactions of the Royal Society*, B 335, 23–30.

Perrett, D. I., M. W. Oram, M. H. Harries, R. Bevan, J. K. Hietanen, P. J. Benson, and S. Thomas (1991). Viewer-centred and object-centred coding of heads in the macaque temporal cortex. *Experimental Brain Research* 86, 159–173.

Pohl, W. (1973). Dissociation of spatial discrimination deficits following frontal and parietal lesions in monkeys. *Journal of Comparative and Physiological Psychology* 82, 227–239.

Pribram, K. H. (1967). Memory and the organization of attention. In D. B. Lindsley and A. A. Lumsdaine, eds., *Brain Function.* vol. 4, UCLA Forum in Medical Sciences 6, 79–112. Berkeley: University of California Press.

Richmond, B. J., and T. Sato (1987). Enhancement of inferior temporal neurons during visual discrimination. *Journal of Neurophysiology* 58, 1292–1306.

Sakai, K., and Y. Miyashita (1992). Neural organization for the long-term memory of paired associates. *Nature* 354, 152–155.

Sakata, H., M. Taira, S. Mine, and A. Murata (1992). Hand-movement-related neurons of the posterior parietal cortex of the monkey: Their role in visual guidance of hand movements. In R. Caminiti, P. B., Johnson, and Y. Burnod, eds., *Control of arm movement in space: neurophysiological and computational approaches*, 185–198. Berlin: Springer-Verlag.

Sayner, R. B., and R. T. Davis (1972). Significance of sign in an S-R separation problem. *Perceptual Motor Skills* 34, 671–676.

Servos, P., M. A. Goodale, and G. K. Humphrey (1993). The drawing of objects by a visual form agnosic: Contribution of surface properties and memorial representations. *Neuropsychologia* 31, 251–259.

Sparks, D. L., and L. E. Mays (1990). Signal transformations required for the generation of saccadic eye movements. *Annual Review of Neuroscience* 13, 309–336.

Stein, J. F. (1992). The representation of egocentric space in the posterior parietal cortex. *Behavioral and Brain Sciences* 15, 691–700.

Stein, J. F., and M. Glickstein (1992). Role of the cerebellum in visual guidance. *Physiological Reviews* 72, 967–1017.

Taira, M., S. Mine, A. P. Georgopoulos, A. Murata, and H. Sakata (1990). Parietal cortex neurons of the monkey related to the visual guidance of hand movement. *Experimental Brain Research* 83, 29–36.

Tanaka, K. (1992). Inferotemporal cortex and higher visual functions. *Current Opinion in Neurobiology* 2, 502–505.

Tarr, M. J., and Pinker, S. (1989). Mental rotation and orientation dependence in shape recognition. *Cognitive Psychology* 21, 233–282.

Ungerleider, L. G., and B. A. Brody (1977). Extrapersonal spatial orientation: The role of posterior parietal, anterior frontal, and inferotemporal cortex. *Experimental Neurology* 56, 265–280.

Ungerleider, L. G., and M. Mishkin (1982). Two cortical visual systems. In D. J. Ingle, M. A. Goodale and R. J. W. Mansfield, eds., *Analysis of visual behavior*, 549–586. Cambridge, MA: MIT Press.

Vallar, G., and D. Perani (1986). The anatomy of unilateral neglect after right-hemisphere stroke lesions. A clinical/CT-scan correlation study in man. *Neuropsychologia* 24, 609–622.

Wald, G. (1950). Eye and camera. *Scientific American* 183, 32.

Walsh, V., S. R. Butler, D. Carden, and J. J. Kulikowski (1992). The effects of V4 lesions on the visual abilities of macaques: Shape discrimination. *Behavioural Brain Research* 50, 115–126.

Walsh, V., D. Carden, S. R. Butler, and J. J. Kulikowski (1993). The effects of V4 Lesions on the visual abilities of macaques: Hue discrimination and colour constancy. *Behavioural Brain Research* 53, 51–62.

Chapter 6

Eye Movements

Eileen Kowler

Before reading any further, take a few minutes to thread a needle. Choose a needle with a fairly small eye so the task won't be too simple. Keep at it until you're successful, and then go on to the material below.

It is likely that you tried to look steadily at the eye of the needle for the several seconds it took to complete the task. You surely tried to hold the needle as still as possible and also to keep your head still, probably choosing to sit rather than stand. In short, you did all the things that should allow you to see the needle and thread as clearly as possible: you used movements of your eyes, with some assistance from movements of your head and arms, to establish conditions optimal for seeing.

It might seem odd to begin a chapter on eye movements by describing a task that requires you to hold the eye still. Nevertheless, this is a good way to illustrate that the purpose of eye movements is to serve the needs of vision. We use eye movements either to bring the line of sight to objects of interest or to keep images of these objects stable on the retina by compensating for any perturbations produced by motions of the object or motions of the head.

We accomplish these tasks so well and with so little effort or awareness that it is easy to take oculomotor skill for granted, seeing it as a low-level sensorimotor reflex under the control of the motion or the position of a stimulus on the retina. But a bit of reflection should tell us that no low-level sensorimotor reflex can do the job by itself. Even something as straightforward as threading a needle would be impossible if we were not able to direct our line of sight wherever we want and to change its position whenever we want. We, rather than the properties of the outside world, control these behaviors.

The mystery is how volition can play such an important role in eye-movement control when we seem to pay so little attention to what our

Preparation of this chapter was supported by grants from the U.S. Air Force Office of Scientific Research, Life Sciences Directorate (91-0342 and F49620-94-3-0333). I thank Eric Anderson, Dan Bahcall, and Chris Araujo for their critical comments on earlier drafts of this chapter.

eyes are doing. Our attention is not devoted to controlling movements of our eyes but to thinking about what we're looking at or planning future activities. The solution to this mystery is that it is precisely by tapping into all of this ongoing cognitive activity that eye movements are able to do their job so well with so little additional effort.

This chapter tells the story of the cognitive control of eye movement, contrasting cognitive influence with the influence of such sensory cues as the position or the velocity of the retinal image. The fundamental role played by cognition in eye-movement control will be shown by making the simple needle-threading task increasingly more complex and demanding of oculomotor skill. The complexities include increasing the contribution of head movements, moving the needle rather than holding it still, providing a structured visual background, searching for a missing needle, and adding concurrent visual, cognitive, or motor tasks. We will see that as the task becomes more complex, and more representative of the ways in which eye movements are used in everyday life, cognition becomes a more central determinant of oculomotor behavior by intervening in and shaping what had traditionally been believed to be low-level, reflexive aspects of eye-movement control. Appreciation of the importance of cognition is due in part to recent methodological developments that allow eye movements to be studied during performance of increasingly more complex and naturalistic tasks.

Below I introduce a few basic concepts needed to understand this chapter. I also outline in some detail the organization of the chapter.

Eye movements fall into two broad classes (Steinman, Kowler, and Collewijn 1990). *Saccadic eye movements* are very rapid jumps of the line of sight made periodically—at most about four times each second but usually much less frequently. *Saccades* are used to bring images of chosen objects to the fovea, the central region of the retina where resolution of fine visual detail is at its best. *Smooth eye movements* are slower, continuous movements designed to track smooth motion of retinal images produced either by the motion of objects or by motion of the eye itself. Smooth eye movements serve to prevent rapid motion of the retinal image, which would blur the visual array and impair acuity.

Consider the nature of smooth and saccadic eye movements in more detail by returning to the needle-threading example. You probably found it easy to hold your eyes still enough to allow you to see the needle and thread quite clearly. Oculomotor research using instrumentation capable of measuring movements as small as one minute of arc—the width of a letter on this page at a distance of seven meters—has shown that such *maintained fixation* is accomplished primarily by low-velocity, smooth eye movements. You would also need to use these same smooth eye movements if the needle had been moving while you were trying to thread it. (Try it.) You would have to move your eye in the same direction and at the same speed as the needle to keep the needle's image relatively stationary on your retina. (Imagine trying to thread a needle while riding on a bus!) Smooth eye movements require sensory cues—particularly the motion of images on the retina—to track the motion of objects in the outside world. However, for reasons to be developed later, tracking based solely on sensory cues would be poor. Instead, cognitive processes—especially expectations and selective attention—account to a large extent for the ability of smooth eye movements to maintain a stable retinal image in the presence of object motion.

If you had the bad luck to drop the needle onto the carpet during your attempts at threading it, you would need to launch a pattern of saccadic eye movements, probably in conjunction with movements of your head or body, to find it. These saccadic movements are different from smooth eye movements in that their goal is to bring selected images to the fovea. Saccades, unlike smooth eye movements, can be directed at will almost anywhere you like, regardless of what happens to be in your visual environment. Yet, unlike many other sequences of voluntary movements (typing or speaking, for example), making accurate saccades seems to effortless. After all, we make errors when we speak or type, but when was the last time you recall *looking* in an undesired location? Saccades, like smooth eye movements, maintain their effectiveness by tapping into ongoing cognitive activity to guide them.

I will begin justifying these claims by discussing eye movements made during the laboratory version of the needle-threading task, in which subjects try to look steadily at a small point target for many seconds on end. This maintained-fixation task, the focus of much oculomotor research during the 1950s and 1960s, serves to introduce the way eye-movement control depends on such sensory cues as the position or the velocity of the retinal image. As we move to descriptions of more complex situations, like those we typically meet in the natural world—when objects move, when we move, or when we search patterned environments—things become less tame. Eye movements do not always work as the early laboratory studies using simple tasks in very sparse visual environments led us to believe. Cognitive processes—expectations, selective attention, and learning—become as or more important than sensory cues. Moreover, the operation of even apparently low-level and involuntary aspects of oculomotor control begin to depend on the cognitive demands of the task as a whole, be it reading, searching, or reaching for objects in space.

6.1   Maintained Fixation

6.1.1   A Brief History

The study of maintained fixation was begun in 1950 by visual scientists interested in the importance of retinal-image motion in maintaining clear vision (Ratliff and Riggs 1950; Barlow 1952; Ditchburn and Ginsborg 1953). Concern with image motion was timely then, in part because of physiological observations showing that many neurons in the retina respond better to transient changes in the image (e.g., onsets or offsets) than to steady presentations (Hartline 1938; 1940), and in part because of visual theories proposing that retinal-image motion can improve visual acuity by means of a neural averaging process (Marshall and Talbot 1942).

Early studies of maintained fixation required subjects to look for many consecutive seconds at a small target (a tiny point or the intersection of a crosshair) located in an otherwise featureless field. These studies revealed a pattern of relatively slow, continuous, irregular oscillations of the eye interrupted once or twice each second by small saccadic jumps. An example of the typical pattern of saccades and slow oscillations during fixation of a small point target is shown in Figure 6.1a. This typical fixation pattern is stable enough to keep the image well within the central twenty minutes of arc of the fovea (the "foveal bouquet"; Polyak 1941) but not so stable that all image motion is eliminated. In fact, if all image motion were to be eliminated—which can be done by moving the stimulus in the same spatial and temporal pattern as the eye—the image would fade from view (Riggs et al. 1953).

An early and influential model of fixational eye movements proposed by Cornsweet (1956) tried to account for the stability of fixation within the formal language of control theory. Control theory was developed in electrical engineering and had been applied to pursuit and saccadic eye movements some time earlier by Westheimer (1954) and to arm movements by Craik (1947). The basic idea behind the application of control theory is that the system monitors how well it is doing by sensing one or more error signals and then responding in ways that correct the error and keep itself in as close to an optimal state as possible. In the application of control theory to fixational eye movements, the error signal was assumed to be caused by the drift of the image from the center of the fovea. *Drift* was a slow eye movement, which Cornsweet assumed was uncontrolled *noise*. Errors caused by such drifts were corrected by the small saccades, which occurred whenever the errors exceeded some threshold value of about three to six minutes of arc.

Although supported by a correlational analysis of the horizontal eye movements of two subjects, it became clear by the mid-1960s that Corn-

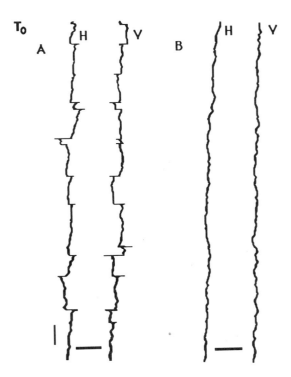

Figure 6.1
(A) Horizontal (H) and vertical (V) eye movements during fixation of a point target. Records were made with a contact lens-optical lever. The record begins at the top. The horizontal black bar at the bottom represents a 15-minute of arc rotation, the vertical bar a 1-second time interval. The abrupt changes in eye position are saccades. (B) Same, except the subject had elected not to make saccades. (Reprinted with permission from R. M. Steinman, G. M. Haddad, A. A. Skavenski, and D. Wyman, Miniature eye movement, 1973, *Science* 181, 810–819.)

sweet's model had it the wrong way around. Stable fixation is achieved not by saccades but by the slow oscillations of the eye—by *slow control* (Steinman, Haddad, Skavenski, and Wyman 1973). The most compelling way to demonstrate slow control is simply to ask someone to stop making saccades (Steinman, Cunitz, Timberlake and Herman, 1967). As the example in Figure 6.1b shows, saccades will disappear and slow eye movements will keep the line of sight quite stable. Typical standard deviations of eye position are about two to three minutes of arc. Stability deteriorates dramatically in total darkness (Figure 6.2), showing that slow control is a genuine response to visual error signals.

What kind of visual error signals drive slow control? One possibility is that slow control is driven by a position error, defined as the distance

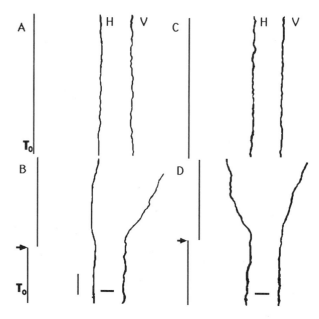

Figure 6.2
(A,C) Slow control in the presence of a fixation target and (B,D) when the fixation target was removed at the time indicated by the arrows to the left of the eye traces. Records begin at the bottom (T_0). The vertical bar represents a 1-second and the horizontal bar a 15-minute of arc rotation. (Reprinted by permission from R. M. Steinman, G. M. Haddad, A. A. Skavenski, and D. Wyman, Miniature eye movement, 1973, *Science* 181, 810–819.)

between the position of the image and the center of the fovea (i.e., the kind of error that Cornsweet had assumed was used to control the saccades during fixation). But if you ask a subject to look to one side of a target, rather than directly at it, smooth eye movements do not correct the positional error introduced by such eccentric fixation. Instead, smooth eye movements keep the line of sight relatively stable and keep it in almost the same place (Epelboim and Kowler 1993). This means that slow control responds to velocity error signals, not to position error signals. Here's how it works: As the eye drifts one way or the other due to some inherent neural or muscular instabilities, retinal-image motion is generated, which in turn is canceled by smooth eye movements going in the opposite direction. Velocity correction works well even for very eccentric target images (i.e., greater than 4 or 5 degrees), although there is some loss in effectiveness as the distance from the fovea increases (Epelboim and Kowler 1993), just as there is a loss in the precision of the perceptual coding of velocity with increasing eccentricity (McKee and Nakayama 1984).

The small saccades, which appear occasionally during fixation even when you are trying not to make any, seem to serve no useful purpose. In fact, they are only prominent during the laboratory version of the fixation task, in which subjects are told to do nothing but simply look carefully at the target. During needle threading—our prototypical natural fixation task—saccades drop out of the pattern and the line of sight is maintained exclusively through slow control (Winterson and Collewijn 1976).

6.1.2   Unrestrained Heads

All the foregoing observations were made with the head supported by either a chinrest or a biteboard. Most eye-movement recording devices accurate enough and precise enough to study maintained fixation cannot be used (at least, not without introducing serious artifacts) when the head is free to move. There are different reasons for this, depending on the properties of the instrument. We describe one reason in some detail here because, until very recently, it constituted a major technical obstacle to studying eye movements under natural conditions.

One of the defining characteristics of an accurate eye monitor is the ability to distinguish rotations of the eye in its orbit from translations of the head. The smooth eye movements and saccades I have described fall into the category of eye rotations. Because each rotation of the eye moves the retinal image by the same amount, rotations are valid indicators of changes in the point of regard. Translations, on the other hand, are produced mainly by moving the head from side to side. Translations change image position too but only for nearby targets and by an amount that depends on target distance. So, the detection of a translational move-ment does not necessarily mean that the point of regard has changed at all. An instrument that confounds rotations with translations—and many do—does not provide an accurate indication of where the subject is looking.

Until relatively recently, nearly all the eye-movement monitors de-signed to be insensitive to translations worked properly only when the head remained within a prescribed area. These instruments include the contact-lens optical lever (used in the early studies of maintained fixation described above) and both the magnetic-field sensor coil and the SRI Dual Purkinje Image Tracker; both the latter came into widespread use begin-ning in the 1960s and 1970s. These three instruments work on very different principles. The contact-lens optical lever measures the reflection of a narrow beam of light from a plane mirror attached to a large, tightly fitting scleral contact lens. The magnetic-field sensor coil works by placing the subject inside a weak alternating magnetic field and measuring the voltage induced in a coil of wire attached to the subject's eye by way of a silicone annulus shaped so as to stick to the anesthetized eye. The Dual Purkinje Image Tracker treats the eye itself as a mirror and measures the separation between the position of infrared reflections from the front surface of the cornea and the back surface of the lens inside the eye. With each of these instruments firm head support is necessary (biteboards or chinrests) to prevent translational artifacts.

No one worried much about the consequences of relying on chinrests or biteboards, at least not until the late 1970s. It was assumed that any motion of the head would be compensated for sufficiently well so that fixation would be as stable with the head unrestrained as it was while the head was supported. (See Questions for Further Thought for an example of how you might demonstrate this for yourself.) This seemed a reasonable assumption because years of prior research testing eye movements during large-amplitude rotations of the head had demonstrated that eye movements compensated for about 95 percent of the rotation. Compensation for head movement is governed by control systems that use both non-visual error signals, which originate in the vestibular system (the vestibulo-ocular response, or VOR), and visual error signals, which originate in the retina. During the 1970s the story changed. Novel instrumentation developed by Collewijn and by Steinman made it possible to produce accurate recordings of eye and head rotations without contamination by translations and with high enough spatial and temporal resolution to study fine-grain properties of eye movements. Steinman and Collewijn (1980) then set out to test the assumption that compensation for head movements is nearly perfect during maintained fixation. In the early experiments, targets were so far away that any translational movement of the head would not affect the position of the retinal image. In the later work (discussed below in the section on saccades), targets were nearby and separate instrumentation was used to measure head translation.

Surprisingly, it turned out that the level of compensation observed previously for large head rotations was not adequate to compensate for the small head rotations that occur even when trying to sit as still as possible. Freeing the head produced considerable motion of the eye during fixation, thus resulting in less image stability. This additional eye motion can be seen in Figure 6.3, which compares fixational eye movements with and without biteboards, and in Figure 6.4, which shows that eye motions are even greater when deliberate, yet modest, motions of the head are made. Figure 6.4 also shows that the motion of the two eyes is not the same, which means that the image of the target rarely, if ever, falls on corresponding retinal points of the two eyes. (When looking at Figures 6.3 and 6.4, keep in mind that all the eye-movement traces represent movements of the eyes in space, which means that all the traces are also valid indicators of movement of the image on the retina.)

Although the average velocity of the retinal image increased to 4 degrees per second during even the modest head rotations shown in Figure 6.4, and the image rarely fell on corresponding retinal points, it was still

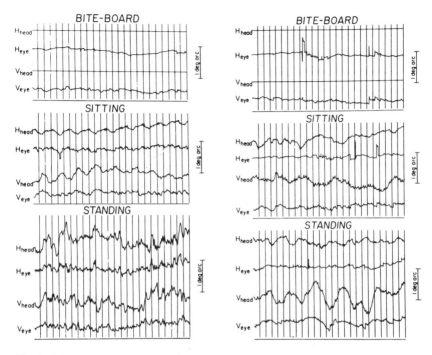

Figure 6.3
Movements of eye and head of two subjects fixating on a distant target while their heads were supported by a biteboard or while sitting or standing as still as possible without head support. Records begin on the right and repetitive vertical stripes indicate 1-second intervals. The vertical bars on the right represent 1-degree rotations. Upward changes in the traces signify rightward movements in (H) and upward movements in (V). (Reprinted by permission from A. A. Skavenski, R. H. Hansen, R. M. Steinman, and B. J. Winterson, Quality of retinal image stabilization during small natural and artificial body rotations in man, 1979, *Vision Research 19*, 675–683.)

possible to see a single, fused, stationary world quite clearly. Studies of visual acuity, visual contrast sensitivity (a measure of the minimum contrast necessary to detect the bars of a grating pattern) and stereopsis in the presence of image motion confirm that vision is surprisingly insensitive to image motion as fast as 2 or 3 degrees per second (Steinman, Levinson, Collewijn, and Van der Steen 1985; Steinman and Levinson 1990; Westheimer and McKee 1975). Image motion as slow as that found when the head is stabilized by a biteboard can even be detrimental in some visual tasks (Riggs et al. 1953).

How could such tolerance, even preference, for image motion be explained? Does the visual system use its supply of neural motion detectors to capture the details of images moving at relatively brisk rates of speed

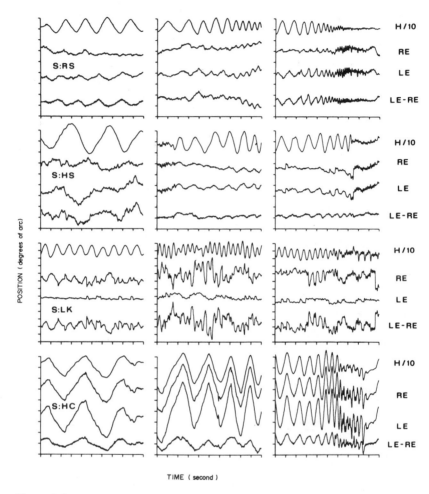

Figure 6.4
Horizontal head and eye movements of four subjects while they fixed on a distant target as they moved their heads. Records begin on the left. The time scale marks signify 1-second intervals. The head trace is scaled to 1/10 of its actual value. Movements of the right eye (RE), left eye (LE) and the difference between the two eyes (i.e., vergence eye movements, LE-RE) are shown with respect to space, so that all movements represent motion of the retinal image. Upward changes in the traces signify rightward movements. Upward changes in the vergence traces signify convergence. (Reprinted by permission from A. A. Steinman and H. Collewijn, Binocular retinal image motion during active head rotation, 1980, *Vision Research* 20, 415–429.)

(2 or 3 degrees per second)? Usually we think of these motion detectors as responsible for allowing us to see moving things; but in fact their real function may be to allow us to see stationary things that move on the retina. Or could it be that high-velocity image motion impairs vision more when it is produced by motion of objects than when it is produced by motion of the eye? This would mean that the motor commands used to control the movements of eye and head are also tapped into by the visual system to improve spatial resolution (Steinman et al. 1985). Answers to these questions are still unknown. Whatever its explanation, tolerance for image motion is a sensible option because it avoids placing excessive, unrealistic demands on oculomotor skill. The visual system appears to have evolved so as to prefer the moderately high image speeds present in natural viewing rather than the artificial, extreme level of stability demonstrated in early studies of fixation done with heads on biteboards.

Summary

This section on fixational eye movements illustrated two things. First, in a very simple oculomotor task—looking at a single, small target present in darkness—sensory cues, specifically the velocity of images moving on the retina, determine how the eye moves. Second, as we introduce what might seem like a trivial modification to the standard laboratory task, namely taking the head off the biteboard, the pattern of eye movements changes, along with ideas about how and why the fixational-eye-movement pattern has the characteristics it does. This is the first of several examples of the way in which making standard laboratory tasks more naturalistic, often in seemingly innocuous ways, can change our conception of how eye movements are controlled. As we proceed, we will see how more naturalistic experiments point to a central role for cognition in oculomotor control.

6.2   Smooth Pursuit

First, we make the fixation task more difficult by setting the needle and thread in motion.

Threading a moving needle is like threading a stationary needle in that the goal remains one of confining the image to the central region of the retina, keeping image velocity within tolerable limits. The same smooth eye movements, which used velocity error signals to track image motion generated by movements of the eye itself, should continue to work in much the same way when image motion is produced by movements of the object. When the object is moving, the smooth eye movements are called *smooth pursuit*.

6.2.1   Motion as a Stimulus for Pursuit

To a certain extent, tracking moving things is much like tracking stationary things. Smooth pursuit can be used to track a relatively simple pattern of target motion with an eye velocity closely approximating but not exactly matching the velocity of the target. The small mismatch between eye velocity and target velocity provides residual retinal-image motion that can serve as the velocity error signal to stimulate subsequent pursuit (Collewijn 1969; Puckett and Steinman 1969), just as velocity error signals are used by slow control when targets are stationary (see above).

One reason we know that motion signals are important in human smooth pursuit is that most people cannot make the eye move smoothly in one or another direction in the absence of stimulus motion. A few rare individuals can make some voluntary smooth-pursuit movements, but the vast majority of us attempting to do so end up producing a sequence of saccades. Voluntary effort is equally useless for suppressing smooth pursuit when the visual field is limited entirely to moving objects. The absence of voluntary control does not, however, mean that smooth pursuit is driven exclusively by velocity error signals and is immune to cognitive control. To see why this is the case, consider the role of prediction in smooth pursuit.

6.2.2   Prediction

When you pursued the movement of needle and thread, you might have developed an awareness of the pattern of motion as a whole, including the immediate position of the target and your expectation of its position at the next instant of time. This sort of prediction of the target's future position need not be limited to the motions you produce when you move needle and thread yourself. Try pursuing a needle moved by someone else. You should be able to predict the path accurately if the pattern of motion is fairly simple, though there may be some uncertainty about precisely when changes in direction will occur. Of course, the person controlling the motion of needle and thread could choose to make sudden or erratic changes in direction (similar to the evasive movements of insects being pursued by predators). If this happens, my guess is that you would find such changes surprising and annoying and have considerable difficulty keeping your eye on the moving needle.

These simple observations provide a fundamental clue to the importance of prediction in human smooth pursuit. Such clues were ignored, or their significance minimized, for many years because the involuntary nature of pursuit made it implausible to consider it was anything other than a reflexive response to visual error signals. In fact, prediction similar to the sort just demonstrated was initially explained, not by reference to cognitive events, but within the domain of control theory by introducing a role for *positive feedback*.

6.2.3   Positive Feedback

Craik (1947), in one of the earliest attempts to use control theory in engineering to explain human motor behavior, introduced positive feedback into models of tracking. He wrote about the tracking movements of the arm and how they might be simulated by mechanical or electrical devices. He noted that a real problem in designing systems that correct tracking errors by monitoring error signals is compensating for the time taken to detect the error and initiate the response. By the time the corrective response occurs, the target has already moved to a new place. To compensate for the lag, he proposed that the tracking system could be set to "go on doing whatever it was doing at the moment," even in the absence of any external input, and to make alterations in the pre-set response only when errors were detected. This system could be simulated mechanically or electrically by allowing responses to be triggered by positive feedback signals, that is, internal replicas of the most recently issued response.

Craik's ideas were incorporated over the years into different models of human smooth-pursuit eye movements, most prominently into models that tried to explain how pursuit can occur in the absence of any retinal motion at all. You can illustrate smooth pursuit in the absence of retinal-image motion by viewing a brief intense flash of light from a conventional camera flashgun. You should see an afterimage, which results from persisting visual excitation left over from the flash. The afterimage will seem to be moving, and the perceived motion will be about the same as the pattern of eye movements. Mach (1906) and Helmholtz (1925) both explained the perceived motion of afterimages by proposing that the accurate registration of object motion results from a combination of signals representing retinal motion with signals representing the motor commands sent to the eye. Whenever the motion on the retina is equal and opposite to the motion of the eye, the signals cancel each other out and objects are perceived as stationary. Any mismatches, on the other hand, are perceived as genuine object motion. Afterimages produce such a mismatch because there is no motion on the retina regardless of how the eye moves. The afterimage moves with the eye. As a result, any perceived smooth motion of the afterimage becomes a subjective indicator of the drift of the eye. Some people's eyes tend to drift steadily in one direction when viewing an afterimage, much as the eye drifts in the dark (Figure 6.2); others can make the eye move one way or the other at will (Cushman et al. 1984).

The existence of *any* directed smooth pursuit with afterimages shows that pursuit is possible even in the absence of retinal velocity signals. Thus, the velocity error signal need not be solely retinal but could come from the same combination of retinal-motion and eye-motion signals responsible for the perceived motion. This suggests that Craik was right in that a positive feedback signal, representing the motion of the eye itself, is part of the input driving pursuit.

A system that keeps replicas of its own activity in order to keep issuing the same response over and over again can appear to be predicting future target motion, provided that the pattern of target motion does not change. In that case, the response, once started, would eventually fall into perfect synchronization with the target and even run on a bit after the motion stops. Others developed the synchronization idea further, proposing that the pursuit system predicts periodic, repetitive patterns of target motion by learning these patterns (Dallos and Jones 1963). However, neither a predictive system that learns repetitive patterns nor the modern engineering counterparts of such models derived from adaptive control theory (Pavel 1990) can explain an unexpected feature of human smooth pursuit: namely, the appearance of the response *before* the start of any stimulus motion at all.

6.2.4   Anticipatory Pursuit

An example of anticipatory pursuit, in which the eye starts out in the direction of stimulus motion before the stimulus begins to move, is shown in Figure 6.5. Anticipatory pursuit was discovered by Raymond Dodge (1927), one of the first to undertake a comprehensive examination of human eye movement. Realizing the enormity of the task of modeling a system in which the response is issued before the stimulus begins, Dodge chose to concentrate on *phase leads*—instances in which the eye changed direction before a target did—during the pursuit of periodic target motion, where the opportunity for learning abounds (Dodge, Travis, and Fox 1930). He attributed phase leads to learning and offered no explanation for the anticipatory response he occasionally observed before target motion had begun.

There have been many studies of anticipatory pursuit and predictive eye tracking since Dodge's time (see Kowler 1990 and Pavel 1990 for reviews). I will concentrate here on one carried out recently to resolve a basic question about the nature of these anticipatory pursuits: was the anticipatory pursuit due to habit or to genuine cognitive prediction?

The stimulus was a disk that moved smoothly down a hollow tube shaped like an inverted letter Y. Figure 6.6 shows a drawing of this stimulus taken from Kowler (1989). As soon as the disk reaches the oblique arms

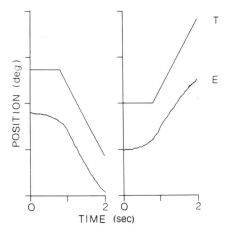

Figure 6.5
Horizontal eye position (bottom traces) as a function of time during smooth pursuit of predictable constant velocity motion (top traces) to the left (left-hand graph) and to the right (right-hand graph). Note the anticipatory pursuit beginning 300 msec before target motion began. (Reprinted by permission from E. Kowler, Cognitive expectations, not habits, control anticipatory smooth oculomotor pursuit, 1989, *Vision Research* 29, 1049–1057.)

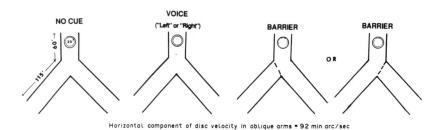

Figure 6.6
Stimulus display used to study anticipatory pursuit. The disk moved down the inverted Y-shaped tube and was equally likely to travel down the right or the left oblique branch. The branch was either undisclosed before each trial (no cue), disclosed by a voice cue, or by a visible barrier cue blocking access to the untraveled branch. (Reprinted by permission from E. Kowler, Cognitive expectations, not habits, control anticipatory smooth oculomotor pursuit, 1989, *Vision Research* 29, 1049–1057.)

of the Y, it continues to move down either the left-hand or right-hand paths. In one set of experimental conditions ("no cue") the disk had a 50 percent chance of traveling down either path, and the subjects did not know in advance which path it would be. The graphs at the far left in Figure 6.7 show average horizontal eye velocity over time for two different subjects (myself and a "naive" subject). The point labeled 0 on the abscissa represents the time when the disk entered one of the oblique arms and the horizontal component of stimulus motion began. Figure 6.7 shows that before time 0 the eye was not stationary but moved smoothly to the left or the right. These are the anticipatory smooth eye movements. The direction of the anticipatory pursuit depended on the target motion on the prior trial. Pursuit was, on average, biased to the right when the prior motion was rightward and to the left when the prior motion was leftward.

This tendency to repeat past tracking movements does not sound like true anticipation but like another example of Craik's rule of "going on doing what you have been doing" until a change is detected. In other words, a habit had been established. Two other conditions were included to test this proposition. In these conditions, illustrated in Figure 6.6, an auditory or visual cue disclosed the direction of target motion on the upcoming trial. A predictive system simply wired to parrot the past would not change its response, regardless of what the auditory or visual cue revealed.

The response did change however. Anticipatory pursuit in the direction disclosed by the cue was prominent, regardless of the target motion on the preceding trial (Figure 6.7, middle and right-hand columns). This shows that the anticipatory pursuit was sensitive to cognitive cues about the direction of future target motion and was not extrapolation or positive feedback masquerading as anticipation. The eye was truly anticipating future motion, using what the subjects knew about upcoming events based on interpretation of both auditory and visual symbolic cues.

The intriguing thing about such an anticipatory response is that it was neither initiated nor could it be suppressed by voluntary effort. The pursuit system took advantage of knowledge about the environment and the objects it contained without asking the subject's permission to do so. This produces an excellent way of keeping up with the target, one that avoids the large tracking errors associated with time delays at a cost of small and surely harmless mismatches between eye and target velocity. (See the section on fixation for a discussion of the tolerance of vision to small amounts of retinal-image motion.)

No one knows the neural basis for such anticipatory pursuit, although investigations of relevant cortical areas are underway (Heinen 1993). One way of producing anticipatory pursuit on a neural level would be to have two separate pursuit systems, one based on the immediate sensory signals

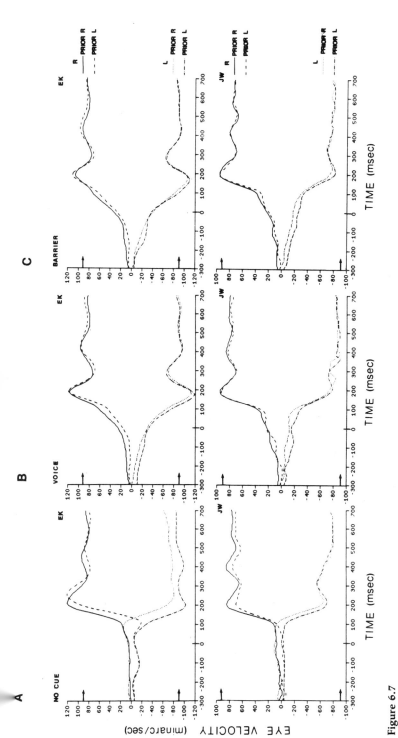

Figure 6.7
Results using the experimental display illustrated in Figure 6.6. Mean horizontal eye velocity over time for two subjects (EK and JW) when either (A) no cue; (B) voice cue or (C) barrier cue signaled the direction of disk motion. Time 0 on the abscissa is when the disk entered of the oblique branches, defining the start of horizontal disk motion. Arrows on the ordinate show horizontal disk velocity; negative values denote leftward motion. The top pair of functions in each graph shows eye velocity when the disk moved rightward; the bottom pair when it moved leftward. Solid lines show eye velocity when the disk motion in the prior trial was to the right; dashed lines when it was to the left. (Reprinted by permission from E. Kowler, Cognitive expectations, not habits, control anticipatory smooth oculomotor pursuit, 1989, *Vision Research 29*, 1049–1057.)

about motion and the other based on cognitive cues about future motion. The response would depend on some rule for combining the outputs of the two systems. The alternative solution would be to have a single pursuit system driven by a motion signal that combines both the sensory registration of target motion and the anticipated motion extrapolated 100 or 200 milliseconds into the future. The signal could simply be the time average of the motion signals over this interval extending from the present into the future. This type of model is quite plausible, for we know human beings can make accurate and precise predictions about the future path of moving objects (Pavel, Cunningham, and Stone 1992; Freyd 1987). The system controlling smooth eye movements may be structured to take fortuitous advantage of this cognitive predictive ability; or it may have provided the evolutionary pressure to develop it.

The prediction of future motion is not the only way cognitive factors influence pursuit. Return once more to needle threading.

6.2.5   Selective Attention

It is likely that your moving needle and thread were not the only objects in your field of view. They were surely seen against a detailed visual background replete with all sorts of objects and contours, any one of which might grab the line of sight at any time. If the visual background you used was not so detailed, try threading the moving needle against such a background. My guess is that the background will not interfere with pursuit at all.

Dodge and Fox (1928) studied the ability of people to look steadily at a small stationary target superimposed on a large pattern of moving stripes. They expected to find subjects' eyes pursuing the stripes in the background, given their greater size and contrast. The fact that they did not indicates that human smooth eye movements do not behave like low-level visuomotor reflexes, despite their involuntary nature.

Maybe backgrounds have such little influence simply because they are imaged on eccentric retina, whereas the target is imaged on the fovea (which is more important for most visual judgments and the natural locus of attention). Kowler et al. (1984) demonstrated that retinal position was not the critical factor. Their stimulus consisted of two, full-field, superimposed patterns of random dots. The patterns were identical except that one was stationary and the other moved to the left at about 1 degree per second (a velocity low enough so that acuity for details remained high even when the pattern was not tracked). The density of the dot patterns was great enough to make it impossible to isolate dots from one pattern on the fovea without dots from the other pattern continuously passing across the fovea at the same time.

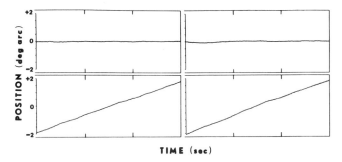

Figure 6.8
Horizontal eye movements under instructions to maintain the line of sight on the stationary (top graphs) or moving (bottom graphs) field of random dots. In the lefthand graphs, only one field was present; in the right both were presented superimposed. (Reprinted by permission from E. Kowler, J. Van der Steen, E. P. Tamminga, and H. Collewijn, Voluntary selection of the target for smooth eye movement in the presence of superimposed, full-field stationary and moving stimuli, 1984, *Vision Research* 24, 1789–1798.)

Subjects had no trouble tracking the selected pattern accurately; there was virtually no interference from the pattern designated as the background (see Figure 6.8). Such near-perfect selection (with only 2 to 4 percent of influence from the background) could not have been achieved by giving greater weight to one or another retinal position because dots from both patterns were present everywhere in the visual field. Neither was selection achieved by paying attention to one or another type of motion (i.e., stationary or leftward). The latter is ruled out because of a quirk in the perception of the patterns tested: namely, the stationary pattern never appeared stationary. Instead, it always appeared to be moving to the right (opposite to the direction of the moving pattern). Yet even though subjects perceived this illusory motion vividly, this perceptual property did not show up in what the eye did; the eye remained stationary when the stationary pattern was the target. The only remaining explanation of the results is that the selectivity of the eye-movement responses was achieved by choosing one or the other of the two fields of dots. In other words, subjects selected the object, not its spatial position nor its pattern of motion. Apparently while voluntary selection determines *what* we pursue, the pursuit system itself determines an appropriate eye velocity that is independent of what we perceive.

6.2.6   One Attentional System or Two?

Is the voluntary selection described above equivalent to selective attention? Attention is usually discussed in purely psychological contexts, as a

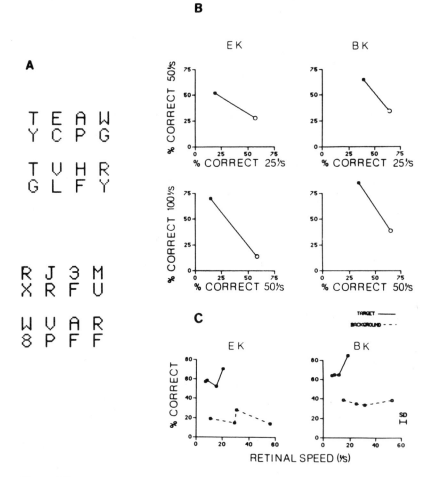

Figure 6.9
(A) An example of a character array containing 16 letters (top) or 14 letters and 2 numerals (bottom). (B) Visual search performance. Percent correct reports for the slower pair of rows in shown on the abscissa, for the faster pair on the ordinate. The slower (open symbols) or faster (filled symbols) rows were pursued. (C) The same data in (B) plotted as a function of measured retinal speed. Performance was always better for the pursued rows and regardless of retinal speed. (Reprinted by permission from B. Khurana and E. Kowler, Shared attentional control of smooth eye movement and perception, 1987, *Vision Research* 27, 1603–1618.)

way to enhance the perceptibility of some objects in contrast to others (see Chapter 2). Models often treat attention as an internal resource to be allocated to one or another place or object (Sperling and Dosher 1986). Is this the sort of attention that determines the target for smooth pursuit? Or, do smooth eye movements have their own, independent, selective filter that allows one to maintain the line of sight on one target while the "mind's eye" explores elsewhere?

Khurana and Kowler (1987) distinguished between these possibilities by testing smooth eye movements and perception at the same time. The stimulus, shown in Figure 6.9a, consisted of four rows of letters, with one pair of rows (the top and third) moving at a velocity that was either twice as great or half as great as the velocity of the other pair (rows two and four). The task was to keep the line of sight in the blank region between rows two and three while matching horizontal eye velocity with either the faster or slower pair of rows. Pursuit was started with the line of sight positioned at the left-hand edge of the display; at about the time it reached the center of the display, the identities of all the characters were briefly— 200 msec—changed. The display then contained two numerals along with the letters; one numeral was located somewhere in the slower rows and the other was somewhere in the faster rows. The subject had to identify both.

The main result was that the numeral in the pair of rows tracked was identified much more accurately than the numeral in the other pair (Figure 6.9b). An important aspect of these results was the demonstration that the superior identification of numerals in the target rows was due to attention and not to any differences in retinal speed of the target and background rows. By analyzing performance as a function of the speed of the retinal image of both the target and background rows, Khurana and Kowler were able to show that even when the retinal speeds were the same (which occurred because, as noted above, eye velocity rarely matched the velocity of the target), performance remained better for the target rows (Figure 6.9c). Separate experiments, in which subjects tried their best to track one set of rows while attending to the other, still showed an advantage for the tracked rows. Perceptual attention cannot be fully dissociated from pursuit. Selection of the pursuit target involves the allocation of perceptual attention to one of the many objects in the visual field.

By linking the selection of the target for pursuit to perceptual attention, the task of maintaining an appropriately stable retinal image in a natural environment becomes vastly simpler. We need no separate set of attentional decisions to control the movements of the eye. We need only decide what is interesting or important; the smooth oculomotor subsystem takes care of the rest.

Summary

Smooth pursuit, because it cannot be initiated or suppressed voluntarily, has always appeared to be a sensorimotor reflex under the control of stimulus motion on the retina. But we have seen that, despite its involuntary nature, pursuit characteristics depend on the ability to predict the future path of a moving target and to attend to the target and ignore its background. Thus, pursuit is effective because it takes advantage of the ongoing cognitive activity that quite naturally accompanies taking an interest in any moving object within your field of view.

6.3   Saccades

Smooth eye movements are used to keep targets relatively stable on the retina, moving at just the right velocity to ensure optimal resolution. Saccades, by contrast, are high-velocity eye rotations used to bring targets to the fovea. Figure 6.10 illustrates a sequence of saccades made while subjects scanned an array of five stationary points configured as a pentagon.

6.3.1   Dropping the Needle

You need saccades in our needle-threading task only for the initial movement of the line of sight to bring the needle and thread to the center of the fovea. They would continue to play a role, however, if you dropped the needle on the floor and needed to search about to find it. In this case— particularly if the floor was so cluttered you had a hard time spotting the needle out of the corner of your eye—you would need to move your line of sight to a number of different locations (search) in hopes that the small target would eventually land on the fovea. You would probably move your head at the same time and, eventually, move your arm to investigate suspicious specks or glints and to pick up the needle when it is (hopefully) finally found. Before continuing with this chapter, try searching for a dropped needle (or, a needle someone has hidden) in a cluttered environment and note how you go about this frustrating kind of search. I will continue to refer to different aspects of this quite common kind of experience throughout this section.

Effective search requires that you make sensible decisions about where the needle might be. An analysis of your eye movements should provide some clues about the search plan. Later I will discuss attempts to infer search and other mental strategies from sequences of saccades. For now, concentrate, not on your decisions, but rather on the oculomotor mechanisms used to carry them out.

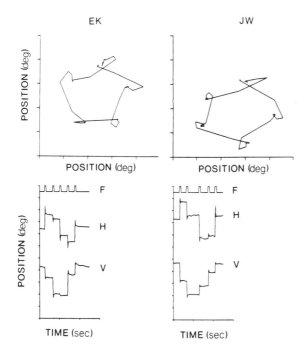

Figure 6.10
Examples of saccadic eye movements made to look at five targets arranged as a pentagon. Top graphs show horizontal versus vertical eye position. Bottom graphs show horizontal (H) and vertical (V) eye position as a function of time. Upward deflections in the eye traces indicate movements to the right or up. The top trace (F) indicates the onset and offset of the saccade detected by a computer algorithm. (Reprinted by permission from C. M. Zingale and E. Kowler, Planning sequences of saccades, 1987, *Vision Research* 27, 1327–1341.)

6.3.2   Target-Step Tracking

An effective saccade is, obviously, one that takes the line of sight to the selected target as accurately as possible in the shortest possible time. The duration of the saccade itself is brief enough for us to ignore the time it takes, at least for now. (For example, a saccade 10 degrees in amplitude lasts only about 50 msec.) The more significant delay in scanning is produced by the intersaccadic interval, which has a minimum duration of about 150 msec. Intersaccadic intervals of 200 to 300 msec are more characteristic of active visual search or of reading (Viviani 1990; O'Regan 1990).

Much of what we know about saccades is based on laboratory studies in which subjects used saccades to follow a single, small point target as it abruptly jumped from a central to an eccentric retinal position (i.e., a *target step* in engineering jargon). This unnatural stimulus has been popular because it specifies precisely what the subject is supposed to do: namely, use a saccade to mimic the jump as closely as possible with as little time delay as possible. By contrast, in natural scenes the viewer decides where and when to direct the eye, so it is hard to distinguish aspects of performance derived from cognitive decisions from those reflecting properties of the sensory or motor mechanisms that carry out these decisions. This distinction is easier to make during target-step tracking, but cognitive factors can certainly intervene even here (Kowler 1990; Westheimer 1989).

Numerous versions of the step-tracking task have been tried, varying the number, size, location, and timing of steps in an attempt to better understand saccades. I mention only a few results here—those related to saccadic accuracy and precision—and then describe very recent work that tries to bridge the gap between the artificial target steps and more natural scanning of complex visual scenes.

6.3.3   Accuracy and Precision

When tracking a target step with saccades, the landing position of the saccade often falls short of the target, requiring one or more additional catch-up saccades to correct the error (Becker and Fuchs 1969). It is not clear, however, that such saccadic inaccuracies actually occur during natural scanning. Even in the traditional sparse laboratory displays containing single-point targets on blank backgrounds, accuracy can be improved dramatically by substituting two stationary points for the jumping target (Lemij and Collewijn 1989; Collewijn, Erkelens, and Steinman 1988), by increasing the time available to evaluate target position (Aitsebaomo and Bedell 1992), by encouraging the subject to expect a larger target jump than will actually be presented (Kapoula and Robinson 1986) or simply by asking the subject to try to be more accurate, increasing latency if necessary in order to do so (Kowler and Blaser 1995).

Saccades to target displacement are also remarkably precise (Kowler and Blaser 1994). Differences in target eccentricity of only 4 percent can produce reliable differences in saccadic landing positions. This level of discriminability is comparable to the ability of people to estimate the distance between two stationary targets, one of our most acute perceptual judgments (Westheimer 1979; Burbeck 1987).

6.3.4   Saccades to Objects

In natural scenes (such as the cluttered environment you searched when you looked for the needle on the floor) targets are not small points but objects of a certain size or shape. When targets are objects, you might expect substantially more scatter of saccadic landing positions than when targets are points, because the precise landing position within such objects is not specified.

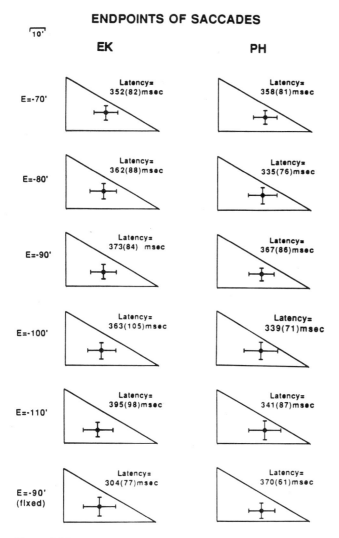

Figure 6.11
Mean endpoints of saccades (±1 SD) made to triangles at different random (70–110 minutes of arc) or fixed (90 minutes of arc) eccentricities. (Reprinted by permission from P. He and E. Kowler, Saccadic localizations of eccentric forms, 1991, *Journal of the Optical Society of America* A, 8, 440–449.)

Surprisingly, there is virtually no change in the spatial precision of saccades when targets increase in size from small points to simple shapes a few degree in diameter. Figure 6.11 shows that when subjects are asked to make a saccade to an eccentric triangle, without trying to choose any particular landing position, but instead look at the form as a whole, the average landing position with respect to the contour remains near the center of gravity of the triangle, despite variation in the triangle's eccentricity. Moreover, the variability of saccadic landing positions is as small as the variability of landing positions of saccades made to single-point targets.

It is doubtful that these precise landing positions are selected deliberately. You can readily demonstrate this for yourself. As you read this text, you may be aware of looking from one word to the next (which is, to a first approximation, what the eye does during reading), but you are surely not aware of choosing a particular letter to land on. The same phenomenon applies to searching for the dropped needle: you think about looking at things, not about pinpointing locations within them. The saccadic system seems to find its own landing position by a spatial pooling process that computes a central reference position within objects. (The same pooling process used to guide saccades may also be used to perceive the location of an object; see Morgan, Hole, and Glennerster 1990.)

We can make saccades to the center of objects with no particular effort or loss of precision, but in natural scanning (for example, searching for the missing needle), we may rarely do so. In a recent study of saccades during reading using fairly long stretches of text (entire paragraphs) and extremely accurate and precise instrumentation, Epelboim, Booth, and Steinman (1994) found that there is considerably more scatter of the landing position of saccades during reading than would be predicted by the small scatter of saccades to eccentric objects described earlier (He and Kowler 1991; Kowler and Blaser 1995). What seems a discrepancy between capacity and performance may be another indication of the congenial relationship between eye movements and cognition. The saccadic system *can* take the line of sight to the precise center of an object, with little variability in landing position. Having such low levels of variability in the simplest sort cf saccadic task, with no important cognitive component, means that there is considerable opportunity for cognitive decisions to influence landing position in more complex and cognitively demanding tasks (such as reading). Decisions would have no noticeable effect on landing position if the saccadic system were "noisy." They can influence the line of sight significantly because there is little variability contributed by other stages of processing.

How might we voluntarily alter the landing position of a saccade within an object? One way is to change the input to the spatial-pooling process by differentially weighing one portion of the object with respect to others. This differential weighting could be carried out by changing how much attention is paid to different portions of the object. The role of attention in saccadic programming is considered next.

6.3.5   Saccades and Attention

Return once more to the search for the needle. If you searched through a sufficiently cluttered environment, you would have noticed that objects were not seen in isolation but against a background of detail and complexity. Backgrounds present the same problem for saccades that they did for smooth eye movements; that is, how are we able to guide movements of the eye on the basis of information from the chosen target object with minimal interference from irrelevant objects? We will arrive at the same solution that worked for smooth eye movements: eye movements are guided by means of selective attention.

Recall that the link between smooth eye movements and attention was demonstrated by asking a subject to pursue and identify targets concurrently (Khurana and Kowler 1987). An analogous experiment for saccades would require the subject to identify an eccentric target while simultaneously preparing either to look at it, or to look elsewhere. If saccades require shifts of attention, then perceptual identification should be better for the target located at the goal of the saccade.

When we set out to do such experiments in my laboratory, we became aware of the methodological difficulties encountered when trying to study how people perform concurrent saccadic and perceptual tasks. We do these tasks outside the laboratory all the time, yet it is surprisingly difficult to study them inside the laboratory. The main problem is deciding what to tell the subject to do, which becomes particularly dicey when you want someone to try to look one way and shift attention elsewhere. If the subject decides to identify the target before programming the saccade, perceptual identification of a target could be accurate, regardless of where the target is with respect to the saccadic goal. Such a strategy would make saccades seem to be independent of attentional shifts, when in fact they might not be. Equally troublesome would be a strategy of confining attention to the saccadic goal, without making sufficient effort to attend to other eccentric locations. This strategy could create the false impression that shifts of attention are necessary to control saccades, when, in fact, the truth might be that sufficient efforts to dissociate saccades and attention might not have been made.

To deal with such problems we (Kowler, Anderson, Dosher, and Blaser 1995) adopted the "dual-task" methodology developed in work on attention (Sperling and Dosher 1986). The basic idea behind these techniques is

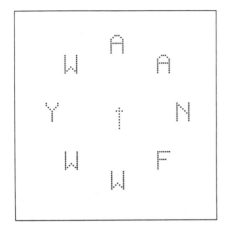

Figure 6.12
Stimulus in experiment on saccades and attention. The task was to look in the direction of the arrow (downward in this case) and report the letter on the right (N). Display duration was 130 msec. (Reprinted by permission from E. Kowler, E. Anderson, B. Dosher and E. Blaser, The role of attention in the programming of saccades, 1995, *Vision Research*, 35, 1897–1916.)

to measure the trade-off between performance of two tasks when subjects are asked to devote different proportions of their attention or effort to accomplishing one or the other.

In our experiment we presented a display containing eight letters and a central arrow cue that disclosed which letter was to be the target of the saccade (see Figure 6.12). The task was to make a saccade in the direction of the arrow and, in the same trial, to identify the letter in the rightmost location. Our goal was to find out whether both these tasks—looking and perceiving—could be done simultaneously as well as each could be done alone.

We told the subject to do the task in three different ways: (1) identify the letter correctly, prolonging saccadic latency if necessary to do the job; (2) keep saccadic latency as short as possible, sacrificing perceptual performance if necessary; and (3) adopt a strategy intermediate between these two extremes. We ran additional experimental sessions in which the saccadic and the perceptual tasks were done alone and experimental sessions in which we tried our best to do both tasks simultaneously as well as we had done each alone.

Results from this experiment are shown in Figure 6.13. These *attentional operating characteristics* (AOCs) show saccadic latency on the abscissa (notice the inverted axis) and perceptual performance on the ordinate. There are two functions in each graph: one for an experiment in which the

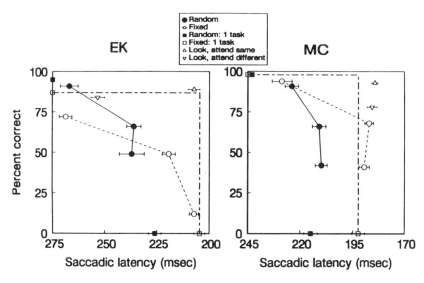

Figure 6.13
Attentional operating characteristics (AOCs) showing saccadic latency (abscissa) and the proportion of correctly identified letters (ordinate). The location of the saccadic target was either selected at random (filled symbols) or remained fixed throughout the session (open symbols). Instructions were to give priority to the saccadic task (lower circles), to the perceptual task (upper circles), or to adopt an intermediate strategy (middle circles). Squares, plotted on the axes, show performance when doing either the saccadic or the letter identification tasks alone. The intersection of the dashed lines emanating from the open squares represents the "independence point," i.e., the point at which there would be no interference in the performance of the two concurrent tasks. The triangles represent attempts to achieve the independence point by trying simultaneously to minimize latency and maximize letter identification; saccadic and perceptual targets were either at the same (upright triangle) or different (inverted triangle) locations. (Reprinted by permission from E. Kowler, E. Anderson, B. Dosher, and E. Blaser, The role of attention in the programming of saccades, 1995, *Vision Research, 35,* 1897–1916.)

direction of the arrow indicating the goal of the saccade varied randomly from trial to trial; and one in which the direction of the arrow remained the same across all trials. Results were similar for both. The functions show that saccades and attention are not independent: performance on one task was sacrificed in order to improve performance on the other. Moreover, subjects never succeeded in doing both saccadic and perceptual tasks together as well as they could do either task alone. The only time in which perceptual and saccadic tasks could be done simultaneously without cost to either is when the saccadic and perceptual targets were in the same location (shown by the upright triangle in each graph). The trade-off between saccadic and perceptual performance shows that we cannot make a saccade to one target while paying full attention to another target located somewhere else. Some attention must be allocated to the goal of the saccade.

An unexpected aspect of the results was the ceiling on the attentional demands of saccades. Substantial improvement in perceptual performance was observed when the subject switched from the instruction to emphasize the saccadic task to the instruction to emphasize both tasks. Yet there was little or no increase in saccadic latency. It was only when near-perfect perceptual performance was required that latency had to be prolonged. Evidently paying too much attention to saccades is a waste of the resources that could be better spent identifying and thinking about what you are looking at. This is clearly a sensible arrangement, protecting us against the need to devote all our attentional resources to motor control.

The two AOCs in each graph (one for random and the other for fixed saccadic direction) are quite similar in shape, the main difference being that latencies were longer for random saccades. This similarity tells us that the attentional demands were connected to the imminent execution of the saccade. The attentional demands of saccades could not be avoided by knowing the saccadic target long in advance and having the opportunity to practice the saccades on many consecutive trials. Even highly practiced saccades require attention to the goal before they are executed. In other words, you can decide that you want to look at a hat, you can use shifts of attention to determine where in your visual field the hat is located, you can think about looking at the hat for a long time, and you can even practice looking at the same hat over and over again, *but* still you will need to shift attention to the hat before you make each saccade.

The link we observed between saccades and attention could be explained by either of the models shown in Figure 6.14. In the spatial model attention is divided between perceptual and saccadic targets during the saccadic latency period. Performance is assumed to depend on the proportion of attention devoted to each. But a problem with the spatial model is that it has no natural way to specify which of the two attended locations is the saccadic goal. For the saccade to be accurate, which they virtually always were in our experiment, a second selective system would be needed—to tell the saccadic subsystem whether to prepare a movement to look down or a movement to look to the right. Attention, by itself, is insufficient.

The temporal model does not require a second selective system. In this model, attention initially resides at the perceptual target and then moves to the saccadic target after sufficient information has been acquired for the perceptual target to be identified. The saccade is initiated right after the attentional shift. In the temporal model, the spatial locus of attention defines the goal of the saccade; saccades would, therefore, be inaccurate if they were initiated too early, before the shift of attention is complete.

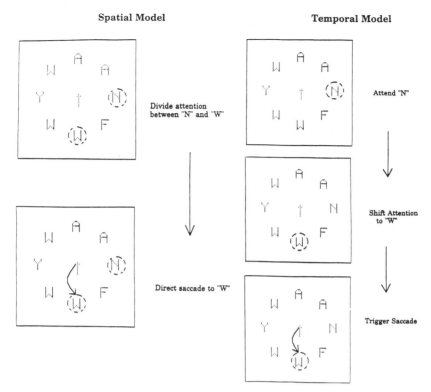

Figure 6.14
Spatial and temporal models of the relationship between attention and saccades. In the spatial model, attention is divided between two locations during the entire saccadic latency period. A saccade is eventually made to one of the locations. In the temporal model (right), attention is initially located at the perceptual target, then shifts to saccadic target. At this time, a saccadic "go" signal is issued, and a saccade is made.

For the temporal model to work, we would need to distinguish between a system that specifies the endpoint of the saccade ("where") and one that initiates the saccade ("when"). The "where" system is the same spatial attentional system that serves perception. There are several candidates for the "when" system in the recently discovered neural areas that inhibit saccades when active and facilitate saccades when silent (see, for example, Munoz and Wurtz 1993a,b).

This distinction between "where" and "when" systems could allow for highly efficient scanning, provided that the two systems communicate with one another. Actually, all that would be necessary would be to pre-set the "when" system to trigger a saccade whenever any transient change in the attentional locus is detected. Scanning would be fast, accurate, and demand nothing more from the observer than to decide which object in the visual field to attend to next. Laboratory studies of attention invariably require that we shut down this saccadic trigger to allow the eye to remain stationary while attention moves to some eccentric location. But it is quite possible that in natural viewing we rarely shut down the saccadic trigger, a speculation that will be supported near the end of this chapter.

The goal of this section on saccades has been to introduce the role of cognition in saccadic control by moving from traditional laboratory studies of saccades, which test very simple and very artificial tasks, to a more naturalistic scanning task, which has been illustrated by the search for the dropped or missing needle. So far, I have discussed what happens when we switch from a sparse laboratory environment, which often contains nothing more than a point target, to a patterned field containing objects of interest located in the midst of distracting background objects. Now that the visual scene has become more natural it is time to take the second step toward natural scanning and take the head off the biteboard.

6.3.6   The Problem of Coordinating Eye and Head

During your search for the missing needle you undoubtedly noticed that movements of your head contributed a lot to shifting the line of sight. Indeed, it would seem absurd to try to search any wide region of space while holding the head still. Yet, until very recently, oculomotor researchers have concentrated on the burdens that movements of the head impose, rather than on the benefits obtained by allowing the head to move.

The burden created by allowing free movement of the head is the need to coordinate head and eye movements so that *gaze shifts* remain accurate. (The term *gaze shift* is typically used to denote shifts of the line of sight in space, carried out when the head is free to move, while *saccade* is reserved for the rotation of the eye within the head.) Suppose you are directing both head and eyes straight ahead and the object you want to look at is 45 degrees to your right. With the head on the biteboard, you would need to rotate the eye 45 degrees, but when the head is free to move, a part of the gaze shift is accomplished by rotating the head in space and a part by rotating the eye in the orbit. Figure 6.15 shows the contribution of eye and head to a gaze shift made by a subject looking between two stationary targets separated by 66 degrees. This gaze shift was quite accurate. How was it possible to specify how much of the gaze shift should be contributed by the head and how much by the eye while ensuring accuracy?

A simple and workable solution was proposed by Morasso, Bizzi, and Dichgans (1973). Their idea was that the saccadic system programs a rotation of the eye by the amount that would bring the line of sight to the target if the head were stationary. To prevent any movements of the head from adding to the movements of the eye and taking the line of sight well

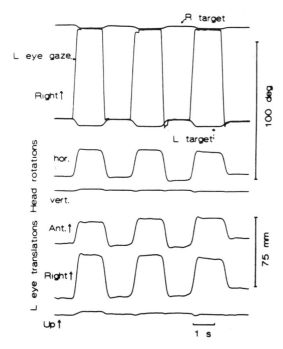

Figure 6.15
The top panel shows gaze shifts between two stationary targets made with the head free. The oscillations of the target trace indicate the perturbation of the retinal image of the stationary target due to head movements. Gaze shifts were accurate, despite the perturbations. The middle panel shows head rotations, and the lower panels head translations. (Reprinted by permission from H. Collewijn, R. M. Steinman, C. J. Erkelens, Z. Pizlo, and J. Van der Steen, The effect of freeing the head on eye movement characteristics during 3-D shifts of gaze and tracking, 1992, in A. Bertolz, W. Graf, and P. P. Vidal, eds., *The head-neck sensory motor system*, New York: Oxford University Press.)

beyond the target, the vestibulo-ocular response (VOR) automatically reduces the amplitude of eye rotation in the head by an amount equal to the rotation of the head in space. The VOR (discussed earlier in the section on maintained fixation with a freely moving head) uses the vestibular signals originating from the semicircular canals to compensate for rotations of the head by generating rotations of the eye in the opposite direction. (If you were staring straight ahead and the VOR were working perfectly, the eye would remain in the same place in space during head rotation.) As we saw in the section on fixation with the head free to move, the VOR does *not* work perfectly for head rotations, leading to considerable motion of the retinal image. (There is, as we shall soon see, considerable controversy about how well the VOR works, and whether it works at all, during shifts of gaze.)

Depending on the VOR to ensure accurate gaze shifts becomes more difficult when targets are nearby. This is because head translations, which are inevitable unless special helmets are used to restrict movements to rotations, displace the retinal image by an amount that depends on target distance. For example, a head translation of only 3 centimeters (cm) moves the retinal image of an object located 20 cm away by nearly 10 degrees. To compensate for the translations we might need a different kind of VOR, triggered by signals representing head translations (Paige 1989). For these vestibular compensatory systems to ensure accurate shifts of gaze they would have to work fast, since gaze shifts are typically too fast for visual signals to play much of a helpful role. For a linear VOR to be effective when targets are nearby, it would have to be sensitive to target distance (evidence suggests it might be; Paige 1989) because translations produce larger retinal displacements the closer the target is to the eye.

The VOR could produce accurate gaze shifts, but a number of investigators have reported that the VOR does not function during shifts of gaze! The whole system is believed to be turned off (Laurutis and Robinson 1986; Guitton and Volle 1987). Accurate shifts of gaze are instead attributed to a system that allows the gaze shift to continue until the monitored error between the current eye position and the target position falls to zero. At that point, the gaze shift stops and the VOR comes back on again. An alternative to both the models described above is that accurate gaze shifts are programmed from the outset, based on visual estimates of target position, with no contribution from the vestibular feedback systems while the gaze shift is in progress.

Any or all of these three models of gaze control are plausible, given the existing evidence. In an attempt to begin to sort things out, let us find out how eye and head are coordinated during a natural scanning task, much like the one you might perform in your search for the missing needle.

6.3.7   Gaze Shifts While the Head Is Free to Move

As discussed in the section on fixational eye movements, a significant obstacle to studying eye movements while the head is free to move is that instruments with the capacity to make accurate recordings of even small eye movements require a stabilized head. This problem was solved by new instrumentation used initially to study fixational eye movements accurately with the freely moving head (see 6.1.2). By adding a separate measuring system for translations, it became possible to determine the position of the line of sight with respect to nearby targets inspected while the head was free to move.

Initial studies had subjects looking back and forth between two nearby, stationary targets (Collewijn, Steinman, Erkelens, Pizlo, and Van der Steen

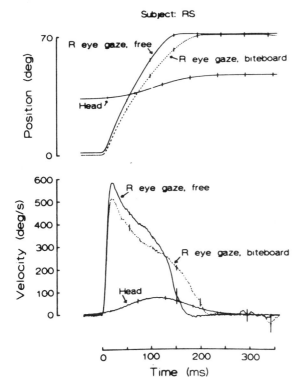

Figure 6.16
Position (top) and velocity (bottom) traces of a single gaze shift with head stabilized
(dotted lines) or free to move (solid lines). (Reprinted by permission from H. Collewijn, R.
M. Steinman, C. J. Erkelens, Z. Pizlo, and J. Van der Steen, The effect of freeing the head on
eye movement characteristics during 3-D shifts of gaze and tracking, 1992, in A. Bertolz,
W. Graf, and P. P. Vidal, eds., *The head-neck sensory motor system*, New York: Oxford
University Press.)

1992). The sample gaze shift in Figure 6.16 shows that gaze shifts while
the head was free were not only accurate (top traces) but also faster than
those observed while the head was stabilized by a biteboard. This result
was not due simply to the velocity of the moving head adding to the
velocity of the eye in the orbit (bottom traces) (which would happen if the
VOR were turned off during the gaze shift) because peak gaze velocity
was reached well before the head began to pick up speed. Evidently,
something about holding the head in place impairs the generation of
saccades. Natural conditions, with freely moving heads, allow more effective performance.

6.3.8   Gaze Shifts During a Visuomotor Task

In the natural world, gaze shifts between targets are rarely, if ever, made for no reason at all. So, a subsequent experiment studied gaze shifts while a subject had to tap at a sequence of vertical rods (Epelboim et al. 1995). The addition of the tapping task added a grab bag of new visual, cognitive, and motor activities to the traditional laboratory gaze shifts. Nevertheless, the gaze shift itself—the movement of eye and head made to look between stationary targets—should remain much the same as it is when gaze shifts are studied in isolation. This prediction was *not* borne out by the results of the experiment described below.

A drawing of the experimental apparatus is shown in Figure 6.17. The task was to tap a sequence of two, four, or six rods placed in wells on a

The Maryland Revolving Field Monitor

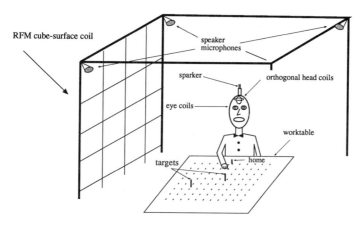

Figure 6.17
Stimulus and recording instruments for experiments on looking and tapping. The subject starts with the finger at the home position and proceeds to tap sequences of 2, 4, or 6 vertical rods located in randomly selected wells on the worktable. The MRFM (Maryland Revolving Field Monitor) cube-surface coil generates a homogeneous, three-dimensional rotating magnetic field. Detection of the phase of the voltages induced in the coils on the eyes and head allow recording of rotational movements to an accuracy of 1 minute or arc. Samples were taken 488 times/second. Translations are measured by recording the arrival time of the wavefronts of sounds (sparks) generated by electrodes mounted on top of the head. The wavefronts of these sounds are detected by four microphones mounted above the subject's head at the corners of the cube-surface coil arrangement. (Reprinted by permission from Epelboim, J., R. M. Steinman, E. Kowler, M. Edwards, Z. Pizlo, C. J. Erkelens and H. Collewijn, The function of visual search and memory in sequential looking tasks, 1995, *Vision Research*, in press.)

large tabletop. Rods were placed in randomly selected wells out of the view of the subject, who had closed his eyes; thus, the only time he saw the stimulus pattern was during the actual trials. Ten consecutive trials with each sequence were recorded. The performance in this tapping task was contrasted with performance when the same sorts of sequences were simply looked at, with no tapping or any other visual, cognitive, or motor requirements added to the task.

There were striking differences between tapping and looking. For one thing, it took much longer for the rod locations to be learned when only looking at them. This result can be seen by comparing the graphs in Figure 6.18 (for tapping) and 6.19 (for looking). Figure 6.18 shows what arm and eye were doing over time for each of the ten trials in a representative sequence. The abscissa shows time, while each horizontal line represents one of the six targets in a sequence. (Targets were color coded so that subjects would know the order in which to proceed.) Each filled circle on a horizontal line shows when a particular target was tapped. This subject tapped all six in the correct order (a typical result; tapping errors were rare). The unfilled boxes show where gaze was over time, with box width indicating how long gaze remained on any of the six targets. Gaze hops about all over the place during the first two or three repetitions while the subject is searching for and learning the location of each rod. Things settle down by the fourth repetition. Throughout all ten trials a tight link between arm and eye is revealed: targets were never tapped without looking at them first. This outcome was rarely violated throughout the entire experiment.

By the time things settled down in repetition four, the pattern of head and eye movements became quite stereotypical from one trial to the next (Figure 6.18). By contrast, it took about nine repetitions to settle on a stereotypical eye-movement pattern when only looking at the rods (Figure 6.19). And, even after all the practice, the time required to complete a "looking-only" trial remained longer than that needed for a "tapping" trial (see Figure 6.20). Although gaze errors were a bit larger in tapping (2.5 degrees) than while looking-only (2 degrees), this does not, by itself, explain the faster tapping times; the time to complete a sequence increased with practice while tapping, but gaze errors remained constant across the ten repetitions. In short, tapping, the nominally harder task because it involves coordination of at least three effector systems (head, eye, and arm), took less time and benefitted more from practice than did the supposedly less-demanding but more artificial task of only looking at targets. Additional analyses performed by Epelboim et al. (1995) suggested that differences between tapping and looking-only lie in different strengths of memory for the target locations.

TAP

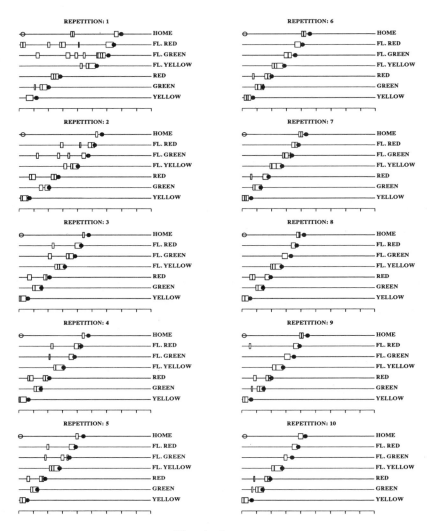

Time (sec)

Figure 6.18
Eye and hand positions over time during tapping of a sequence of 6 rods. Ten repetitions with the same sequence are shown. Rods were color coded to disclose the order in which they should be tapped. The filled circle on each graph shows when each rod was tapped. Open rectangles show where the subject was looking as a function of time. (Reprinted by permission from Epelboim, J., E. Kowler, M. Edwards, H. Collewijn, C. J. Erkelens, and R. M. Steinman, Natural oculomotor performance in looking and tapping tasks, 1994, *Proceedings of the Cognitive Science Society*, 16, 272–277.)

LOOK

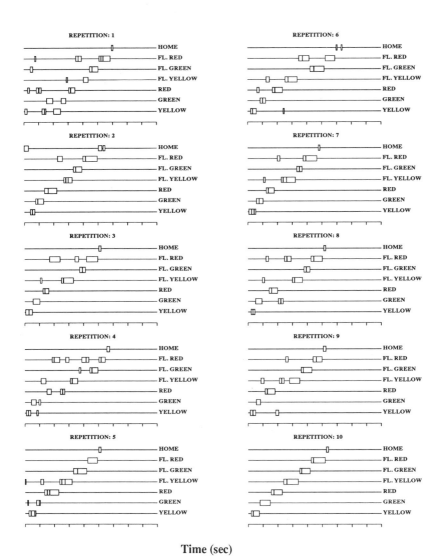

Time (sec)

Figure 6.19

Eye position over time when looking at a sequence of 6 rods. Ten repetitions with the same sequence are shown. Rods were color coded to disclose the order in which they should be looked at. Open rectangles show where subject was looking as a function of time. (Reprinted by permission from Epelboim, J., E. Kowler, M. Edwards, H. Collewijn, C. J. Erkelens, and R. M. Steinman, Natural oculomotor performance in looking and tapping tasks, 1994, *Proceedings of the Cognitive Science Society*, 16, 272–277.)

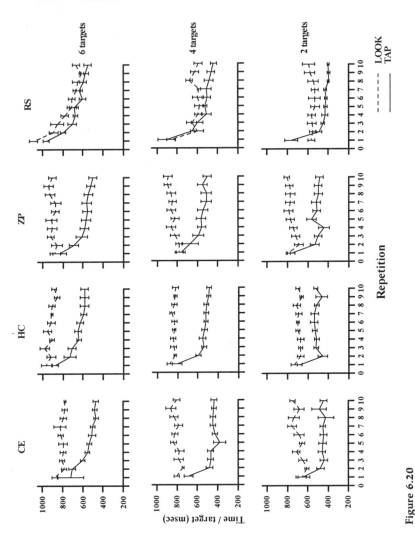

Figure 6.20
Average time per target taken to complete a tapping or looking trial as a function of repetition number. Graphs show performance for different subjects and different sequence lengths. (Reprinted by permission from Epelboim, J., E. Kowler, M. Edwards, H. Collewijn, C. J. Erkelens, and R. M. Steinman, Natural oculomotor performance in looking and tapping tasks, 1994, *Proceedings of the Cognitive Science Society*, 16, 272–277.)

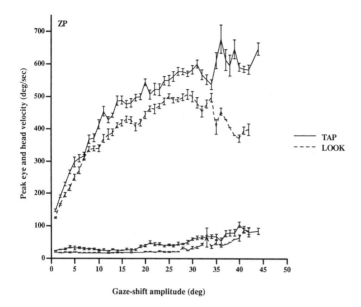

Figure 6.21
Peak velocity of gaze and head as a function of gaze-shift amplitude for tapping and looking. Gaze velocity (top two traces) was much faster than head velocity.

Differences between tapping and looking persist when we examine the velocity of the gaze shifts themselves (Epelboim et al. 1994b). Figure 6.21 compares gaze velocities during looking and tapping by plotting peak gaze velocities as a function of gaze-shift amplitude. Gaze velocities are faster during tapping, thus enhancing the improvement in gaze velocities (described earlier) created simply by allowing free head movement. Once again, just as in freeing the head, the improvement is not caused by the simple addition of eye and head velocities, for gaze velocities during tapping were faster than during looking, even when the movements of the head were similar (as shown in Figure 6.21). Simple addition of eye and head velocities can also be ruled out by the observation that the velocity of the eye within the head was faster during tapping than during looking (Figure 6.22).

What accounts for the faster gaze while tapping? Epelboim et al. (1994b) speculated that gaze velocity depends on the effectiveness of the VOR. They proposed that there is a functioning VOR during gaze shifts that is by no means perfect. Increasing the "gain" of the VOR (where "gain" is the ratio of eye movement to head movement) should improve the accuracy of the gaze shift, but at the expense of a reduction in gaze velocity. If, on the other hand, speedy shifts of gaze are what you want, then the best

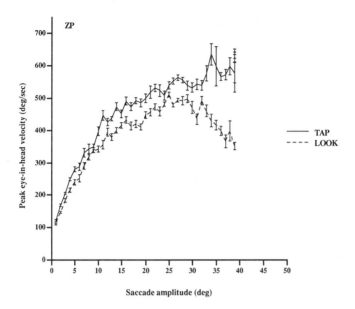

Figure 6.22
Peak velocity of saccade (eye-in-head) as a function of saccade amplitude for tapping and looking.

tack is to decrease the VOR gain, thus allowing the eye to ride along with the head a bit more, and tolerate any resulting increase in gaze errors that might occur. This may well have been the strategy adopted during tapping, when gaze shifts were faster and gaze errors were larger.

It is doubtful that any of the subjects performing these tasks had any idea that they were adjusting their VOR gain or making decisions about the kinds of gaze errors they wanted to tolerate. They were concentrating on tapping targets rapidly and in the correct order (just as you were concentrating on finding the missing needle). All the rules for controlling head, eye, and arm could have been built into a package of instructions formulated at a high, task-based level at which typical tasks are things like tapping, searching, or reading (Epelboim et al., 1995).

The existence of task-level rules implies that the influence of concurrent cognitive activities on eye movements is more profound and basic to the operation of eye movements than had been previously supposed. Up to this point, in most of the examples presented in this chapter, the influence of cognition was confined to the formation and definition of the effective input to the oculomotor control system. The differences between gaze shifts in looking and tapping tasks implies that the way the oculomotor system operates on the input is also constrained by the cognitive events accompanying the task as a whole. We do not know which cognitive events might be important, nor how their influence is exerted. Research allowing accurate measurement of head and eye movements during natural tasks is still at an early stage. By continuing to study eye movements and head movements in the context of natural performance, we may ultimately arrive at a clear understanding of how these complex systems do the jobs they were intended to do.

6.3.9   Inferring Mental States from Saccades

A related attempt to understand coordination of head, eye, and arm during the performance of a natural task is described by Ballard, Hayhoe, and Pelz (1995). Their focus is less on the mechanisms of oculomotor control than on the way the sequences of movements provide clues about underlying cognitive strategy.

The experimental task was to copy a random pattern of eight colored blocks generated on a computer screen. To the right of the pattern to-be-copied (the *model*) was a *source* area containing an assortment of blocks. The subject had to retrieve blocks from the source area and copy the model pattern in a third region termed the *workspace*. A sample display is shown in Figure 6.23.

This simple task, like the tapping described above, was performed with a continual sequence of eye and head movements, and subjects tended to follow a very stereotypical pattern. The most striking characteristic of the performance was how little the subjects relied on memory for the pattern as a whole. Instead, they began each sequence by looking at the model, choosing one of the blocks, then moving eye and arm to the source to find a block of the appropriate color, glancing back at the model to confirm the block's location, and then finally moving eye, head, and arm to the workspace to put the block down. This "model-pickup-model-block" strategy was followed in the majority of the trials, particularly when copying the first two blocks in a sequence. Although the memory load for the colors and locations would seem to be fairly light, subjects chose instead to use the display itself as an external memory. They also, apparently, chose to make saccades rather than relying on perceptual attention to sample information from eccentric retina. This outcome is consistent with the suggestion made earlier (see subsection 6.3.5) that attentional shifts without saccades may be rare and that it may be more efficient to operate in a mode in which saccades are set to be triggered by any transient change in the locus of attention.

The use of the display as an external memory is known as a *diectic strategy*, which has played a prominent role in many algorithms used to control robot vision. The virtue of such a strategy, argue Ballard et al.

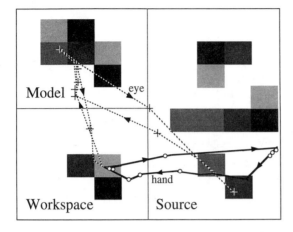

Figure 6.23
Stimulus display used in the block-copying task showing the model, workspace, and source. The eye position trace is shown by the cross and the thin line. The cursor trace is shown by the arrow and dark line. A single cycle is shown, from dropping off block two to dropping off block three (in the experimental trial the blocks were colored). (Reprinted by permission from D. H. Ballard, M. M. Hayhoe and J. B. Pelz, *Memory representations in natural tasks*, 1995, *Journal of Cognitive Neuroscience*, 7, 66–80.)

(1995), is that eye movements are inexpensive ways to retrieve information. Why use up internal memory storing elaborate representations of space when the answer to the immediate question can be obtained by a quick glance? This logic provides a convincing rationale for using eye movements as an online record of cognitive decisions about search and problem-solving strategies. Ballard et al.'s (1995) approach to inferring cognitive states from saccades may succeed because the material they are using (random assortments of blocks) has no obvious structure or pattern, and so mnemonic or grouping strategies for reducing memory load would prove difficult. In such cases, eye movements may be the best bet for retrieving information and the subject is forced to produce this overt indication of the retrieval strategy. (For an alternative approach to inferring mental states from saccades with cognitively complex material, see Suppes 1990.)

6.4   Summary and Conclusions

The goal of this chapter was to tell the story of the integral role played by cognitive processes in eye movement control. We have seen that cognitive processes are involved not only in high-level aspects of target selection, but in numerous stages of oculomotor programming, including those formerly believed to fall into the domain of involuntary sensorimotor events immune to cognitive influence of any sort. Research on oculomotor control in the context of natural tasks has begun to reveal that eye movements work differently, indeed, actually better, when performed concurrently with "real" tasks that require us to recognize objects, reach for targets, or plan sequences of actions. Eye movements evolved to serve the needs of vision and, at the same time, to take advantage of the visual and cognitive activities made possible by the optimal retinal conditions that the eye movements themselves create.

Suggestions for Further Reading

Several comprehensive reviews of eye-movement research provide excellent surveys and critical evaluations of behavioral and physiological research, as well as treatments of models (most of which are derived from engineering). For example, two noteworthy and timeless monographs are Collewijn's (1981) review of eye movements and oculomotor physiology in the rabbit and Yarbus's (1976) classic treatment of eye movements and vision. More recent and more broadly based reviews are Carpenter's book (1988) and edited volume (1990) and Hallett's chapter in the *Handbook of Perception and Human Performance* (1986). The series of edited books published by Elsevier, *Reviews of Oculomotor Research*, contains comprehensive reviews of topics related to, for example, oculomotor adaptation (Berthoz and Melvill Jones 1985), saccades (Wurtz and Goldberg 1989), and visual motion (Miles and Wallman 1993). For reviews of topics most directly related to vision and cognition, see volume 4 of the Elsevier series (Kowler 1990), which contains detailed reviews and critical analyses of both the way eye movements are influenced by cognitive processes and how vision and cognition (in tasks ranging from acuity to reading and mental arithmetic) depend on eye movements. This volume includes critical reviews of attempts to infer cognitive events from oculomotor patterns in chapters by O'Regan, Viviani, and Suppes.

Many innovative methods for measuring eye movements have been developed over the past century, going back to Delabarre (1897), who put a plaster ring around his cornea and recorded movements by means of a protruding wire that made marks on a rotating smoked drum. Carpenter (1988) and Hallett (1986) survey the most popular modern methods, including discussion of relevant geometrical issues and sources of measurement artifacts. For a good introductory-level discussion of the problem of translational artifacts, see Cornsweet (1976); and for a critical evaluation of some modern methods and discussion of the methods for measuring eye movements with a freely moving head, see Steinman and Levinson (1990), Erkelens et al. (1989a, 1989b), Epelboim, Booth, and Steinman (1994) and Edwards et al. (1994).

Several topics in eye-movement research have been the focus of heated controversy. One is the merits of engineering models; for a good debate of this issue, see Robinson (1986) versus Steinman (1986a, 1986b). Another, more long-standing debate concerns the contribution of eye movements to visual localization. Much of the controversy revolves around the role of internal representations of eye position ("efferent copies") in both perceptual localization and motor control. See Skavenski (1990) for a particularly clear and thoughtful treatment of the history and controversies, Van Gisbergen and Van Opstal (1989) for discussion of oculomotor models incorporating efferent copy and Matin and Li (1994) for a new slant on the problem of how vision affects representations of eye position.

Problems

6.1   Here is a classical demonstration that is often described by researchers who study both vestibular and visual compensatory eye movements. Using this page of text as your stimulus, rotate your head back and forth, first at a fairly low velocity, then gradually making your movements faster and faster. Take general note of how fast the head movements have to be before the text becomes too blurry to read. Now observe the appearance of the text again, but this time moving the page in the same oscillatory pattern while keeping your head as still as possible. The usual outcome of this demonstration is that moving the text produces considerably more blur than moving your head. Offer at least two explanations for this phenomenon.

6.2   Professor L. Dell'Osso, author of many papers on an oculomotor disorder known as *congenital nystagmus* (CN) (e.g., Dell'Osso et al., 1992), has CN himself and has made many observations, both formal and informal, on its consequences for vision. CN is characterized by a periodic, high-velocity oscillation of the eye, with only very brief instances (60 milliseconds) during each cycle of the oscillatory movement in which the eye is relatively stationary (velocity about 2 degrees per second). I asked Professor Dell'Osso how well he threads needles, expecting that the high-velocity image motion would be a major impediment. He said he can thread a needle faster than anyone else he knows. The only visual judgment that is problematical for him is perceiving the motion of slowly moving objects (such as the motion of a deer walking in the forest). He often does not notice objects moving slowly against structured backgrounds. Given the above description of the CN pattern, and assuming that the effects of retinal-image motion in someone with CN are comparable to the effects of image motion in others, explain why the characteristics of the CN pattern as described above might predict that threading needles would be easy, while judging low-velocity motion would be difficult.

6.3   A frequent question that comes up about the experiment on anticipatory pursuit eye movements made in response to symbolic cues (see Figures 6.6 and 6.7) is whether the cues determine the direction of anticipatory pursuit on the very first trial of the experiment, or whether it takes time for cues to become effective. What are the specific implications of finding, and not finding, effects of the cues on the very first trial for theories of smooth eye tracking?

6.4   One result of the experiments on gaze shifts during looking and during tapping was that the time between successive shifts of gaze increased with the number of targets to be looked at or the number of targets to be tapped. What are some of the possible explanations of the effects of sequence length on timing between responses?

Questions for Further Thought

6.1   There is a long tradition of using saccadic patterns to infer cognitive events in tasks such as search or reading. Suppose that you have a record of the pattern of eye movements made by a subject counting the number of beads on a tabletop and that, in addition, you know whether the answer he or she gave to the counting problem was correct. Your job is to analyze the eye-movement pattern and use it to infer the subject's counting strategy. What eye-movement characteristics would you look for? What specific assumptions would you have to make in order to infer the counting strategy from eye-movement characteristics?

6.2   One of the reasons that eye movements (particularly saccades) are so important for human beings is that we have a fovea, a small retinal region where visual resolution is best. Those studying natural vision in humans and animals, and those designing artificial vision systems for robots, often speculate about whether the system would perform better as a whole if there were no fovea and if, instead, a high degree of resolution were available across large areas of the retina. Given what you now know about the mechanisms controlling gaze shifts and saccades, what might be the various advantages and disadvantages of abandoning our fovea (with the accompanying need for saccades) in favor of a homogeneous retina with uniformly high levels of visual resolution?

References

Aitsebaomo, A. P., and H. E. Bedell (1992). Psychophysical and saccadic information about direction for briefly presented visual targets. *Vision Research* 32, 1729–1737.

Andre-Deshays, C., A. Berthoz, and M. Revel (1988). Eye-head coupling in humans. I. Simultaneous recording of isolated motor units in dorsal neck muscles and horizontal eye movements. *Experimental Brain Research* 69, 399–406.

Ballard, D. H., M. M. Hayhoe, and I. B. Pelz (1995). Memory use during hand-eye coordination. *Journal of Cognitive Neuroscience* 7, 66–80.

Barlow, H. B. (1952). Eye movements during fixation. *Journal of Physiology* 116, 290–306.

Becker, W., and A. F. Fuchs (1969). Further properties of the human saccadic system: Eye movements and correction saccades with and without visual fixation points. *Vision Research* 9, 1247–1258.

Berthoz, A., and G. M. Jones, eds. (1985). *Adaptive mechanisms in gaze control.* Amsterdam: Elsevier.

Burbeck, C. A. (1987) Position and spatial frequency in large-scale localization judgments. *Vision Research* 27, 417–427.

Carpenter, R. H. S. (1988). *Movements of the eyes,* 2nd ed. London: Pion Press.

Carpenter, R. H. S., ed. (1990). *Eye movements,* vol. 9, *Vision and visual dysfunction.* Boca Raton, FL: CRC Press.

Collewijn, H. (1969). Optokinetic eye movements in the rabbit. *Vision Research* 9, 117–132.

Collewijn, H. (1981). The *oculomotor system of the rabbit and its plasticity.* Berlin: Springer-Verlag.

Collewijn, H., C. J. Erkelens, and R. M. Steinman (1988). Binocular co-ordination of horizontal saccadic eye movements. *Journal of Physiology* 404, 157–182.

Collewijn, H., R. M. Steinman, C. J. Erkelens, Z. Pizlo, and J. Van der Steen (1992). The effect of freeing the head on eye movement characteristics during 3-D shifts of gaze and tracking. In A. Berthoz, W. Graf, and P. P. Vidal, eds., *The Head-neck sensory motor system.* New York: Oxford University Press.

Cornsweet, T. N. (1956). Determination of the stimuli for involuntary drifts and saccadic eye movements. *Journal of the Optical Society of America* 46, 987–193.

Cornsweet, T. N. (1976). The Purkinje image method of recording eye position. In R. A. Monty and J. W. Senders, eds. *Eye movements and psychological processes.* Hillsdale, NJ: L. Erlbaum.

Craik, K. (1947). Theory of the human operator in control systems. *British Journal of Psychology* 38, 56–61.

Cushman, W. B., J. F. Tangney, R. M. Steinman, and J. L. Ferguson (1984). Characteristics of smooth eye movements with stabilized targets. *Vision Research* 24, 1003–1009.

Dallos, P. J., and R. W. Jones (1963). Learning behaviour of the eye fixation control system. *IEEE Transactions on Automatic Control* AC8, 218–227.

Delabarre, E. B. (1897). A method of recording eye movements. *American Journal of Psychology* 9, 572–574.

Dell'Osso, L. F., J. Van der Steen, R. M. Steinman, and H. Collewijn (1992). Foveation dynamics in congenital nystagmus. I: Fixation. *Documenta Ophthalmologica* 79, 1–23.

Ditchburn, R. W., and B. L. Ginsborg (1953). Involuntary eye movements during fixation. *Journal of Physiology* 119, 1–17.

Dodge, R. (1927). *Elementary conditions of human variability*, New York: Columbia University Press.

Dodge, R., and J. C. Fox (1930). Optic nystagmus. *Archives of Neurology and Psychiatry* 20, 812–823.

Dodge, R., R. C. Travis, and J. C. Fox (1930). Optic nystagmus III. Characteristics of the slow phase. *Archives of Neurology and Psychiatry* 24, 21–34.

Edwards, M., Z. Pizlo, C. J. Erkelens, H. Collewijn, J. Epelboim, M. R. Stepanov, E. Kowler, and R. M. Steinman (1994). *The Maryland revolving field monitor: Theory of the instrument and processing its data*. Center for Automation Research Technical Report, CAR-TR-711. College Park, MD: University of Maryland.

Epelboim, J., J. R. Booth, and R. M. Steinman (1994a). Reading unspaced text: Implications for theories of reading eye movements. *Vision Research* 34, 1735–1766.

Epelboim, J., H. Collewijn, M. A. Edwards, C. J. Erkelens, E. Kowler, Z. Pizlo, and R. M. Steinman (1994b). Coordinated movements of the arm and head increase gaze-shift velocity. *Investigative Ophthalmology and Visual Science Supplement* 35, 1550.

Epelboim, J., E. Kowler, M. Edwards, H. Collewijn, C. J. Erkelens and R. M. Steinman (1994c). Natural oculomotor performance in looking and tapping tasks. *Proceedings of the Cognitive Science Society* 16, 272–277.

Epelboim, J., R. M. Steinman, E. Kowler, M. Edwards, Z. Pizlo, C. J. Erkelens and H. Collewijn (1995). The function of visual search and memory in sequential looking tasks, *Vision Research*, in press.

Epelboim, J., and E. Kowler (1993). Slow control with eccentric targets: Evidence against a position-corrective model. *Vision Research* 33, 361–380.

Erkelens, C. J., R. M. Steinman, and H. Collewijn (1989a). Ocular vergence under natural conditions I. Continuous changes of target distance along the median plane. *Proceedings of the Royal Society of London* B236, 417–440.

Erkelens, C. J., R. M. Steinman, and H. Collewijn (1989b). Ocular vergence under natural conditions II. Gaze shifts between real targets differing in distance and direction. *Proceedings of the Royal Society of London* B236, 441–465.

Freyd, J. F. (1987). Dynamic mental representations. *Psychological Review* 94, 427–438.

Guitton, D., and M. Volle (1987). Eye-head coordination during orienting movements to targets within and beyond the oculomotor range. *Journal of Neurophysiology* 58, 427–459.

Hallet, P. E. (1986). Eye movements. In K. R. Boff, L. Kaufman, and J. P. Thomas, eds., *Handbook of perception and human performance* vol. 1, *sensory processes and perception*, Chapter 10. New York: John Wiley.

Hartline, H. K. (1938). The response of single optic nerve fibers of the vertebrate eye to illumination of the retina. *American Journal of Physiology* 121, 400–415.

Hartline, H. K. (1940). The receptive field of the optic nerve fibers. *American Journal of Physiology* 130, 690–699.

He, P., and E. Kowler (1991). Saccadic localization of eccentric forms. *Journal of the Optical Society of America* A8, 440–449.

Heinen, S. J. (1993). Characteristics of predictive smooth-pursuit neurons in the dorsomedial frontal cortex. *Investigative Ophthalmology and Visual Science Supplement* 34, 1500.

Helmholtz, H. (1925). *Treatise on physiological optics*, transl. by J. P. Southall. New York: Dover.

Kapoula, K., and D. A. Robinson (1986). Saccadic undershoot is not inevitable: Saccades can be accurate. *Vision Research* 26, 735–743.

Khurana, B., and E. Kowler (1987). Shared attentional control of smooth eye movement and perception. *Vision Research* 27, 1603–1618.

Kowler, E. (1989). Cognitive expectations, not habits, control anticipatory smooth oculo-motor pursuit. *Vision Research* 29, 1049–1057.

Kowler, E., ed. (1990). *Eye movements and their role in visual and cognitive processes.* Amsterdam: Elsevier.

Kowler, E. (1990). The role of visual and cognitive processes in the control of eye movement. In E. Kowler, ed., *Eye movements and their role in visual and cognitive processes.* Amsterdam: Elsevier.

Kowler, E. (1991). The stability of gaze and its implications for vision. In R. H. S. Carpenter, ed., *Eye movements. Vision and visual dysfunction,* vol. 9. Boca Raton, FL: CRC Press.

Kowler, E., E. Anderson, B. Dosher, and E. Blaser (1995). The role of attention in the programming of saccades. Vision Research, 35, 1897–1916.

Kowler, E., E. Blaser (1995). The accuracy and precision of saccades to small and large targets. *Vision Research,* 35, 1741–1754.

Kowler, E., and S. McKee (1987). Sensitivity of smooth eye movement to small differences in target velocity. *Vision Research* 27, 993–1015.

Kowler, E., J. van der Steen, E. P. Tamminga, and H. Collewijn, (1984). Voluntary selection of the target for smooth eye movement in the presence of superimposed, full-field stationary and moving stimuli. *Vision Research* 24, 1789–1798.

Kowler, E., Z. Pizlo, G. L. Zhu, C. J. Erkelens, R. M. Steinman, and H. Collewijn (1992). Coordination of head and eye during the performance of natural (and unnatural) visual tasks. In A. Berthoz, W. Graf, and P. P. Vidal, eds., *The head-neck sensory motor system.* New York: Oxford University Press.

Laurutis, V. P., and D. A. Robinson (1986). The vestibulo-ocular reflex during human saccadic eye movements. *Journal of Physiology* 373, 209–233.

Lemij, H. G., and H. Collewijn (1989). Differences in accuracy of human saccades between stationary and jumping targets. *Vision Research* 29, 1737–1748.

Mach, E. (1906/1959). *Analysis of sensations,* New York: Dover.

Marshall, W. H., and S. A. Talbot (1942). Recent evidence for neural mechanisms in vision leading to a general theory of sensory acuity. *Biological Symposium* 7, 117–164.

Matin, L., and W. Li (1994). The influence of a stationary single line in darkness on the visual perception of eye level. *Vision Research* 34, 311–330.

McKee, S. P., and K. Nakayama (1984). The detection of motion in the peripheral visual field. *Vision Research* 24, 25–32.

Miles, F. A., and J. Wallman, eds. (1989). *Visual motion and its role in the stabilization of gaze.* Amsterdam: Elsevier.

Miles, F. A., and J. Wallman (1993). *Visual motion and its role in the stabilization of gaze.* Amsterdam: Elsevier.

Morasso, P., E. Bizzi, and J. Dichgans (1973). Adjustment of saccadic characteristics during head movements. *Experimental Brain Research* 16, 492–500.

Morgan, M. J., G. J. Hole, and A. Glennerster (1990). Biases and sensitivities in geometrical illusions. *Vision Research* 30, 1793–1810.

Munoz, D. P., and R. H. Wurtz (1993a). Fixation cells in monkey superior colliculus I. Characteristics of cell discharge. *Journal of Neurophysiology* 70, 559–575.

Munoz, D. P., and R. H. Wurtz (1993b). Fixation cells in monkey superior colliculus II. Reversible activation and deactivation. *Journal of Neurophysiology* 70, 576–589.

O'Regan, J. K. (1990). Eye movements and reading. In E. Kowler, ed., *Eye movements and their role in visual and cognitive processes.* Amsterdam: Elsevier.

Paige, G. D. (1989). The influence of target distance on eye movement responses during vertical linear motion. *Experimental Brain Research* 77, 585–593.

Pavel, M. (1990). Predictive control of eye movement. In E. Kowler, ed. *Eye movements and their role in visual and cognitive processes.* Amsterdam: Elsevier.

264   Kowler

Pavel, M., H. Cunningham, and V. Stone (1992). Extrapolation of linear motion. *Vision Research* 11, 2177–2186.

Polyak, S. L. (1941). *The retina.* Chicago: University of Chicago Press.

Puckett, J. de W., and R. M. Steinman (1969). Tracking eye movements with and without saccadic correction. *Vision Research* 9, 695–703.

Ratliff, F., and L. A. Riggs (1950). Involuntary motions of the eye during monocular fixation. *Journal of Experimental Psychology* 40, 687–701.

Riggs, L. A., F. Ratliff, J. C. Cornsweet, and T. N. Cornsweet (1953). The disappearance of steadily fixated test objects. *Journal of the Optical Society of America* 43, 495–501.

Robinson, D. A. (1986). The systems approach in the oculomotor system. *Vision Research* 26, 91–99.

Skavenski, A. A. (1990). The role of visual and cognitive processes in the control of eye movement. In E. Kowler, ed., *Eye Movements and Their Role in Visual and Cognitive Processes*, pp. 263–287. Amsterdam: Elsevier.

Skavenski, A. A., and R. M. Steinman (1970). Control of eye position in the dark. *Vision Research* 10, 193–203.

Skavenski, A. A., R. H. Hansen, R. M. Steinman, and B. J. Winterson (1979). Quality of retinal image stabilization during small natural and artificial body rotations in man. *Vision Research* 19, 675–683.

Sparks, D. L., and L. E. Mays (1983). Spatial localization of saccade targets. I. Compensation for stimulation-induced perturbations in eye position. *Journal of Neurophysiology* 49, 45–63.

Sperling, G., and B. A. Dosher (1986). Strategy and optimization in human information processing. In K. R. Boff, L. Kaufman, and J. P. Thomas, eds., *Handbook of perception and human performance*, vol. I, *Sensory processes and perception*. New York: John Wiley and Sons.

Steinman, R. M. (1986a). The need for an eclectic, rather than systems, approach to the study of the primate oculomotor system. *Vision Research* 26, 101–112.

Steinman, R. M. (1986b). Eye movement. *Vision Research* 26, 1389–1400.

Steinman, R. M., and H. Collewijn (1980). Binocular retinal image motion during active head rotation. *Vision Research* 20, 415–429.

Steinman, R. M., R. J. Cunitz, G. T. Timberlake, and M. Herman (1967). Voluntary control of microsaccades during maintained monocular fixation, *Science* 155, 1577–1579.

Steinman, R. M., G. M. Haddad, A. A. Skavenski, and D. Wyman (1973). Miniature eye movement. *Science* 181, 810–819.

Steinman, R. M., E. Kowler, and H. Collewijn (1990). New directions for oculomotor research. *Vision Research* 30, 1845–1864.

Steinman, R. M., and J. Z. Levinson (1990). The role of eye movement in the detection of contrast and spatial detail. In E. Kowler, ed., *Eye movements and their role in visual and cognitive processes*. Amsterdam: Elsevier.

Steinman, R. M., J. Z. Levinson, H. Collewijn, and J. Van der Steen, (1985). Vision in the presence of known natural retinal image motion. *Journal of the Optical Society of America* A, 2, 226–233.

Suppes, P. (1990). Eye-movement models for arithmetic and reading performance. In E. Kowler, ed. *Eye movements and their role in visual and cognitive processes*. Amsterdam: Elsevier.

Van Gisbergen, J., and J. Van Opstal (1989). *Models.* In R. H. Wurtz and M. E. Goldberg, eds. *The neurobiology of saccadic eye movements*. Amsterdam: Elsevier.

Viviani, P. (1990). Eye movements in visual search: Cognitive, perceptual and motor control aspects. In E. Kowler, ed., *Eye movements and their role in visual and cognitive processes*. Amsterdam: Elsevier.

Walls, G. L. (1962). The evolutionary history of eye movements. *Vision Research* 2, 69–80.

Westheimer, G. (1954). Eye movement response to a horizontally moving visual stimulus. *Archives of Ophthalmology* 52, 932–941.

Westheimer, G. (1979). The spatial sense of the eye. *Investigative Ophthalmology and Visual Science* 18, 893–912.

Westheimer, G. (1989). History. In R. H. Wurtz, and M. E. Goldberg, eds., *The Neurobiology of Saccadic Eye Movements*. Amsterdam: Elsevier.

Westheimer, G., and S. P. McKee (1975). Visual acuity in the presence of retinal image motion. *Journal of the Optical Society of America* 65, 847–850.

White, J. M., D. M. Levi, and A. P. Aitsebamo (1992). Spatial localization without visual references. *Vision Research* 32, 513–526.

Winterson, B. J., and H. Collewijn (1976). Microsaccades during finely guided visuomotor tasks. *Vision Research*, 1387–1390.

Wurtz, R. H., and M. E. Goldberg, eds. (1989). *The neurobiology of saccadic eye movements*. Amsterdam: Elsevier.

Yarbus, A. L. (1967). *Eye movements and vision*. New York: Plenum Press.

Zingale, C. M., and E. Kowler (1987). Planning sequences of saccades. *Vision Research* 27, 1327–1341.

Chapter 7
Mental Imagery
Stephen M. Kosslyn

What seems to happen when you try to decide which is higher off the ground, the tip of a racehorse's tail or its rear knees? Or when you think about how a new arrangement of furniture would look in your living room? Many people report that in performing these tasks, they "see" with their "mind's eye" the horse's tail and the furniture. Visual mental imagery is "seeing" in the absence of the appropriate immediate sensory input; imagery is a "perception" of remembered information, not new input. But if all one is doing is "mentally perceiving" what has already been perceived, what is the use of imagery? And in what way does it make sense to talk about "seeing," "hearing," and the like without actually perceiving? Indeed, the very idea of mental images is fraught with puzzles and possible paradoxes. What, exactly, is being "perceived"? Surely images cannot be actual pictures in the head; there is no light in there, and who or what would look at the pictures, even if they were there? And, given that there are no hands in the brain, how do we "move things around" in images?

At first glance (and even at second glance) these are knotty problems indeed. In this chapter we will explore recent attempts to answer these and related questions. Because most research has focused on visual imagery, we will concentrate on that type of imagery here, although many people report experiencing imagery in all sensory modalities ("hearing" with the "mind's ear," "tasting" with the "mind's tongue," and so on).

7.1 Purposes and Problems

It is helpful to begin by considering the purposes of imagery. Once we have a sense of what imagery is used for, we can ask about the nature of the problems that must be solved by a system (including a brain) to accomplish these purposes.

7.1.1 What Is Imagery For?

All characterizations of imagery rest on its resemblance to perception. One way to consider the purposes of imagery is to explore the parallels between imagery and perception in the same modality. For example, we use vision primarily for two types of purposes: to identify objects, parts, and characteristics (such as color and texture); and to track moving objects, to navigate, and to reach appropriately. Similarly, one purpose of imagery is to identify properties of imaged objects, which allows us to retrieve information from memory. For example, consider how you answer the following questions: What shape are a beagle's ears? Which is darker green, a Christmas tree or a frozen pea? Which is bigger, a tennis ball or a 100-watt light bulb? Most people claim that they visualize the objects and "look" at them in order to answer these questions. In these cases, we apparently call on some of the machinery used in perception to classify parts or properties that previously were noticed (and stored in memory) but were not explicitly categorized or labeled at the time.

Imagery is used to retrieve information from memory in a variety of circumstances, but primarily when (1) the information to be remembered is a subtle visual property; (2) the property has not been explicitly considered previously (and hence labeled); and (3) the property cannot easily be deduced from other stored information (for example, from information about a general category to which the object belongs). To get a feel for these principles, try the following experiment on a friend (and "watch" what happens as you think about it as well). Ask your friend to describe the shape of Snoopy's ears. After she answers, ask her again. And then again (begging patience!). After about the third time, your friend will probably report no longer using imagery; at this point she will simply remember the answer (say, "rounded"). The change was in condition 2; the property became categorized, and hence imagery was not needed (see Kosslyn and Jolicoeur 1980). Can you think of ways of testing the other two conditions?

A second purpose of imagery parallels the role of vision in allowing us to navigate, track, and reach: imagery is a way of anticipating what will happen if one's body or a physical object moves in a particular way. For example, we can visualize a container and "see" whether there is room for it on the top shelf in the refrigerator; or we can mentally project an object's trajectory, "seeing" where it will land. Imagery is used when we reason about the appearance of an object when it is transformed, especially when we want to know about subtle spatial relations.

These two functions of imagery allow us to use imagery in a wide variety of ways. Kosslyn et al. (1990) asked people to keep a diary in which they recorded the kinds of images they formed and the purpose of each image. As one would expect from the previous discussion, some respondents reported using imagery in memory retrieval and visual thinking. For example, one subject reported that she visualized what she had worn the day before when getting dressed in the morning. In addition, people used imagery to help produce descriptions, to help understand descriptions, as part of "mental practice" (one subject visualized a "mental scenario" of what he would say when asking a professor about a grade change), and to induce emotional or motivational states (one subject imaged what she would look like thinner to help her stay on her diet). These reports are consistent with Shepard and Cooper's (1982) descriptions of cases in which scientists used "imaged models" as aids to reasoning and Paivio's (1971) evidence that memory can be improved if one visualizes the material and then encodes the images into memory. Surprisingly, however, the most often reported imagery arose during "free association," for no specific purpose. It is possible that the major use of imagery is to prompt one to think about previously neglected aspects of objects or events.

7.1.2   Problems To Be Solved

If you were going to program a computer to mimic these uses of human imagery, the system would have to be able to solve several types of problems. Perhaps the most fundamental problem the program would have to solve is the *generation* of images. That is, we do not have a given image all the time; it only comes to mind when we are faced with a specific situation. We store in long-term memory the information necessary to form images, but some process or processes must use that information to create the image per se. In addition, once the program generated an image, it would have to be able to interpret the shape and other properties of the imaged object. To "see" that Snoopy has rounded ears, we must be able to *inspect* the object in the image, classifying it in a new way.

Many of the uses of imagery also require solving two other problems. In many imagery tasks we must be able to *maintain* the image over time. Unless the property is very salient and immediately obvious, careful inspection may be necessary; a process or processes must be available for hanging onto the image as long as necessary. Finally, in many situations we want to *transform* the object in an image in some way. A large number of such transformations are possible, and there must be some means of carrying them out.

There are many other problems that must be solved to use imagery, such as those involved in deciding what object to visualize, how to transform an imaged object, and so on. The processes that solve these problems appear to be part of a more general system that is used in thinking more generally. Thus, we will not consider these processes further here (see Volume 1).

7.2   The Human Processing System

A considerable amount is now known about how the brain generates, inspects, maintains, and transforms objects in images. We consider each ability in turn.

7.2.1   Image Generation

Logically, there are only two ways a visual mental image can be formed. One can retain perceptual input online, or one can activate information stored in long-term memory. The fact that images can be generated from stored information does not imply that there is a single mechanism that generates images; nor does it imply that there is a mechanism dedicated to generating images. That is, several mechanisms working together could generate images, and these mechanisms may have other roles as well. By analogy, a car can slow down if one simply takes one's foot off the gas, which does not activate a separate "slowing-down" system. Similarly, it now appears that there are several ways images can be generated; and at least two—and probably more—subsystems are used in each type of image generation.

First, consider images of objects, which may include color and texture as well as shape. One subsystem seems to be used to activate visual-memory representations. This subsystem is apparently also used to "prime" the visual system so that one can more easily encode an expected object or part. In recent work in my laboratory, we have shown that imagery does prime perception. We found that people can more quickly determine whether an picture is an object or not if they had visualized that object than if they visualized another object or formed no images. Similarly, if subjects had recently read a word printed in lowercase font, they could later more easily visualize it than if they had not recently seen the word. It is possible that the mechanisms that prime one to see an object can form an image of that object merely by priming the representation more strongly.

The subsystem that activates stored visual information may be sufficient to form images in tasks that require only a global form of an image. For example, such images are all that is needed to decide from memory whether a mug is higher than it is wide, or whether Carter has a rounder face than Clinton. But other tasks require that high-resolution parts be added to the image. For example, such images are necessary to decide whether Nixon had a longer nose than George Washington. Images of complex objects apparently are built up on the basis of distinct parts, each of which is activated individually. Reed (1974) showed subjects patterns such as those in Figure 7.1, then presented them with parts of patterns and asked whether any of the previous patterns had contained the parts. Subjects

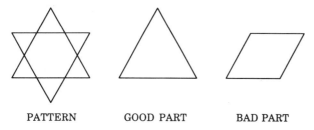

PATTERN   GOOD PART   BAD PART

Figure 7.1
Examples of types of stimuli used by Reed (1974). Some parts are more difficult to see in a pattern than others, both in perception and in imagery.

were more successful with "good" parts than with "bad" ones (in the Gestalt sense). Presumably, the pattern was initially broken down perceptually into units, which were stored and later used to generate the image. If the probed part corresponded to a unit or units initially encoded, well and good; if not, it was difficult to make the comparison. Consistent with this interpretation, the time needed to generate images increases with each additional part to be included. Indeed, the time needed to generate images changes when the number of parts is varied in a host of ways, including simply asking people to construe a pattern in one way or another. For example, a Star of David can be seen as relatively few overlapping figures (two triangles) or as relatively many adjacent figures (a hexagon with six small triangles). If subjects "see" it the latter way, they require considerably more time later to form an image of the pattern (for a review of such findings, see Kosslyn 1980).

A second subsystem (or group of subsystems) apparently is used to position individual parts in the image. This inference is consistent with the observation that the brain processes location information separately from shape information (Ungerleider and Mishkin 1982; Mishkin and Ungerleider 1982). If the two types of information are entered into memory separately, it is reasonable to assume that separate subsystems use each type of information when the system is running in the other direction—that is, forming an image rather than encoding one. Many researchers (for example, Pylyshyn 1973) noticed that parts could be visualized in the wrong location, which is consistent with the idea that shape and location are processed by separate mechanisms, even in imagery. Indeed, the simple fact that people can arrange imaged objects according to a description suggests that locations are processed separately from shapes. For example, can you visualize George Washington floating above a horse? Or George Bush riding on an elephant?

Some findings that support the distinction between a subsystem that activates visual memories and subsystems that arrange component parts were reported by Farah et al. (1985) and Kosslyn et al. (1985). They tested patients who, for medical reasons, had had their corpus callosum—the major connection between the two hemispheres of the brain—surgically severed. Thus, in these people the left and right halves of the brain no longer communicate effectively. These researchers found that the left hemisphere could perform imagery tasks that require putting parts together. For example, one patient was asked whether animals such as cats and apes have ears that protrude above the top of the skull. This task requires imaging the ears in the correct relationship to the head. The left hemisphere could also perform tasks that required forming a global image and assessing overall shape. For example, the patient was asked whether an animal such as a hog is larger (as seen from the side) than a goat. This task does not require putting parts together in a high-resolution image; the general shape will do. In contrast, the right hemisphere could only effectively perform the latter kind of imagery task; it could form images but had great difficulty performing tasks that require putting parts together. Thus, there is evidence for a distinction between two subsystems: one activates individual stored perceptual units well in both hemispheres, and another juxtaposes parts effectively only in the left hemisphere. Such a dissociation would not occur if the mechanisms were not distinct. Consistent with the inference that the left hemisphere is critically involved in generating multipart images of the sort described here, Farah (1984) found that most patients who could not generate such images had a lesion in the (posterior) left cerebral hemisphere (see also Trojano and Grossi 1994).

The subsystems that arrange parts into images must use a stored representation of how the parts are arranged. Kosslyn, Maljkovic, Hamilton, Horwitz, and Thompson (in press) provide evidence that there are two ways in which this can be done: parts can either be positioned on the basis of "categorical" representations of spatial relations, such as "connected to" or "above," or they can be arranged on the basis of metric spatial information. The left cerebral hemisphere appears to be better at using categorical spatial-relations representations, whereas the right appears to be better at using metric information (see also Sergent 1989 and 1990).

As if this is not complicated enough, there is evidence that people can form images of different types. So far we have been discussing images of objects formed by activating visual memories; but people can also form images of patterns that do not require activating visual memories. For example, look at the tile pattern of a bathroom floor. Can you "see" letters and other geometric patterns as if they were formed by some of the rows and columns of tiles? When people perform this sort of task, the visual-memory parts of their brains need not be activated (Kosslyn, Alpert, Thompson, Maljkovic, Weise, Chabris, Hamilton, Rauch, and Buonanno 1993). In this case, people appear to form images by positioning and fixating their attention selectively. Indeed, parts of the brain (e.g., the thalamus) that are involved in fixating attention are very active in this type of imagery task. In contrast, when people visualize familiar objects and cannot easily fixate attention, parts of the brain that appear to store visual memories are activated, whereas parts of the brain involved in attention are not as strongly activated.

In short, images can be formed by activating visual memories of global patterns, by activating visual memories of individual parts and then arranging them (probably using at least two different methods), or by selectively allocating attention. The brain is a complicated mechanism, and it would not be surprising if these are only a few of its possible methods for generating mental images (for additional possibilities, see Kosslyn 1994).

7.2.2   Image Inspection

Given the apparent parallels between the uses of imagery and those of like-modality perception, it is not surprising that imagery apparently shares some of the same processing mechanisms used in recognition, navigation, and tracking. In fact, objects seem to be "inspected" in imagery in much the same way that they are in actual perception. Once a representation is formed, it apparently is treated the same way regardless of whether it arose from the senses or from memory. There are numerous sources of evidence that imagery shares processing mechanisms with like-modality perception (for an excellent review, see Finke and Shepard 1986). One is the finding that imagery selectively interferes with like-modality perception. For example, Segal and Fusella (1970) asked subjects to hold an image in mind while trying to detect a faint auditory or visual stimulus. They found that holding a visual image (say, of a flower) impairs visual perception more than auditory perception, but that holding an auditory image (say, the sound of a telephone ringing) has the reverse effect. Thus, they found modality-specific interference, which suggests that imagery and like-modality perception share common mechanisms that are not used in other-modality processing (see also Craver-Lemley and Reeves 1987; Farah 1985). Indeed, the process of "looking" at objects in images shares many properties of actual perception. As another example, image a honeybee at a very small size, and then try to decide what color its head is. Many people report having to "zoom in" in order to "see" the head, which is not necessary if the object is visualized at a larger size. And in fact, more time is required to "see" parts of objects when they are imaged at smaller sizes (Kosslyn 1980).

Neuropsychological data also support the claim that image inspection is accomplished by the same mechanisms used in perceptual recognition. For example, patients who have suffered damage to the right parietal lobe sometimes show "unilateral visual neglect": they ignore objects to their left side (the right side of the brain receives input from the left side, and vice versa). Bisiach and Luzzatti (1978) asked such patients to image a scene that was very familiar prior to the stroke that caused their brain damage. In one experiment, subjects were asked to image standing on one side of a plaza and report what they could "see." These patients were very familiar with the plaza prior to the stroke and had no difficulty forming the image. However, they described only buildings to their right, ignoring buildings to their left (just as they would do in perception). Bisiach and Luzzatti then asked the patients to image standing at the opposite side of the plaza looking toward the spot where they had previously stood and, again, report what they could "see." They again mentioned only buildings to their right, which led them to name the buildings they had previously ignored and ignore those they had previously mentioned!

Levine, Warach, and Farah (1985) present additional evidence that perceptual recognition mechanisms are also used in imagery. They identified patients who had lost the ability to perceive either shape or location and then tested their ability to visualize the same properties. As expected, patients who could not recognize faces perceptually also could not interpret faces in imagery (for instance, when asked whether George Washington had a beard); similarly, patients who could not register locations perceptually could not do so in imagery (for instance, when asked to describe how to get from one place to another).

The fact that perceptual mechanisms are used in inspecting images probably explains why it is difficult to reorganize complex patterns in images. Chambers and Reisberg (1985) found that subjects could not reinterpret an ambiguous figure in an image, "seeing" the alternative interpretation. Perceptual mechanisms organize the input into units and spatial relations among them, and reorganizing these units requires time. Images can be retained only with effort and apparently often cannot be retained long enough to reorganize them. Finke, Pinker, and Farah (1989) showed, however, that if images are simple enough, subjects can in fact reorganize and discover new patterns in them. For example, they asked subjects to visualize the uppercase letter D, rotate it 90 degrees counterclockwise, and then to image a J under it. Many subjects had no difficulty seeing that the result resembled an umbrella.

In addition, Hyman and Neisser (1991) and Peterson, Kihlstrom, Rose, and Glisky (1992) showed that under certain circumstances images of ambiguous objects could be reinterpreted. For example, Hyman and Neisser found that people can mentally reorganize even images of complex objects, provided that they understand how they are to reorganize the image. They told their subjects that the front of the animal they were visualizing was the back of another animal, which allowed over half of the subjects to reverse the image of the ambiguous duck/rabbit figure. Moreover, Brandimonte and Gerbino (1993) showed that subjects are much better at reversing imaged figures if they are prevented from verbally encoding the figure when studying it. By verbally describing a figure, one encodes descriptions of specific parts and properties, which later may be used to reactivate those perceptual units while the image is maintained—thereby preserving the initial perceptual organization of the figure and making it difficult to reinterpret.

Finally, as will be discussed shortly, many of the brain areas that are activated when we recognize and identify objects are also activated during visual mental imagery (for a review, see Kosslyn 1994). Farah (1988) reviews additional neuropsychological results that demonstrate the use of perceptual mechanisms in image inspection.

7.2.3   Image Maintenance

Ask a friend to participate in the following informal experiment. His or her job is to listen to a set of directions and form an image of a one-inch line segment for each direction, connecting it to the end of the previously imaged segment. If you read "North," your friend should visualize a vertical one-inch line; if you then read "West," he or she should connect the right end of a horizontal line to the top of the first line, and so on. Ask your friend to image the path described by the following directions: North, Northeast, North, West, South, West, West, South, West, Northeast, West, Southwest, East. Ask him to report back the series. A normal human, you will find, will not recall all of the directions but will recall non-random combinations of segments. For example, your friend is apt to recall "North, West, South," because it forms a simple visual pattern but may not recall "West, West"; he may forget the length of the line, only remembering that there is a line in that direction.

The typical results from this simple experiment illustrate the most important aspects of what we know about image maintenance. First, we can retain relatively little information in an image at once (see also Weber and Harnish 1974). Second, the critical measure is the number of *chunks*, the number of perceptual units that are present. For example, we can remember roughly as many squares as individual, unorganized line segments: what is important is the number of groups of units.

The fact that images fade quickly and require effort to maintain makes sense if imagery relies on mechanisms that are also used during visual perception. In vision, we do not want sensory input to stick around; if it did, we would get blurring whenever we shifted our eyes to a new stimulus. What is a virtue in vision—the fast fade rate, as it were—is a drawback in imagery. When we form an image, it fades rapidly and effort is required to keep it in place. Many people report that holding an image is a little like a juggling act: each part is refreshed while other parts are fading. Presumably, the total amount that can be kept in mind at once depends on the speed with which the parts are refreshed and the speed with which they fade.

7.2.4   Image Transformation

Examine the pairs of stimuli in Figure 7.2. Are the members of each pair identical, regardless of their individual orientations? If we had measured the time you required to make each judgment, we would have found that the time increases as the disparity in orientation between the members of a pair increases (Shepard and Metzler 1971). We can "mentally rotate" imaged objects, and the time to do so increases linearly with increasing amounts of rotation (Shepard and Cooper 1982). If you think about it, this finding is remarkable: Mental images are not actual objects that must obey the laws of physics. That is, real objects must pass through intermediate points along the trajectory as they change orientation, but objects in images need not obey the laws of physics; they are not real, rigid entities. At least one part of our transformation mechanism seems to be built to

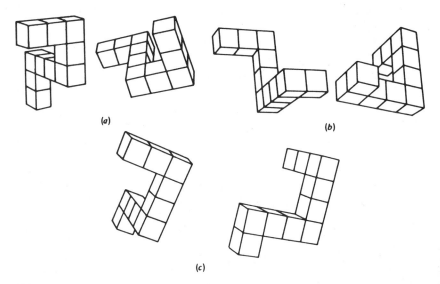

(a)

(b)

(c)

Figure 7.2
Examples of stimuli used by Shepard and Metzler (1971). The time to decide whether the figure in each pair is identical increases with the amount of mental rotation necessary to align them. (Reprinted by permission of the publisher and author from R. N. Shepard and J. Metzler, Mental rotation of three-dimensional objects, 1971, *Science* 171, 701–703. Copyright 1971 by the AAAS.)

Figure 7.3
Examples of types of stimuli used by Bundesen and Larsen. The time to decide whether the forms are identical increases with the degree of size scaling necessary to align them.

mimic perceptual processes. Shepard and Cooper (1982) argue that this makes good sense from an evolutionary point of view, given that one of the purposes of imagery is to mimic what would happen in actual physical situations. It is also of interest that motor areas of the brain are active when subjects mentally rotate objects (e.g., Deutsch, Bourbon, Papanico-laou, and Eisenberg 1988). It is possible that one forms moving images by priming the visual system as if expecting to see the results of physically manipulating an object. If so, then the incremental nature of transforma-tions may occur, at least in part, because our movements must traverse trajectories (see Kosslyn 1994).

In addition to being rotated, imaged objects can be expanded or re-duced in size. For example, consider the stimuli in Figure 7.3. Are they the same or different? The time to decide increases linearly with the disparity in sizes (Bundesen and Larsen 1975). Furthermore, it is possible to perform such esoteric transformations as mentally folding cubes. Shepard and Feng (1972) asked subjects to view stimuli like those illustrated in Figure 7.4 and to decide whether the arrows would meet when the sheet is folded into a cube (the shaded side is the base). The time to decide increased linearly with the number of sides that had to be shifted in the image to fold the cube.

At least one of the processes used in image transformations is more effective in the right hemisphere of the brain (e.g., Ratcliff 1979). Re-searchers have found that subjects with damage to the right parietal lobe have difficulty with mental rotation tasks and other image transformation

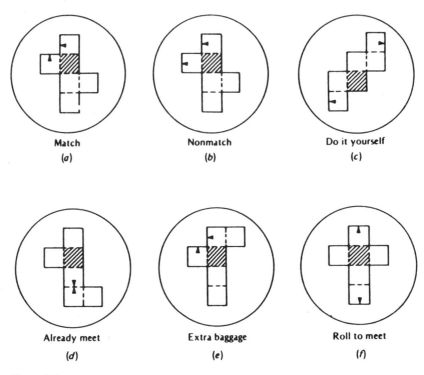

| Match | Nonmatch | Do it yourself |
| (a) | (b) | (c) |

| Already meet | Extra baggage | Roll to meet |
| (d) | (e) | (f) |

Figure 7.4
Examples of the unfolded cubes used by Shepard and Feng (1972). The time to decide whether the arrows will meet when the cube is folded (using the shaded square as the base) increases with the number of sides that must be mentally moved. (Reprinted by permission of the publisher and author from R. N. Shepard and C. Feng, A chronometric study of mental paper folding, 1972, *Cognitive Psychology* 3, 228–243.

tasks such as cube folding. Consistent with these findings, the isolated right hemisphere of a split-brain patient has been found to be better than the isolated left hemisphere at performing spatial-manipulation tasks. However, other studies have provided evidence that the left hemisphere also plays a role in image transformations (for discussions, see Kosslyn 1987; 1994).

7.3   Individual Differences

The scientific study of imagery has always been intimately bound to the study of individual differences. Indeed, Sir Francis Galton, credited as one of the founders of the scientific study of individual differences in intelligence and personality, was one of the first to study imagery. In 1883 Galton asked people to visualize their breakfast table and report what was on it that morning. To his surprise, many of his initial subjects professed not to know what he was talking about when he asked about their use of imagery in this task. As it turned out, he initially queried his friends, who tended to be professional men; as he broadened his sample, he found that about 12 percent of the people questioned reported not having the experience of mental imagery. In a more recent study, however, only about 3 percent of respondents reported not experiencing visual imagery; what makes this result particularly puzzling is that the more recent respondents were members of Mensa, a society for people who score high on IQ tests, who presumably were similar to the professional people who made up Galton's nonimaging 12 percent (McKellar 1965). Even small differences in the wording of the two questionnaires may have led the subjects to think about the phenomenology differently.

In addition to the apparent unreliability of simple introspective reports, there is a deeper reason for the disparity in the two studies. Imagery ability is not all-or-none; a given person is not generally good or bad at imagery. Consider the following study: a wide range of people were given thirteen different imagery tasks, including ones that required rotation, retention, generation, and so on, in different combinations (Kosslyn, Brunn, Cave, and Wallach 1984). If imagery ability is a single trait, then those who did well on one task should have done well on the others. This did not occur. Rather, there was a wide range of correlations in the performance of the tasks. Indeed, for some pairs of tasks, doing well on one implied doing poorly on the other. These results make sense if different aspects of imagery are accomplished by using separate subsystems, which are invoked in different combinations in different tasks. A person who is poor at a process, such as shifting the image during rotation (because one or more necessary subsystems are ineffective), will be poor at all tasks that require it—but not necessarily poor at tasks that do not require it. The precise nature of the task is an important variable in assessing imagery ability, and different tasks will not necessarily produce similar results. This discovery is consistent with what we know about how imagery is carried out in the brain. As noted earlier, the generation of many types of multipart images apparently involves left-hemisphere processes, and image rotation and inspection involve right-hemisphere processes (probably in addition to left-hemisphere ones). In general, imagery is not carried out by a single "center," or even one cerebral hemisphere (Erlichman and Barrett 1983; Farah 1984; Kosslyn 1987; 1994). Rather, imagery appears to depend on mechanisms located in various parts of the brain, and the combination of mechanisms that will be recruited depends on the task. Hence, questionaires that require different combinations of processes may well produce different results.

7.4   What Is a Mental Image?

We have been discussing mental images rather blithely, relying heavily on introspection and intuition. Not surprisingly, much effort has been expended in recent years to understand what a mental image is. We experience mental images as fleeting, ethereal entities; how should we conceptualize them in a way that explains not only how they can represent information about the world but also how mental images relate to the physical brain itself? This issue has plagued theorists at least since the time of Plato, who in his *Theaetetus* likened images to patterns etched on a block of wax (individual differences in imagery ability were explained by differences in the hardness and purity of the wax). The brain is an organ of the body, and like other organs of the body it can be described at numerous distinct levels of analysis. We can divide these levels into two general classes, the physical and the functional. For example, stomachs can be described either in terms of cell types, enzymes, and so on, or in terms of their role in digestion. Similarly, brains can be described in terms of their physical composition—cells, anatomical connections, neurotransmitters, and so on—or in terms of their functions. The primary function of brains is to store and process information. A mental representation is a description at the functional level of analysis of how the brain stores information.

The study of the nature of mental-image representations entered a new stage in the early 1970s. No one denies that people experience "seeing with the mind's eye," but there is controversy over what this experience reveals about how the brain actually stores most information. Two means of representation have been proposed for mental images, one that confers a special status on images and one that treats them as no different in kind from the representations of linguistic meaning. The two alternatives are called *depictive* and *propositional* representation (see Kosslyn 1980). These are different *formats*, different types of codes. Every type of code is distinguished in part by a specific syntax. The syntax is characterized by (1) the elementary, or "primitive," symbols, and (2) the set of rules for combining the symbols. Symbols usually belong to different *form classes* (e.g., "noun," "verb," "determiner," and so on), and the rules of combination are defined in terms of these classes—which allows them to generalize over an infinite number of distinct symbols. A format is also defined in part by the semantics of a code. The semantics is determined by how meaning is conveyed by individual symbols and combinations of symbols. For example, the symbol *A* can be interpreted as a part of speech if read as a word, or as a configuration of birds in flight (as seen from above) if interpreted as a picture. The same marks are used in both formats, but what differs is how they are interpreted as conveying meaning. The rules of semantics assign a meaning (or sometimes more than one, if the symbol is ambiguous) to a specific symbol.

In contrast, the *content* of a representation is the specific information conveyed. The same content can be conveyed in numerous different formats. For example, the information in this sentence could be conveyed by speaking it aloud (where the symbols are composed of sound waves), by writing it down in Morse code (where the symbols are dots and dashes), and so on. Palmer (1978) offers a more elaborate treatment of these distinctions, but this overview is sufficient for present purposes. In order to see how the results from experiments can distinguish between the use of propositional and depictive representations, we need to go into more detail in characterizing these representational types.

7.4.1   Propositional Representations

Consider a propositional representation of a simple scene that contains a ball sitting on a box. We can write the representation using the notation "ON (BALL, BOX)." This kind of notation is close to the way propositions are represented in computers and also serves to make it clear that we are not talking about sentences in a natural language (like English). A propositional format can be characterized (roughly) as follows:

Syntax.   (1) The symbols belong to a variety of form classes, corresponding to relations (e.g., on), entities (e.g., ball, box), properties (e.g., red, new), and logical relations (e.g., if, not, some). (2) The rules of symbol combination require that all propositional representations have at least one relation (e.g., "BALL, BOX" does not assert anything). (3) Specific relations have specific requirements concerning the number and types of symbols that must be used (e.g., ON(BOX) is unacceptable because ON relates one object to another, and hence at least two symbols must be used).

Semantics.   (1) The meaning of individual symbols is arbitrarily assigned, which requires the existence of a lexicon (as is true for words in natural languages, whose meanings must be looked up in such a mental dictionary). (2) A propositional representation is defined to be unambiguous, unlike words or sentences in natural languages. A different propositional symbol is used for each of the senses of ambiguous words (e.g., "ball" as dance versus "ball" as sphere). (3) A propositional representation is abstract. That is, (a) it can refer to nonpicturable entities, such as sentimentality; (b) it can refer to classes of objects, not simply individual ones (such as boxes in general); and (c) it is not tied to any specific modality (a propositional representation can store information seen, acquired through language, felt, and so on). (4) Some theorists add another characteristic to the semantics of propositions: they are either true or false (see, for example, Anderson and Bower 1973). However, this seems to me to be not a property of the representation per se but a relation between the representation and a specific state of the world (see Palmer 1978).

7.4.2 Depictive Representations

Now consider a drawing of the same scene, of a ball on a box. The drawing is an example of a depictive representation. Depictive representations differ from propositional ones on almost every count. There is no explicit symbol that stands for the relation ("on" is not represented separately but only emerges from the juxtaposition of the symbols standing for the ball and the box). The rules of combination are not defined over form classes. Indeed, the rules are very lax; any dot can be placed in any relationship to any other dot—as surrealist painters have delighted in demonstrating to us. The semantics are not arbitrarily assigned, and depictions are inherently ambiguous because they are interpreted as resembling an object (and a picture can be seen as resembling more than one object). Depictions are not abstract: they cannot directly refer to nonpicturable concepts (although they can do so indirectly, by forming a separate association); they represent individual instances (not classes); and they are visual.

Thus, depictions are not propositions. But what are they? We can characterize depictive representations as follows:

Syntax.   (1) The symbols belong to two form classes: points and empty space. (2) The points can be arranged so tightly as to produce continuous variations or so sparsely as to be distinct (like the dots in a comic strip). (3) The rules for combining symbols require only that points be placed in spatial relation to each another.

Semantics.   (1) The association between a representation and what it stands for is not arbitrary; rather, depictions "resemble" the represented object or objects. That is, (a) each part of the representation must correspond to a visible part of the object or objects, and (b) the represented "distances" among the parts of the representation must correspond to the distances among the corresponding parts of the actual object (as they appear from a particular point of view). Thus, a pattern in an array in a computer can be a depiction because the points can correspond to points on the surface of an object with the corresponding distances on the object being preserved by the number of cells (filled or empty) between the points in the array. Similarly, there need be no actual picture in the brain to have a depiction: all that is needed is a *functional space* in which distance can be defined vis-à-vis the interpretation of the information. (For a more detailed treatment of the concept of a functional space, see Kosslyn 1980.) However, although all that is required in order to have a depiction is a functional space, such representations may occur in brain regions that are physically—as well as functionally—spatial, as will be discussed shortly.

7.5   The Imagery Debate

The modern debate about mental imagery has gone through two distinct phases and is just now entering a third phase. The first began in 1973 with the publication of Pylyshyn's paper "What the Mind's Eye Tells the Mind's Brain: A Critique of Mental Imagery" and Anderson and Bower's book *Human Associative Memory*. Pylyshyn's critique of mental imagery focused on arguments that the very idea of imagery is paradoxical (Who looks at the images?) or muddled (In what ways are images like pictures? Why can't you "see" the number of stripes on an imaged tiger?). The thrust of the critique of imagery was that a depictive representation does not occur in the brain when we experience mental images; instead, propositional representations are used for all forms of cognition—including imagery. The depictive features of images evident to introspection were thus taken to be epiphenomenal: these features have nothing to do with the representation used to perform the task, just as the lights flashing on the outside of a mainframe computer have nothing to do with carrying out the internal processing (if the lights are removed, it keeps working just as well).

The imagery debate provides a nice case study of one kind of research in cognitive science. In the following section, we consider one example of the sort of data gathered to address this issue and see how these sorts of data led to elaboration of the propositional arguments, which in turn led to more experimentation, which in turn led to the second phase of the debate. Following this, we consider briefly how a different sort of data is now being marshaled to address the issue, perhaps more conclusively.

7.5.1   Scanning Visual Mental Images

By their very nature, depictions embody space (recall that "distance" is an intrinsic part of the representation). Thus, if depictive representations underlie the experience of "having an image," then the spatial nature of the representation should affect how images are processed. On the other hand, if the underlying representation is propositional, we have no reason to expect distance to affect processing times (given that the description of an object's appearance would be stored in a list or network of some kind, just as in language).

Different Mechanisms? The First Phase of the Debate

In this section we consider a series of experiments carried out largely by my colleagues and me; these experiments illustrate how one can make abstract ideas concrete and how one can grasp a conceptual issue by the horns, so to speak.

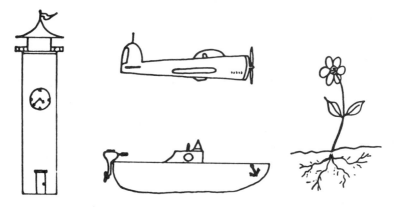

Figure 7.5
Examples of the drawings used by Kosslyn (1973) to study image scanning.

We reasoned that one way to discover whether image representations embody space is to determine whether it takes more time to shift attention greater distances across an imaged object. If subjects take more time to scan a longer distance across an imaged object, we would have evidence that distance is indeed embodied in the representation of the object. The first experiment began by asking subjects to memorize a set of drawings (Kosslyn 1973). Half the drawings were vertical and half were horizontal, as illustrated in Figure 7.5. After the subjects had memorized the drawings, they closed their eyes, heard the name of one object (say, *speedboat*), and visualized it. Once it was imaged, the subjects were asked to mentally focus ("stare" with the "mind's eye") at one end of the object in the image. Then the name of a possible component of the object (say, *motor*) was presented on tape. On half the trials the name labeled part of the drawing, and on the other half it did not. The subjects were asked to "look for" the named component on the imaged object. An important aspect of this experiment was that the probed parts were either at one end or the other of a drawing or in the middle. The subjects were told that we were interested in how long it took to "see" a feature on an imaged object (the word *scan* was never mentioned in the instructions). They were to press the true button only after "seeing" the named component and the false button only after "looking" but failing to find it. I reasoned that if image representations depict information, then it ought to take more time to locate the representations of parts located farther from the point of focus. And in fact this is exactly what occurred.

At first glance, the results from this experiment seemed to show that depictive representations are used in imagery. But it soon became clear that a propositional explanation could easily be formulated. Bobrow (per-

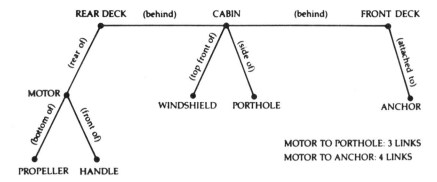

Figure 7.6
A propositional representation of the drawing of a speedboat illustrated in Figure 7.5. The greater the distance between two parts of the drawing, the larger the number of links between them in the network.

sonal communication) suggested that the visual appearance of an object is stored in a propositional structure like that illustrated in Figure 7.6. This representation is a series of linked hierarchies of propositions, with each hierarchy describing the relations among parts of a portion of the object. Note that we could rewrite the propositions illustrated here as BOTTOM-OF (PROPELLER, MOTOR), REAR-OF (MOTOR, REAR DECK), and so on. That is, each link is a relation that combines the symbols at the connected nodes into a proposition. According to Bobrow's theory, people automatically (and unconsciously) construct these sorts of propositional descriptions when asked to memorize the appearance of drawings. When subjects were asked to focus on one end of the drawing, they would then activate one part of the representation (for instance, for *speedboat*, the node for *motor*). When, subsequently, they were asked about a part, they searched the network for its name. The more links they had to traverse through the network before locating the name, the more time it took to respond. For example, for *speedboat* it took more time to find *anchor* than *porthole* after having been focused on the motor because four links had to be traversed from motor to anchor but only three from motor to porthole. Thus, the effect of "distance" on scanning time may have nothing to do with distance being embodied in an underlying depictive representation but may instead simply reflect the organization of a propositional network (see also Lea 1975). According to such theories, the conscious experience of scanning a pictorial mental image is somehow produced by processing this network, and the depictive aspects of images open to introspection are epiphenomenal.

It should now be clear why it was necessary to go into so much detail in characterizing the differences between the types of representations: We need a reasonably precise characterization of the two formats if we are to perform experiments to discriminate between them. According to the characterization provided earlier, although propositional structures can be formulated to capture the spatial arrangement of the drawings, they are not depictions. Recall that in depictions, in contrast to this sort of propositional representation, the shape of empty space is represented as clearly as the shape of filled space and there is no explicit representation of relations (such as REAR-OF).

The next experiment was designed to eliminate a critical problem with the first one. In this experiment we independently varied the distance scanned across and the number of items scanned over. The results of this experiment were straightforward: both distance and amount of material scanned over affected the reaction times. Time increased linearly with greater distance scanned over, even when the amount of material scanned over was kept constant (for details, see Kosslyn, Ball, and Reiser 1978), as expected if images rely on depictive representations. The notion of depiction leads us to expect that image representations embody distance in at least two dimensions. To test this hypothesis, we asked subjects to memorize the map illustrated in Figure 7.7. On this map were seven objects. The subjects learned to draw the locations of each of the seven objects on the map. These objects were positioned in such a way that the members of each of the twenty-one pairs were a different distance apart. The subjects later closed their eyes, visualized the map, focused on a given location, and then scanned to another location (or "looked" for it and failed to locate it); subjects eventually scanned between all possible pairs of objects on the map. As is evident in Figure 7.8, the time to scan the image increased linearly with greater distance scanned across. This result was as predicted if image representations depict information.

However, it is possible to create a propositional counterexplanation even here. Now the network contains *dummy nodes* that mark off distance. That is, these nodes convey no information other than the fact that an increment of distance (say, 5 centimeters) exists between one object and another; hence, there would be more nodes between nodes representing parts separated by greater distances on the map. By putting enough dummy nodes into a network, the propositional theory developed for the original results can be extended to these results as well. To attempt to rule out this propositional counterexplanation, we conducted a control experiment, which involved a variation on the map-scanning task. In this experiment, subjects again imaged the map and focused their attention on a particular point, but now they were told simply to decide as quickly as possible whether the probe named an object on the map. If the propositional theory were correct, we reasoned, we should find effects of distance here too; after all, we asked the subjects to form the image (which corre-

Figure 7.7
A map that was memorized and later imaged and scanned. The seven objects were placed
in such a way that members of each of the 21 pairs were a different distance apart.

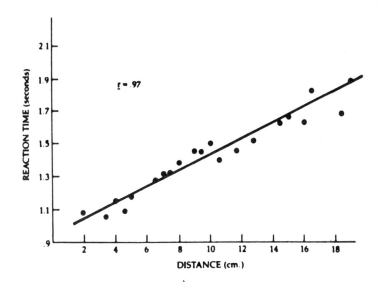

Figure 7.8
The time to scan between pairs of objects on an image of the map illustrated in Figure 7.7.

sponds to accessing the appropriate network). However, there were absolutely no effects on response times of the distance from the focus to target objects. In other experiments we varied the size of the imaged objects being scanned, asking subjects to adjust the size of an object in the image after they memorized it. Not only did time increase with the distance scanned, but more time was required to scan across larger images. The finding of effects of size on scanning time allows us to eliminate yet another nondepictive explanation for the effects of distance on response times: One could argue that the closer two parts are on an object or drawing, the more likely it is that they will be grouped into a single perceptual chunk and stored as a single unit and, hence, the easier it will later be to look up the two parts in succession. Because the size of the image was not manipulated until after the actual drawing was removed, this explanation cannot account for the effects of size on scanning time.

Demand Characteristics? The Second Phase of the Debate

The second phase of the debate began about eight years after the first, when Pylyshyn elaborated his views (Pylyshyn 1981). This phase of the debate focused on the data collected earlier. Whereas the proponents of depictive representation claimed that the data reflected the processing of such representations, the propositionalists now focused on possible methodological problems with the experiments. Two such problems were raised: *experimenter-expectancy effects* and *task demands*.

First, consider Intons-Peterson's (1983) claim that the imagery data could be understood simply as reflecting experimental-expectancy effects. She performed an experiment in which she compared scanning images to scanning physically present displays. Half the experimenters were told that image scanning should be faster and half were told that perceptual scanning should be faster. She found that the experimenters' expectations influenced the results. When experimenters expected faster perceptual scanning, the subjects produced this result; when they expected faster image scanning, there was no difference in overall times. Thus, the experimenters were somehow leading the subjects to respond as the experimenters expected.

Jolicoeur and Kosslyn (1985) investigated further the claim that the increase in times with increasing distance scanned occurs only because subjects respond to experimenter expectancy effects. We performed a series of experiments using Intons-Peterson's methodology. For example, we told one experimenter that we expected a U-shaped function, with the most time being required to scan the shortest and longest distances. The reason for this prediction, we explained, was that the four closest objects should be grouped into a single chunk—because of the Gestalt laws of similarity (the four closest objects were also drawn darker, to make them more similar) and proximity—and so they are cluttered together, making it difficult to scan among them. And the longest distances require more time than the medium ones because more scanning is involved. Like the results from our previous experiments, we found that response times increased linearly with greater distance. In additional experiments, Jolicoeur and Kosslyn varied experimenter expectancy in different ways, none of which affected the increase in time to scan greater distance. Indeed, these experimenters failed to replicate Intons-Peterson's original results.

What could be going on here? Many details of such experiments may differ from laboratory to laboratory (for instance, ensuring that subjects always keep their fingers on the response buttons), and these details could be critical for obtaining experimenter-expectancy effects. The important point is that, whatever caused the experimenter expectancy effect in Intons-Peterson's study, it was not present in the procedures used in the initial studies of image scanning. Thus, these results cannot be explained away as simply reflecting how well subjects can satisfy the expectations of the experimenter.

Taking an alternative tack, Pylyshyn (1981) claimed that the very instruction to scan an image induces subjects to mimic what they would do if they were scanning an actual object—which leads them to take more time to respond when they think they would have taken more time to scan across a visible object. The way the subjects estimate (unconsciously) how long to wait would involve propositional processing of some sort. This potential concern was ruled out by image-scanning experiments that eliminated all references to imagery in the instructions. Finke and Pinker (1982, 1988; see also Pinker, Choate, and Finke 1984) showed subjects a set of random dots on a card, removed the card, and presented an arrow. The question was, if the arrow were superimposed over the card containing the dots, would it point directly at a dot? Subjects reported using imagery to perform this task, and Finke and Pinker found that the response times increased linearly with distance from the arrow to a dot. Furthermore, the rate of increase in time with distance was very similar to what my colleagues and I had found in our earlier experiments. Because no imagery instructions were used, let alone mention of scanning an image, a task-demands explanation seems highly implausible.

Goldston, Hinrichs, and Richman (1985) actually went so far as to tell their subjects the predictions, which is never done in typical psychological experiments. Even when subjects were told that the experimenter expected longer times with shorter distances, they still required more time when longer distances were scanned. Telling subjects different predictions did affect the degree of the increase with distance, but this result is not surprising; given the purposes of imagery, one had better be able to control imaged events. What is impressive is that even when subjects were, if anything, trying for the opposite result, they still took longer to scan longer distances.

Denis and Carfantan (1985) described many types of imagery experiments to naive subjects and asked them to predict the outcomes. Although these subjects were good at predicting many of the effects of imagery (for example, that it will help one to memorize information), they were very poor at predicting the results of scanning experiments and the like. If subjects are using knowledge about perception and physics to "fake" the data in the experiments, it is puzzling that they evince no such knowledge in this situation.

In short, although alternative explanations are possible for each of the individual results, different alternative explanations will be necessary to explain the different effects (such as the effects of size scale or of distance and number of intervening items on scan times). In contrast, the depictive imagery theory is general to all of the results.

Cognitive Neuroscience: The Third Phase of the Debate

The second phase of the imagery debate ended with a fizzle. Most researchers found the arguments about methodology uninteresting, and many felt that the issue could not be resolved without making many difficult-to-defend assumptions (e.g., see Anderson 1978). Thus, it is notable that the critical information for resolving the debate came not from psychology, computer science, philosophy, linguistics, or any of the fields allied with cognitive science. Rather, the key facts came from neuroscience. Specifically, three pieces of information were critical. First, it has long been known that some visual areas of the brain are topographically organized. These regions of cortex preserve the spatial structure (roughly) of the retina; patterns of stimulation of the retina are represented in a functional space that is implemented in physical space in these regions of cortex. Indeed, the macaque monkey brain contains some fifteen distinct areas that are topographically organized in this way (e.g., see Felleman and Van Essen 1991). Second, it has been found that connections between visual areas typically do not simply send information downstream. Rather, these connections usually run in both directions. Third, the areas of the brain that store visual memories are not topographically organized.

These facts are consistent with the notion that visual memories are stored in an abstract (propositional?) format and that an image is formed in order to make accessible information about the local geometry of a shape. An image is formed, presumably, using the backwards connections that run from the areas involved in visual memory to (at least some of) the areas that are topographically organized. The image would make accessible spatial and visual information that was only implicit in the long-term memory representation. If so, then image representations would be depictive in the strongest sense of the term: they would be patterns in a physical, and also functional, space. Hence, they would literally be "pictures in the head." Notice, however, that there need not be somebody in there looking at such pictures. Just as in vision proper, the neural connections to other areas serve to pass the information along for additional processing, eventually leading to the accessing of relevant stored information (and hence interpretation; see Chapters 1, 4, and 5).

To test the hypothesis that images rely on topographically organized visual cortex, my colleagues and I (Kosslyn et al. 1993) conducted a number of studies using positron emission tomography (PET). This technique allows researchers to track which parts of the brain are more active during one task than another. We found that parts of the brain known to be topographically organized in humans are active during visual mental imagery, even when subjects close their eyes. In one study, we asked subjects to visualize letters either at a small size (i.e., so that they seemed to subtend a small visual angle) or at a large size (i.e., so that they seemed to subtend a large visual angle). By comparing brain activation when images were formed at the two sizes, we showed not only that visual areas were activated, but also that the precise location of activation depended on the size of the image. In humans, the fovea (the high-resolution, central part of the eye) projects input to the very posterior portion of several visual areas in the occipital lobe, and more parafoveal parts of the eye (which register stimuli that subtend larger visual angles) project input to more anterior regions of these visual areas. Imagery had the same effect: images of letters formed so that they seemed to subtend very small visual angles activated the very posterior part of visual cortex more than images formed so that they seemed to subtend larger visual angles. In contrast, images of letters that seemed to subtend larger visual angles activated more anterior portions of visual cortex. Moreover, the precise regions of activation were quite close to what one would expect based on the results from visual stimulation studies in the literature.

One could argue, however, that the representation in topographically organized cortex is purely epiphenomenal during imagery. Perhaps the backwards connections automatically engender activation in these areas but this plays no role in information processing. Thus, it is important that Farah, Soso, and Dasheiff (1992) found that damage to these regions disrupts imagery. Indeed, they studied a patient who had half of her occipital lobe—which contains the topographically organized visual areas—removed (for medical reasons) and measured the angle subtended by objects in her images before and after the operation. The horizontal angle shrank, by roughly half, after the operation. The visual areas that are active during imagery clearly play a critical role in the process.

These neuropsychological data are important not simply in their own right but in combination with the behavioral data summarized earlier. We not only have behavioral evidence that distance is an intrinsic part of image representations, but we also have neurological evidence showing that topographically organized parts of the brain play a key role in imagery. These areas are organized to depict the local geometry of an object's shape, and they are not involved in language or other types of information processing. These areas are tailor-made to support depictive representations and lack the appropriate characteristics to support language-like representations. Thus, the evidence is very strong that depictive representations underlie visual mental images.

The imagery debate, evolved over time as new issues arose, and much was learned as a consequence; not only were the empirical issues refined, but the conceptual foundations for a theory of imagery were formed. One important result of the debate is that it is clear that imagery cannot rely solely on depictive representations. Although such representations may underlie the experience of "seeing with the mind's eye," other types of representations also are critical. I noted earlier that categorical spatial relations can be used to store relations among parts and can later be used to arrange parts into an image; such representations are propositional. But more than that, images must be "under a description"; in both perception and imagery, the image must be interpreted at some point, and this interpretation process must remove the ambiguity inherent in depictive representations. It is clear that depictive representations are one component of the ensemble of representations and events that underlie imagery, but they are not the only component.

7.6   Images and Brain

As research on imagery has progressed, it has become increasingly tied to findings about the brain (e.g., see Kosslyn 1994; Tippett 1992). The insight that imagery is not a single entity was critical in beginning to understand how it is carried out in the brain (Farah 1984; Kosslyn, Brunn, Cave, and Wallach 1984). Before investigators had some idea of the component processes, it was not known what kinds of individual functions one should expect to be carried out in a single location. The experiments summarized here illustrate how behavioral data can be combined with neurological results, allowing us to distinguish among alternative theories of mental representation. Such research is cumulative. For example, once we establish that scanning reflects processing the representation, scanning can then be used to study additional issues; Pinker (1980) and others have used scanning to study how images represent information in three dimensions, and

Denis and Cocude (1989; 1992) have used it to study factors that allow people to use verbal descriptions to form images.

The study of mental imagery is interesting in part as a bridge between perception and mental activity. As such, it is perhaps the cognitive faculty "closest to the neurology" because so much is now known about the neural mechanisms of perception. Given its long history, it seems fitting that the study of imagery may end up providing one of the first case studies of how the brain gives rise to the mind.

Suggestions for Further Reading

Although the experimental study of mental imagery has developed in a number of different directions, the most vigorous research in the field has recently focused on two of them. One line of work centers on exploring the relationship between imagery and like-modality perception. Perhaps the best overview of this work is by Finke and Shepard (1986), who point out numerous and varied examples of experiments that demonstrate the existence of common mechanisms in imagery and perception (see also Finke 1989). In addition, Shepard and Cooper (1982) review the evidence that transformations of objects in images are in many ways analogues of corresponding transformations of the actual objects. A particularly intriguing contribution to this line of inquiry is provided by Farah (1988), who reviews previously overlooked evidence for the visual nature of visual mental imagery. This evidence, which springs from neuropsychology, concerns the common neural systems used in imagery and like-modality perception.

The other line of work focuses on questions about the nature of the processing system that underlies imagery. This research has become increasingly neuropsychological, on the one hand, and computational, on the other. Farah (1984) offers a neuropsychological analysis of imagery into separate processes, as does Kosslyn (1987; 1994). Both authors attempt to decompose imagery into underlying processing components. Tippett (1992) provides a review of neuropsychological research on one specific set of processes, namely those involved in image generation. Pinker (1985) reviews the now-classic work taking the computational approach, some of which grows out of research in artificial intelligence.

Finally, Morris and Hampson (1983) explore the role of consciousness in imagery (and vice versa), a topic not reviewed in this chapter; and Paivio's (1971) book plumbs the uses of imagery in learning and memory. The *Journal of Mental Imagery* publishes a very wide range of material on imagery (from case reports of the use of imagery in psychotherapy to experimental studies of individual differences in imaging ability) and provides good illustrations of the different approaches to studying imagery and applying research on imagery.

These sources provide references to a large literature that serves to integrate the study of imagery into the study of cognition more generally.

Problems

7.1 Generate an account for each of the following results based on both a descriptive and a depictive representation:

a. Subjects take more time to inspect objects that are imaged at small sizes than ones that are imaged at large sizes. (Example: Objects are represented as lists of descriptions of properties. When asked to form an image at a small size, subjects recall fewer properties of the object and thus are more likely to have to dig into long-term memory when subsequently asked about a particular property.)

b. Subjects can identify a picture more quickly if they form an image of the object in advance, but this effect is most pronounced if the picture is of exactly the same object (same size, orientation, and so on) as the one being imaged.

c. Subjects require more time to rotate objects greater amounts.

7.2   How could you discover whether experimenter expectancy effects were due to ESP?

7.3   Imagery has been thought to be both a "primitive" form of thought and a very advanced form of thought. What arguments can you offer for each of these extreme positions?

7.4   Captain DeWitt is interested in hiring only good imagers to navigate his yacht. Thus, he gives them the DeWitt Imagery Test (DIT). This test requires people to rate how vividly they can form images of common objects. What is wrong with his intended use of this test?

7.5   The minister of education of a wealthy foreign country is interested in raising the IQ of his population. Thus, he hits on the idea of training everyone to use imagery more in their thinking. Is this a good idea?

References

Anderson, J. R. (1978). Arguments concerning representations for mental imagery. *Psychological Review*, 85, 249–277.

Anderson, J. R., and G. H. Bower (1973). *Human associative memory*. New York: V. H. Winston.

Bisiach, E., and C. Luzzatti (1978). Unilateral neglect of representational space. *Cortex* 14, 129–133.

Brandimonte, M., and W. Gerbino (1993). Mental image reversal and verbal recoding: When ducks become rabbits. *Visual Cognition*, 21 (1), 23–33.

Bundesen, C., and A. Larsen (1975). Visual transformation of size. *Journal of Experimental Psychology: Human Perception and Performance* 1, 214–220.

Chambers, D., and D. Reisberg (1985). Can mental images be ambiguous? *Journal of Experimental Psychology: Human Perception and Performance* 11, 317–328.

Craver-Lemley, C., and A. Reeves (1987). Visual imagery selectively reduces vernier acuity. *Perception* 16, 533–614.

Denis, M., and M. Carfantan (1985). People's knowledge about images. *Cognition* 20, 49–60.

Denis, M., and M. Cocude (1989). Scanning visual images generated from verbal descriptions. *European Journal of Cognitive Psychology* 1, 293–307.

Denis, M., and M. Cocude (1992). Structural properties of visual images constructed from poorly or well-structured verbal descriptions. *Memory and Cognition* 20, 497–506.

Deutsch, G., W. T. Bourbon, A. C. Papanicolaou, and H. M. Eisenberg (1988). Visuospatial experiments compared via activation of regional cerebral blood flow. *Neuropsychologia* 26, 445–452.

Erlichman, H., and J. Barrett (1983). Right hemisphere specialization for mental imagery: A review of the evidence. *Brain and Cognition* 2, 55–76.

Farah, M. J. (1984). The neurological basis of mental imagery: A componential analysis. *Cognition* 18, 245–272.

Farah, M. J. (1985). Psychophysical evidence for a shared representational medium for mental images and percepts. *Journal of Experimental Psychology: General* 114, 91–103.

Farah, M. J. (1988). Is visual imagery really visual? Overlooked evidence from neuropsychology. *Psychological Review* 95, 307–317.

Farah, M. J., M. S. Gazzaniga, J. D. Holtzman, and S. M. Kosslyn (1985). A left hemisphere basis for visual imagery? *Neuropsychologia* 23, 115–118.

Farah, M. J., M. J. Soso, and R. M. Dasheiff (1992). Visual angle of the mind's eye before and after unilateral occipital lobectomy. *Journal of Experimental Psychology: Human Perception and Performance* 18, 241–246.

Felleman, D. J., and D. C. Van Essen (1991). Distributed hierarchical processing in the primate cerebral cortex. *Cerebral Cortex* 1, 1–47.

Finke, R. A. (1989). *Principles of mental imagery.* Cambridge, MA: MIT Press.

Finke, R. A., and S. Pinker (1982). Spontaneous imagery scanning in mental extrapolation. *Journal of Experimental Psychology: Human Learning and Memory* 8, 142–147.

Finke, R. A., and S. Pinker (1983). Directional scanning of remembered visual patterns. *Journal of Experimental Psychology: Learning, Memory, and Cognition* 9, 398–410.

Finke, R. A., S. Pinker, and M. Farah (1989). Reinterpreting visual patterns in mental imagery. *Cognitive Science* 13, 51–78.

Finke, R. A., and R. N. Shepard (1986). Visual functions of mental imagery. In K. R. Boff, L. Kaufman, and J. P. Thomas, eds., *Handbook of perception and human performance.* New York: Wiley.

Galton, F. (1883). *Inquiries into human faculty and its development.* London: Macmillan.

Goldston, D. B., J. V. Hinrichs, and C. L. Richman (1985). Subjects' expectations, individual variability, and the scanning of mental images. *Memory and Cognition* 13, 365–370.

Hyman, I. E., and U. Neisser (1991). Reconstructing mental images: Problems of method. (Emory Cognition Project Tech. Rep. No. 19). Atlanta: Emory University.

Intons-Peterson, M. J. (1983). Imagery paradigms: How vulnerable are they to experimenters' expectations? *Journal of Experimental Psychology: Human Perception and Performance* 9, 394–412.

Jolicoeur, P., and S. M. Kosslyn (1985). Is time to scan visual images due to demand characteristics? *Memory and Cognition* 13, 320–332.

Kosslyn, S. M. (1973). Scanning visual images: Some structural implications. *Perception and Psychophysics* 14, 90–94.

Kosslyn, S. M. (1980). *Image and mind.* Cambridge, MA: MIT Press.

Kosslyn, S. M. (1987). Seeing and imagining in the cerebral hemispheres: A computational approach. *Psychological Review* 94, 148–175.

Kosslyn, S. M. (1994). *Image and brain: The resolution of the imagery debate.* Cambridge, MA: MIT Press.

Kosslyn, S. M., N. M. Alpert, W. L. Thompson, V. Maljkovic, S. B. Weise, C. F. Chabris, S. E. Hamilton, S. L. Rauch, and F. S. Buonanno (1993). Visual mental imagery activates topographically organized/visual cortex: PET investigations. *Journal of Cognitive Neuroscience* 5, 263–287.

Kosslyn, S. M., T. M. Ball, and B. J. Reiser (1978). Visual images preserve metric spatial information: Evidence from studies of image scanning. *Journal of Experimental Psychology: Human Perception and Performance* 4, 47–60.

Kosslyn, S. M., J. Brunn, K. R. Cave, and R. W. Wallach (1984). Individual differences in mental imagery ability: A computational analysis. *Cognition* 18, 195–243.

Kosslyn, S. M., J. D. Holtzman, M. J. Farah, and M. S. Gazzaniga (1985). A computational analysis of mental image generation: Evidence from functional dissociations in split brain patients. *Journal of Experimental Psychology: General* 114, 311–341.

Kosslyn, S. M., and P. Jolicoeur (1980). A theory-based approach to the study of individual differences in mental imagery. In R. E. Snow, P. A. Federico, and W. E. Montague, eds., *Aptitude, learning, and instruction: Cognitive processes analyses of aptitude,* vol. 1. Hillsdale, NJ: L. Erlbaum.

Kosslyn, S. M., V. Malkovic, S. E. Hamilton, G. Horwitz, and W. L. Thompson (in press). Two types of image generation: Evidence for left- and right-hemisphere processes. *Neuropsychologia.*

Kosslyn, S. M., C. Segar, J. D. Pani, and L. A. Hillger (1990). What is imagery for? *Journal of Mental Imagery* 14, 131–152.

Lea, G. (1975). Chronometric analysis of the method of loci. *Journal of Experimental Psychology: Human Perception and Performance* 2, 95–104.

Levine, D. N., J. Warach, and M. J. Farah (1985). Two visual systems in mental imagery: Dissociation of "what" and "where" in imagery disorders due to bilateral posterior cerebral lesions. *Neurology* 35, 1010–1018.

McKellar, P. (1965). The investigation of mental images. In S. A. Barnett and A. McLaren, eds., *Penguin science survey*. Harmondsworth, Eng.: Penguin.

Mishkin, M., and L. G. Ungerleider (1982). Contribution of striate inputs to the visuospatial functions of parieto-preoccipital cortex in monkeys. *Behavioural Brain Research* 6, 57–77.

Morris, P. E., and P. J. Hampson (1983). *Imagery and consciousness*. New York: Academic Press.

Paivio, A. (1971). *Imagery and verbal processes*. New York: Holt, Rinehart and Winston.

Palmer, S. E. (1978). Fundamental aspects of cognitive representation. In E. Rosch and B. Lloyd, eds., *Cognition and categorization*. Hillsdale, NJ: L. Erlbaum.

Peterson, M. A., J. F. Kihlstrom, P. M. Rose, and M. L. Glisky (1992). Mental images can be ambiguous: Reconstruals and reference-frame reversals. *Memory and Cognition* 20, 107–123.

Pinker, S. (1980). Mental imagery and the third dimension. *Journal of Experimental Psychology: General* 109, 354–371.

Pinker, S. (1985). Visual cognition: An introduction. In S. Pinker, ed., *Visual cognition*. Cambridge, MA: MIT Press.

Pinker, S., P. A. Choate, and R. A. Finke (1984). Mental extrapolation in patterns constructed from memory. *Memory and Cognition* 12, 207–218.

Pylyshyn, Z. W. (1973). What the mind's eye tells the mind's brain: A critique of mental imagery. *Psychological Bulletin* 80, 1–24.

Pylyshyn, Z. W. (1981). The imagery debate: Analogue media versus tacit knowledge. *Psychological Review* 87, 16–45.

Ratcliff, G. (1979). Spatial thought, mental rotation and the right cerebral hemisphere. *Neuropsychologia* 17, 49–54.

Reed, S. K. (1974). Structural descriptions and the limitations of visual images. *Memory and Cognition* 2, 329–336.

Segal, S. J., and V. Fusella (1970). Influence of imaged pictures and sounds on detection of visual and auditory signals. *Journal of Experimental Psychology* 83, 458–464.

Sergent, J. (1989). Image generation and processing of generated images in the cerebral hemispheres. *Journal of Experimental Psychology: Human Perception and Performance* 15, 170–178.

Sergent, J. (1990). The neuropsychology of visual image generation: Data, method, and theory. *Brain and Cognition* 13, 98–129.

Shepard, R. N., and L. A. Cooper (1982). *Mental images and their transformations*. Cambridge, MA: MIT Press.

Shepard, R. N., and C. Feng (1972). A chronometric study of mental paper folding. *Cognitive Psychology* 3, 228–243.

Shepard, R. N., and J. Metzler (1971). Mental rotation of three-dimensional objects. *Science* 171, 701–703.

Tippett, L. (1992). The generation of visual images: A review of neuropsychological research and theory. *Psychological Bulletin* 112, 415–432.

Trojano, L., and Grossi, D. (1994). A critical review of mental imagery deficits. *Brain and Cognition* 24, 213–243.

Ungerleider, L. G., and M. Mishkin (1982). Two cortical visual systems. In D. J. Ingle, M. A. Goodale, and R. J. W. Mansfield, eds., *Analysis of visual behavior*. Cambridge MA: MIT Press.

Weber, R. J., and R. Harnish (1974). Visual imagery for words: The Hebb Test. *Journal of Experimental Psychology* 102, 409–414.

Chapter 8

The Development of Object Perception

Elizabeth S. Spelke, Grant Gutheil, and

Gretchen Van de Walle

Although the environment can be described in many ways, we tend to perceive our surroundings as an arrangement of objects: from small bodies such as seeds and marbles, to medium-sized bodies such as dogs and chairs, to large bodies such as trains and houses. This tendency is reflected in our language. The simplest common nouns that apply to the scene in Figure 8.1, for example, refer to the kinds of objects it portrays: *orange*, *pitcher*, and *table*. More elaborate expressions are required to name the collection of objects on the table or one part of an object, such as the right back leg of the table. The tendency to perceive objects is reflected, as well, in our actions. If we attempted to transform the array in Figure 8.1, we would probably do so by manipulating one or more of the objects, not by manipulating the array in its totality or by moving a single object part.

8.1 Perceiving Objects

Perceiving objects makes sense, because a single object usually forms a more stable configuration than does a scene as a whole. If we follow the scene in the figure over time, the bowl and the table may part company, but the objects themselves are likely to persist. Although every momentary scene presents a new perceptual configuration, therefore, we can perceive scenes as familiar and meaningful by dividing them into objects of known kinds. Given these advantages, perhaps it is not surprising that adults perceive objects immediately and effortlessly, even in complex and cluttered environments, and that children appear to focus on objects as soon as they begin to talk about the world (Markman 1989) and act upon it (see below). What is surprising is how difficult a feat object perception turns out to be.

8.1.1 Problems of Object Perception

In apprehending objects, our perceptual systems solve two general problems. One is the problem of *parsing*, of carving up perceptual arrays into

Figure 8.1
"Still Life," by Paul Cezanne. (Reprinted by permission of the National Gallery, Washington D. C.)

regions that correspond to the bodies that the scene contains. The other is the problem of *recognition*, of perceiving each body as an entity of a known kind. The recognition problem has been studied intensively and is discussed in a number of the chapters in this volume (see the chapters by Biederman, Farah, Goodale, Nakayama, He and Shimojo, and Pashler). Here, we focus primarily on the parsing problem, discussing object recognition only insofar as recognition processes enter into the parsing of scenes.

The parsing problem has two aspects. First, perceivers must apprehend the boundaries of objects in the surface layout that surrounds them at any given time, deciding what parts of the layout belong to the same or different bodies. In philosophy, this task forms part of the general problem of *individuation*, determining what in our surroundings counts as a single entity; in computational vision and perceptual psychology, this task forms part of the general problem of *unit formation*, dividing visual arrays into the entities that receive further processing. Second, perceivers must apprehend the persistence of objects over successive encounters, determining when something viewed at one time is the same body as something viewed at another time. Philosophers discuss this task as part of the general problem of *identity*: determining which of our experiences or descriptions of the world pertain to a single entity. For students of psychophysics and computational vision, this task presents one aspect of the general *correspondence problem*: the problem of determining what elements in multiple visual representations pertain to a single entity in the scene represented. At first glance, both these tasks appear simple, because we normally solve them so effortlessly. Even after centuries of study by philosophers and intense recent work by psychologists, computer scientists, and linguists, however, no simple procedure for accomplishing either task has been found.

Imagine first an ideal perceiver with full information about the layout of surfaces in a scene: not just the surfaces visible from one point of observation but all the hidden surfaces and their hidden connections as well. Because each object in the scene rests upon other objects, and because distinct objects can be complexly intertwined, it is extremely difficult to state criteria by which such a perceiver could decide where one object ends and the next begins. Philosophers have puzzled over such everyday cases as that illustrated in Figure 8.2: Why do we describe the array in this figure as "a car and trailer" but not "a bumperless car-trailer and bumper," viewing the front bumper but not the trailer as part of the car? Moreover, how should we describe an event in which parts of the car are replaced? Does the car persist over the replacement of one fender? Does it persist over the replacement of all its original parts? What is the status of a second car, composed entirely of the original parts of the first car? (For a discussion of these and other problems, see Hirsch 1982.)

The problems of object perception are compounded when we consider the task faced by an actual perceiver, who does not have access to complete information about objects and their connections. Objects are not fully visible from any one point of observation: the back of every opaque object is hidden, and the front surfaces of many objects are partly occluded as well. In Figure 8.1, for example, only one lemon and a few oranges present front surfaces that are fully in view. Moreover, the connections and separations among adjacent surfaces in a scene are not themselves visible. No visual information clearly indicates, for example, that the bowl and the stem of the glass are parts of a single connected body whereas the dish and the pitcher are not. Finally, objects are frequently obscured from view by movements—the perceiver's, the object's itself, or those of other objects. In these cases, perceivers have no direct information about what might have occurred while objects were occluded and must decide, from

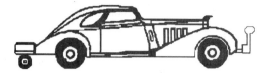

Figure 8.2
A puzzle of individuation: a car and a trailer (after Hirsch, *The concept of identity*, 1982, Harvard University Press.)

highly incomplete information, whether something seen now is the same object as something seen in the past.

Cognitive scientists do not agree about how these problems are solved. Controversies concern such fundamental questions as these: (1) What is an object for a human perceiver? Is there one basic level at which we divide up the world (singling out, perhaps, such entities as a person and a house), or do we divide scenes into objects at multiple levels with equal facility (singling out a hand and a family as well as a person, and a kitchen and a village as well as a house)? (2) At what point or points in perceptual analysis does object perception occur, and on what information is it based? Do perceivers organize arrays into objects early in visual analysis or only after extracting considerable information about perceived scenes? (3) What is the relation between our ability to detect object boundaries over space and our ability to trace the identity of an object through time? (4) What is the relation between our ability to perceive objects in real time and from limited sensory information, and our ability to form judgments about the unity and identity of objects in real or hypothetical situations in which time pressures and perceptual limitations are minimized? Is object perception guided by common principles in all these cases? (5) On what processes does object perception depend? In particular, we may distinguish two different classes of procedures by which arrays could be divided into objects both over space and over time: those employing principles that apply to all objects and those employing knowledge of the properties and behavior of objects of particular kinds.

8.1.2   General and Kind-Based Approaches to Object Perception

It is possible that humans perceive objects in accord with a set of rules or principles that apply to any objects and scenes that we might encounter. The most well-known proposal of this sort comes from Gestalt psychology, a theoretical and experimental approach to perception and other psychological phenomena that developed near the beginning of this century and generated ideas that remain influential today. The Gestalt psychologists suggested that perceivers inherently tend to organize the surrounding layout into the simplest, most regular units. This tendency can be expressed as a set of principles such as *similarity* (surfaces lie on a single object if they share a common color and texture), *good continuation* (surfaces lie on a single object if their edges lie on the same line or smooth curve), *good form* (surfaces lie on a single object if their edges can be joined to form a region with a symmetrical shape), and *common fate* (surfaces lie on a single object if they move together). In Figure 8.1, for example, the principles of similarity, good continuation, and good form serve to group together all the visible regions of the dish and to single out the dish as separate from the oranges. Gestalt principles of organization were also thought to underlie perception of object identity over successive encounters. When an object appears successively in different locations, perceivers were said to perceive a single, persisting body by grouping its appearances into the simplest patterns of motion and change (Michotte 1963).

Considerable evidence appears to favor the Gestalt theory. Presented with meaningless arrays of unfamiliar forms, adults organize the arrays into units that are maximally simple and regular (Figure 8.3a), even when their knowledge about the objects in those arrays contradicts this organization (Figure 8.3b). In addition, this theory provides a natural account of how children develop the ability to perceive objects. If one set of general principles serves to organize the layout into objects, then children who are endowed with some or all of these principles would be able to perceive and direct their actions to objects before they have acquired any specific knowledge about the kinds of objects that surround them. By perceiving and acting on objects, children could learn to recognize, categorize, and talk about the objects they encounter, and they could learn to make inferences about the distinctive properties and behavior of objects of particular kinds.

General theories of object perception, nevertheless, have been subjected to serious criticism within philosophy, psychology, and computational vision for a simple reason: no set of general principles has yet been found that singles out objects under all and only the conditions that human perceivers do. To see the difficulties faced by any theory of object perception based on general principles, consider again the example in Figure 8.2. Why do we perceive a car and a trailer, instead of a front bumper and a bumperless car-trailer? Gestalt principles do not distinguish between these two organizations. Different sets of general principles, focusing perhaps on the material composition or the mode of connection of the car and the trailer (e.g., Palmer and Rock 1994), do not appear more promising, because our perception of the objects in this scene does not appear to change if we imagine that the trailer and the bumper are composed of the same substance and are connected to the car body in the same manner. Examples such as the disassembled car present similar problems for any general

(a)

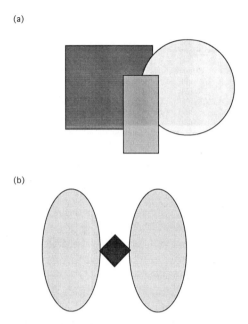

(b)

Figure 8.3
Some illustrations of the Gestalt principles of organization: In (a), we group lines into figures with smooth, symmetrical shapes and uniform textures; in (b), this grouping takes precedence over recognition of the letters *M* and *W* (after Wertheimer 1923).

account of perception of object identity. For example, no general principles readily explain why we are inclined to judge that a car persists when its transmission is replaced, but would be less inclined to judge that a dog persists if its central nervous system were replaced. These examples, and similar examples that have been discussed within computational vision (see Marr 1982), appear to defy any general theory of object perception.

Faced with these problems, many cognitive scientists have proposed that object perception depends on the ability to recognize objects of particular kinds and to apply knowledge about those objects' properties and permissible transformations. Object perception may depend on processes of object recognition, which depend in turn on the perceiver's vocabulary of internal representations, or models, of the kinds of objects that furnish our surroundings.

More specifically, the processes by which perceivers recognize part of a visual scene as a car or a tree may operate on an unparsed representation of that scene. Models of visible objects may be applied everywhere to the scene, irrespective of the actual boundaries of objects (which, by hypothesis, the perceiver does not yet represent). When a sufficiently good match is found between an object model and a region of the scene, an object of the kind that corresponds to the model is recognized. Once an object is recognized, perceivers can apply their knowledge of the properties and the behavior of that kind of object both to perceive the object's boundaries in the visual scene and to trace the object's identity through time. Because we know that cars have bumpers but not trailers as proper parts, we perceive a car but not a bumperless car-trailer in Figure 8.2. Because we know that dogs but not cars have behavioral and mental capacities supported by certain internal structures, we consider certain transformations of dogs to be more radical than other, superficially similar transformations of cars.

Studies in computational vision and perceptual psychology provide support for the plausibility of model-based unit formation. A variety of mechanical recognition schemes have been devised that operate on input representations that have been parsed into elementary edges and features but not into objects (see Ullman 1989). Moreover, recent experiments provide evidence that human perceivers can recognize objects before they have divided a scene into units. Before the visual system determines whether the white regions in Figure 8.4 are objects or a shapeless background, for example, recognition processes appear already to have categorized the region as a person (Peterson 1995). Processes of object recognition, therefore, could provide a feasible basis for object perception.

Nevertheless, kind-based approaches to object perception have certain limitations. First, they do not provide a natural account of our ability, under certain circumstances, to perceive that a "thing" has appeared without perceiving what that thing is, or our ability to change how we categorize an object while perceiving the object to persist over this change (see Kahneman and Treisman 1984). Second, kind-based approaches provide no natural account of the development of object perception in young children. If processes of object recognition underlie perception of the boundaries and identity of objects, then one can perceive objects only if one has a vocabulary of object models. But how do children develop this vocabulary? Because it is hardly likely that humans possess innate knowledge of the visual appearance of cars and trailers, children must have the means to learn about these objects as they encounter them. As the Gestalt psychologists emphasized, however, it is not at all obvious how one can learn about the appearance of a car or trailer if one does not first have the means to single out these entities as units in a visual scene. Because every visual scene presents a novel arrangement of objects, a perceiver who possesses neither a rich store of object models nor a set of general principles for organizing arrays into units seems doomed to experience a meaningless succession of novel arrays. Cars and trailers might populate these arrays and travel through them in a predictable fashion, but children with no

Figure 8.4
A reversible figure-ground display (after Peterson, 1994). When the black region is perceived as figure, the white regions appear to be indefinite in form and to continue behind the figure. Nevertheless, subjects who experience the black region as figure may activate representations of people in the region perceived as background. Evidence for these representations comes from the finding that this figure undergoes a reversal faster than an otherwise comparable figure-ground display in which the background has no meaning, and from the finding that subjects presented with this background are slower to identify a subsequently presented figure with the same shape. (Redrawn by permission from M. A. Peterson, Object recognition processes can and do operate before figure-ground organization, 1994, *Current Directions in Psychological Science.*)

means to perceive such objects would not experience their presence or learn from their predictable behavior.

In summary, neither general theories nor kind-based theories appear to provide a complete account of human perceivers' abilities to apprehend the boundaries and the persistence of everyday objects. Whereas some evidence suggests that object perception depends on general principles that override knowledge of the recognizable objects in a scene (e.g., Figure 8.3b), other evidence suggests the reverse (e.g., Figure 8.2). Moreover, each class of theory appears to have inherent limitations: general theories do not do justice to the richness and specificity of mature object perception, and kind-based theories do not appear to support an account of how object perception develops.

These observations raise the possibility that object perception depends both on a small set of general principles and on a much larger body of kind-specific knowledge. General principles of object perception may serve to organize the perceptible surface layout into bodies in the absence of any knowledge of the kinds of objects that visual arrays contain. These principles may be available to children early in development; they may serve as a basis for the development of kind-specific knowledge; and they may underlie adults' ability to perceive the boundaries and persistence of objects that we fail to recognize or that we categorize differently at different times. Kind-specific knowledge could supplement the general principles of object perception, accounting for our fine-tuned perceptions of and intuitions about the boundaries and persistence of familiar things.

8.2   Infants' Perception of Object Boundaries

If mature object perception depends both on general and on kind-specific processes, then the general process may be difficult to discern in adults, who possess a vast store of knowledge of object kinds. This process may be studied more easily in people whose knowledge is less extensive. Studies of object perception in infancy might reveal what the basic process for perceiving objects is and how it becomes extended and changed by the growth of knowledge.

In the sections that follow, we ask whether and how infants perceive the unity and boundaries of objects in momentary scenes and the persisting identity of objects over successive encounters. We consider both the nature and the limits of infants' perception of objects in hopes of shedding light on the process by which infants organize perceptual arrays. After exploring this process in infancy, we consider its possible relation to processes of object perception in adults. Our primary focus concerns the roles of general principles and kind-specific knowledge in object perception. We will see, however, that studies of early perceptual development also bear on the other questions about object perception with which we began, concerning the nature of the objects perceivers apprehend, the nature of the information on which object perception depends, the relation between perception of object boundaries and perception of object identity, and the relation between immediate perception of objects, and intuitive judgments about them.

8.2.1   Figure-Ground Organization

Adults perceive objects as separate from whatever stands behind them, and we perceive the border between an object and its background as bounding the object, not the ground. Psychologists have studied the development of perception of these figure-ground relationships by presenting infants with simple scenes such as those illustrated in Figure 8.5 and observing infants' patterns of reaching for the displays.

To reach for and pick up an object effectively, the hand must be directed to the object's external boundaries (see Chapter 5). This ability, in turn, depends on the ability to perceive the object as distinct from the objects and surfaces around it. Experiments from a number of laboratories provide evidence that at the age when infants begin reaching for objects (about

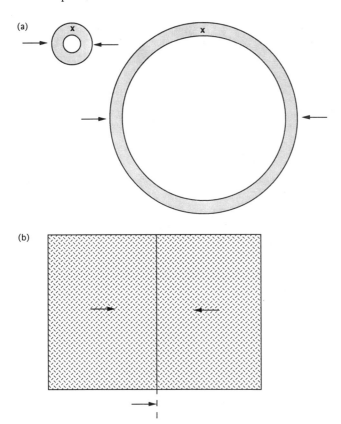

Figure 8.5
Schematic depictions of displays for studies of infants' perception of an object as separate from its background. In (a), the location of a sound source is indicated by the x; frequent points at which the reaching hands contact the object are indicated by arrows (after Clifton et al. 1991). In (b), arrows indicate the direction of motion of each of the two regions at one point in time (over time, the regions reverse direction). Infants reach for the region that moves together with the border—in this case, the region on the left (after M. E. Arterberry, L. G. Craton, and A. Yonas, Infants' sensitivity to motion-carried information for depth and object properties, 1993, in C. E. Granrud, ed., *Visual perception and cognition in infancy*, vol. 23, Carnegie-Mellon Symposia on Cognition, L. Erlbaum.)

four-and-a-half months), they reach for object boundaries. For example, Clifton et al. (1991, Figure 8.5a) presented six-month-old infants, in alternation, with a large and a small ring within reaching distance. Infants reached for each object by directing their hands to its borders, aiming at different spatial locations for the different objects. These reaching patterns suggest that infants perceived each object as separate from the background surface behind it and represented, with some accuracy, the location of each object's boundaries. Interestingly, infants' representations of object boundaries appeared to persist in the absence of visual information, because their reaching was also directed to the borders of the objects when the lights were extinguished and reaching took place in the dark.

What information do infants use to perceive the boundaries of a visible object? Experiments by Yonas and colleagues suggest that infants perceive object boundaries by detecting the relative motion patterns of surfaces and edges (Arterberry, Craton, and Yonas 1993). When a surface and its borders move as a unit relative to surrounding surfaces, infants perceive the surface as a bounded object (see Figure 8.5b). Such motions are normally produced when an object moves against a stationary background or when a stationary object stands in front of its background and is viewed by the baby with a moving head.

According to the Gestalt psychologists, the simplest example of figure-ground organization occurs in displays containing no motion or depth changes: static, two-dimensional displays consisting of two regions of different brightness or color (e.g., Figure 8.4). Perception of these displays as containing a bounded figure in front of an unbounded ground was thought to depend on the same organizational principles as perception of objects in moving, three-dimensional arrays. Research with infants casts doubt on this view: Infants do not appear to reach for the borders between the two regions in a two-dimensional figure-ground display (Spelke 1988; Arterberry et al. 1993). They appear to perceive figure-ground relations by analyzing the three-dimensional spatial arrangements and motions of surfaces but may not organize surfaces into regions of a common brightness and color.

8.2.2   Perception of Object Boundaries in Scenes of Multiple Objects

Infants rarely encounter objects in the simple arrays studied above because the objects in natural scenes stand upon, beside, and in front of other objects. Can infants perceive the boundaries of objects in these more complex situations? Some experiments have approached this question by presenting infants with displays containing two objects that touched or overlapped in depth, and investigating infants' patterns of reaching for the displays.

(a)

(b)

(c)

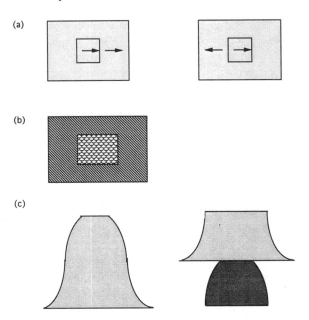

Figure 8.6
Schematic depictions of displays for studies of infants' perception of the boundaries of objects in two-object arrays. (Redrawn by permission from C. von Hofsten and E. S. Spelke, Object perception and object-directed reaching in infancy, 1985, *Journal of Experimental Psychology: General* 114, 198–212; R. Kestenbaum, N. Termine, and E. S. Spelke, Perception of objects and object boundaries by three-month-old infants, 1987, *British Journal of Developmental Psychology* 5, 367–383; and E. S. Spelke, K. Breinlinger, K. Jacobson, and A. Phillips, Gestalt relations and object perception: A developmental study, 1993, *Perception* 22, 1483–1501.)

For example, infants have been presented with one object that was small and close to them and a second object that was larger and more distant (Hofsten and Spelke 1985; Figure 8.6a). A variety of observations indicate that infants tend to reach for the object that is closest (Yonas and Granrud 1985). If infants perceived the two objects as distinct, therefore, they were expected to reach for the borders of the small object. If infants perceived the two objects as a single unit, in contrast, they were expected to reach primarily for the borders of the larger object, because it provided most of the external borders of the display. Infants reached for the borders of the small object when the objects were spatially separated, either vertically or in depth, and when the objects moved relative to one another. In contrast, they reached for the two objects as a single unit when the objects were stationary and adjacent in depth and when the objects moved together. Like the studies of one-object arrays, these studies suggest that infants perceive objects by detecting surface arrangements and motions, grouping together all surfaces that are adjacent and undergo no relative motion. The above studies suggest that infants perceive objects by analyzing the three-dimensional arrangements and motions of surfaces, not by analyzing motions and configurations in the visual representations that precede the recovery of depth information (see Chapter 1). Consider, for example, the display in Figure 8.6a, in which two objects that are arranged in depth undergo a rigid translatory motion. Because the objects are at different distances from the infant, they undergo different two-dimensional displacements in the infant's visual field: The closer object is displaced at a greater speed and to a greater extent than the more distant object. If object perception depends on an analysis of two-dimensional patterns of image displacement, infants should perceive these two objects as separate units, because the images of the two objects undergo different retinal motions. Contrary to this prediction, infants perceived a single object in this display. This and other findings (e.g., Kellman et al. 1987) provide evidence that infants' perception of objects occurs relatively late in visual analysis, after the recovery of information about the arrangements and motions of surfaces in depth.

In all the above studies, infants' perception of object boundaries has been inferred from patterns of reaching. Further studies have investigated object perception in infancy by focusing on infants' visual attention to displays of objects. If infants are presented repeatedly with a single visual display, the time they look at the display tends to decline over successive presentations. If a new display is presented subsequently, their looking time increases. This pattern of preferential looking at novel displays has provided psychologists with a different method for assessing infants' perception of object boundaries. Infants are first familiarized with an array of two objects and then presented with new arrays in which either both objects appear in new locations—but with the same internal arrangement—or one object appears in a new location, changing the objects' arrangement. If infants perceive the objects as separate units, we would expect them to see the array in which both objects are displaced together as more novel and, therefore, to look at it longer. If infants perceive the objects as a single unit, they should look longer at the array in which one object is displaced relative to the other.

The findings of experiments using this preferential-looking method converge closely with the findings of experiments using reaching methods. In particular, preferential-looking experiments provide evidence that infants perceive objects as separate units if the objects are separated visibly or in depth (Kestenbaum et al. 1987; Figure 8.6b). In contrast, the experiments provide evidence that infants perceive two stationary, adjacent objects as a single unit, even if the objects differ in color, texture, and shape and their edges are not aligned (Kestenbaum et al. 1987; Spelke et al. 1993; Figure 8.6c).

The above studies support the Gestalt thesis that perceivers have an early-developing, general process for organizing arrays into objects, but they cast doubt on the Gestalt psychologists' characterization of that process. Infants do not appear to perceive object boundaries by organizing visual scenes into units that are maximally homogeneous in color and texture and maximally smooth and regular in shape. For a young infant, the relatively complex objects in Figure 8.6c appear to be perceived as units just as readily as the simpler objects. Infants' failure to perceive object boundaries in accord with the Gestalt principles cannot be explained by failures to perceive the colors or shapes of surfaces, because studies of infants' sensitivity to color and pattern suggest that these properties are detectable (see, for example, Teller and Bornstein 1987). Further studies suggest that infants use such properties as lightness similarity and good continuation to group together closely spaced elements on a single surface (Quinn, Burke, and Rush 1993; Van Giffen and Haith 1984). Infants appear to detect Gestalt relationships and to use them to perceive surface textures, but they do not appear to use these relationships to perceive the boundaries of objects.

These studies suggest that infants would perceive the display in Figure 8.2a as a single object: a "car-trailer," rather than a car and a trailer. A final preferential-looking experiment has tested this suggestion quite directly (Xu and Carey 1994). Ten-month-old infants were presented with two toys—for example, a yellow rubber duck and a red metal truck—arranged so that one toy stood on top of the other. In one condition, the duck moved back and forth on top of the truck, remaining in contact with the truck throughout the motion; in the other condition, the two toys were stationary. After familiarization with one or the other of these displays, infants viewed two test events in which a hand entered the display, grasped the duck, and lifted it into the air. In one event, the duck moved separately from the truck; in the other event, the duck and truck moved together. Infants who had been familiarized with the duck and truck undergoing relative motion looked longer at the second event, suggesting that they had perceived the two toys as separate objects and expected them to move independently. In contrast, infants who had been familiarized with the duck and truck without motion looked longer at the first event, suggesting that they had perceived the two toys as a single object. This study provides further evidence that infants fail to perceive object boundaries in accord with the Gestalt principles of similarity, good continuation, and good form. In addition, it suggests that ten-month-old infants failed to perceive the object boundaries by recognizing the toy duck and the toy truck as objects of distinct kinds.

All the findings discussed in this section suggest that infants perceive objects through a general process for grouping surfaces into units by analyzing their three-dimensional arrangements and motions, irrespective of their colors, textures, or shapes. Similar conclusions emerge from studies focusing on infants' perception of objects that are partly hidden.

8.2.3   Perception of Objects over Occlusion

Because almost every object in a natural visual scene is partly hidden by parts of itself or by other objects, perceiving object boundaries requires that we represent the connections among parts of objects that are occluded. Research with infants has investigated infants' perception of partly hidden objects under a variety of circumstances in which the visible parts of an object appear at spatially separated places or at different times.

For example, researchers have familiarized infants with an object whose top and bottom are visible but whose center is occluded and tested their perception of the connectedness and the shape of the partly occluded object by comparing looking times to subsequent fully visible displays (Figure 8.7a). If the visible ends of the center-occluded object moved together, four-month-old infants subsequently looked longer at a fully visible display in which the ends were separated by a gap than at a fully visible display in which the ends were connected. Because infants tend to look longer at displays they perceive as more novel, this preference provides evidence that the infants perceived the original center-occluded object as connected behind the occluder (Kellman and Spelke 1983; S. Johnson and Nanez, in press; Slater, et al. 1990). Infants did not show this preference when the entire display was stationary, when the ends of the center-occluded object were stationary and the occluder moved, or when one end of the object moved while the other did not (Kellman and Spelke, 1983). These findings suggest that infants perceive the unity of a center-occluded object by detecting the common motion of its visible surfaces.

Further experiments provided evidence that perception of the unity of a center-occluded object was unaffected by the color and form relations among its visible surfaces (Kellman and Spelke 1983, Figure 8.7b), suggesting that Gestalt relations such as similarity, good continuation, and good form do not influence infants' perception. Moreover, experiments provide evidence that Gestalt relations do not dictate the forms of the objects that infants perceive. Infants' perception of the form of a moving center-occluded object was investigated by familiarizing infants with the occlusion display and then presenting fully visible objects that differed in form (Figure 8.7c). Infants were found to look equally at displays presenting the simple and regular form perceived by adults and at displays presenting

312   Spelke, Gutheil and Van de Walle

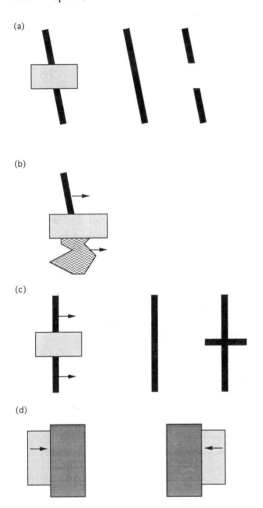

Figure 8.7
Displays for studies of infants' perception of partly occluded objects. (Redrawn by permission from P. J. Kellman and E. S. Spelke, Perception of partly occluded objects in infancy, 1983, *Cognitive Psychology* 15, 483–524; L. G. Craton and R. Baillargeon, personal communication; and G. Van de Walle and E. S. Spelke, L'intégration spatio-temporelle dans la perception des objets chez le bébé, 1993, *Psychologie française* 38, 75–83.)

a more complex form (Craton and Baillargeon, personal communication, December 1991). This finding suggests that whereas infants perceive the connectedness of a moving center-occluded object, they have no definite perception of its shape.

Finally, experiments have investigated infants' perception of objects whose parts become visible only over time (Figure 8.7d). In one study, five-month-olds were familiarized with an object that moved laterally behind one large occluder such that its center was always hidden, and its left and right sides underwent the same motion and were visible in immediate succession. After this familiarization, infants were presented in alternation with a connected object and with the two previously visible parts separated by a gap. Infants looked longer at the latter display, suggesting that they perceived the visible parts in the original occlusion display as a unitary object (Van de Walle and Spelke 1993). In other experiments presenting objects with successively visible parts, however, infants were found not to perceive properties of an object such as its overall form (see Arterberry, Craton, and Yonas 1993). The ability to perceive the form of an object whose parts appear in succession appears to develop over the second half of the first year. At younger ages, infants appear to perceive the unity and boundaries of such an object but not its shape.

Recently, a number of investigators have explored the early development of the ability to perceive the unity of a moving, partly occluded object. Experiments with infants from birth to four months suggest that this ability emerges at about two months of age. Newborn infants who are familiarized with a center-occluded object subsequently look longer at a fully visible display containing one connected object than at a display containing the two visible parts of the original object separated by a gap: the opposite looking pattern from that shown by four-month-old infants (Slater et al. 1990). At two months of age, infants show no preference between these two displays when the occluder is large (S. Johnson and Nanez, 1995); they look longer at the display with the gap when the occluder is reduced in size (Johnson and Aslin, in press). Two-month-old infants, therefore, perceive the unity of a center-occluded object under some but not all of the conditions that are effective at four months.

These findings suggest that a developmental change in object perception occurs over the first four months. At this writing, the nature of this change is not clear. It has been suggested that the change depends on the maturation of the neural systems underlying the perception of coherent motion and the control of attention to spatially separated parts of the visual field (M. Johnson 1990). Alternatively, the change may depend on developmental changes in visual resolution, allowing greater access to information about surfaces and their arrangement (S. Johnson 1994; see Banks and Shannon, 1993). Finally, the change may reflect the emergence, through learning or maturation, of the system for perceiving objects itself (Slater et al. 1990; Smith and Katz, in press).

8.2.4   Principles of Object Perception

Although the above studies cast doubt on the thesis that infants perceive objects in accord with Gestalt principles such as similarity and good continuation, they do suggest that infants organize the three-dimensional surface layout into objects on some basis. Two- to four-month-old infants appear to perceive objects by relating the motions of surfaces to the arrangement of surfaces in the following ways. First, when they are unable to see a spatial separation between two stationary surfaces, infants infer that the surfaces consist of a single object, regardless of differences in color and texture. This inference accounts for infants' tendency to reach for adjacent objects as one unit and to dishabituate to changes that alter the perceived boundaries of such an array. Two surfaces that are spatially separated, in contrast, are perceived as two distinct objects, despite similarities in color and texture. This inference accounts for infants' tendency to reach for surfaces separated by a visible gap as independent units and to generalize habituation to displays in which such objects change position with respect to one another. Third, when infants observe two surfaces that undergo a common rigid motion, they infer that the surfaces are connected, provided that this interpretation is not inconsistent with the perceivable layout (i.e., the surfaces are not separated by a visible gap). This inference accounts for infants' perception of the connectedness of the visible portions of a moving partly occluded object. Finally, when infants observe two surfaces that move relative to one another, they infer that the surfaces belong to two separate objects. This inference accounts for infants' tendency to reach for adjacent objects undergoing independent motion as separate units.

Taken together, infants' inferences about surface arrangements and motions appear to reflect two fundamental properties of inanimate, material objects. First, objects are cohesive: they are internally connected and externally bounded entities that maintain both their connectedness and their boundaries over time and space. Second, objects influence one another's motions if and only if they touch. These object properties have led us to propose that object perception initially accords with two principles: the principles of *cohesion* and *contact* (Spelke and Van de Walle 1993; see Figure 8.8a and 8.8b).

8.3   Infants' Perception of Object Identity

When an object disappears from view and later returns to view, are inexperienced perceivers ever able to recognize that the two appearances in-

A. The principle of cohesion: A moving object maintains its connectedness and boundaries

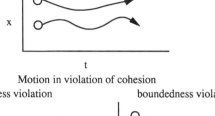

Motion in accord with cohesion

Motion in violation of cohesion

connectedness violation   boundedness violation

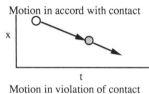

B. The principle of contact: Objects move together if and only if they touch

Motion in accord with contact

Motion in violation of contact

action on contact violation   no action at a distance violation

C. The principle of continuity: A moving object traces exactly one connected path over space and time

Motion in accord with continuity

Motion in violation of continuity

continuity violation   solidity violation

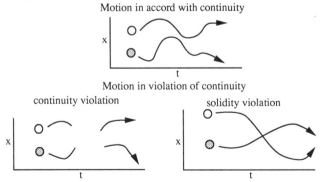

Figure 8.8
Three proposed principles of object perception in infancy.

volve a single, persisting object? Studies of infants have addressed this question through preferential-looking methods similar to those described above.

8.3.1 Perceiving Objects over Successive Encounters

One series of experiments investigated whether young infants perceive the identity or distinctness of objects by analyzing the spatiotemporal continuity or discontinuity of object motion: a factor that strongly influences adults' intuitions about object persistence and change (Spelke et al. 1995). Separate groups of four-month-old infants were familiarized with one of two events involving objects that moved in and out of view (Figure 8.9a). In one event, one object moved at a constant speed behind two spatially separated occluders. Because the occluders were almost as narrow as the object, the motion in this condition appeared to be continuous. In a second event, the same object motions occurred on the far sides of the two occluders, but no object was seen to move in the space between the two occluders. The visible paths of object motion therefore were discontinuous. To assess infants' perception of the identity or distinctness of the objects in these events, experimenters tested all the infants with fully visible displays containing one or two objects. The infants who had been familiarized with the two-screen event involving continuous motion showed a greater preference for the two-object event than those who had been familiarized with the two-screen event involving discontinuous motion. This finding provides evidence that infants' perception of the number of objects in the occlusion events was influenced by the apparent continuity or discontinuity of object motion. Whereas infants presented with continuous motion appeared to perceive a single object that moved in and out of view, those presented with discontinuous motion appeared to perceive two distinct objects.

Following the ideas of the Gestalt psychologists, Michotte (1963) proposed that perceivers apprehend the identity of objects through time by organizing events so as to maximize the smoothness of object motion. Experiments tested whether infants exhibit this tendency by comparing infants' perception of object identity over occlusion events in which objects either undergo smooth motion or motion that changes in speed (Spelke et al., in press). They presented four-month-old infants with events in which two objects moved in succession behind one large occluder (Figure 8.9b). In one event, the objects' visible speed and the duration of occlusion gave rise, for adults, to the impression that one object moved in and out of view at a constant speed. In two further events, the objects moved at speeds and with occlusion times that suggested an abrupt change in object motion behind the occluder. After familiarization with any of these events, infants showed no consistent looking preferences

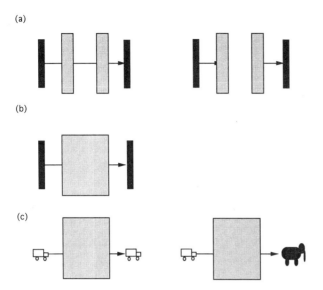

Figure 8.9
Displays for studies of infants' perception of the identity of objects over successive encounters. (Redrawn by permission from E. S. Spelke, R. Kestenbaum, D. Simons, and D. Wein, Spatio-temporal continuity, smoothness of motion, and object identity in infancy (1995), *British Journal of Developmental Psychology*, F. Xu and S. Carey, Infants' metaphysics: The case of numerical identity (in press), *Cognitive Psychology*.

between displays of one versus two objects. These findings cast further doubt on the Gestalt thesis that object perception depends on a general tendency to organize the perceptual world in to the simplest and most regular units. Infants' perception of object identity did not appear to be influenced by the apparent smoothness of object motion.

Further preferential-looking experiments have investigated whether infants perceive the identity of objects over successive appearances by analyzing properties of objects such as surface color and shape or by taking account of object kind (Xu and Carey, in press, Figure 8.9c). In these studies, two objects moved into view, in succession, on the opposite sides of a single, wide box. In one condition, the objects were visually indistinguishable and from the same category. In a second condition, the objects differed in color, texture, and shape, and were members of different categories (for example, a red metal truck and a blue rubber elephant). Infants' perception of object identity was also tested under conditions in which the spatiotemporal discontinuity of object motion specified that the objects were distinct: Two toys appeared on the opposite sides of two narrow, spatially separated boxes, or they were visible on the two sides of the wide box simultaneously. In all cases, perception of object identity was tested by familiarizing ten-month-old infants with one occlusion event and then comparing their looking times to fully visible displays of one versus two objects.

Once again, infants' perception of identity was influenced by information for the spatiotemporal continuity or discontinuity of object motion. Infants familiarized with two spatially separated objects that were visible simultaneously, or with two objects whose successive visible motions were separated by a visible gap, looked longer at the fully visible display containing only one object. In contrast, infants' perception of identity did not appear to be influenced either by such perceptible properties of the objects as their color or shape or by object kind. When objects appeared in succession from within a single box, infants' perception of the identity or distinctness of the objects appeared to be indeterminate, and equally so, regardless of whether or not the objects differed in color, shape, and kind.

In the situation studied by Xu and Carey, ten-month-old infants' perception of object identity appears to differ dramatically from the perceptions of adults. When adults view a toy elephant and a toy truck that emerge from a box in succession, they are strongly inclined to view the second object as distinct from the first. Young infants, in contrast, appear to be uncommitted as to the identity or distinctness of the objects. What accounts for this difference?

One might propose that infants failed to perceive object identity by taking account of the perceptible differences between the elephant and the truck because they failed to detect these differences. The best evidence against this possibility comes from Xu and Carey's own experiments. In one study, they presented infants with an occlusion event in which one object appeared and disappeared on the left side of the box and then a second object (that either was featurally indistinguishable from the first object or came from a different category) appeared on the right side of the box and remained in view for as long as the infant looked at it. The infants who viewed two featurally distinctive objects looked longer at the second object than those who viewed two featurally indistinguishable objects. This preference provides evidence that infants detected the featural differences between the former pair of objects. Nevertheless, infants in both conditions showed the same indeterminate perception of the number of objects participating in the event. Ten-month-old infants evidently detected differences between a toy elephant and a toy truck, but they did not infer from these differences that the elephant and truck were numerically distinct objects.

Following philosophers such as Wiggins (1980) and Hirsch (1982), Xu and Carey suggest that adult perception of the distinctness of the elephant and truck depends on the knowledge that elephants and trucks are different sorts of things. Perception of object identity depends, in their view, on knowledge of object categories. Xu and Carey's research suggests that conceptual categories such as *elephant* do not guide ten-month-old infants' perception of object identity. Instead, infants may view the elephant/truck event in the way that many adults would view temporally separated appearances of a chrysalis of one species and a mature moth of a different species. Although a naturalist might draw on knowledge of insect kinds to perceive the first object as numerically distinct from the second, most of us would note the property differences between the chrysalis and the moth and yet be uncertain as to the identity or distinctness of the objects.

8.3.2   More Principles of Object Perception

In summary, infants appear to share some, but not all, of adults' abilities to perceive the identity of objects over successive encounters. Like adults, young infants perceive object identity by analyzing the positions and motions of objects. When infants view an event in which two successive encounters with objects can be seen to be connected by a single path of motion, they infer that a single object participated in the event. Conversely, when infants view an event in which they can see that no continuous path of motion connects two object appearances, either because two successive object motions are separated by a gap or because two objects are visible simultaneously in distinct locations, then infants infer that two distinct objects participated in the event. Unlike adults, young infants do not appear to perceive object identity by analyzing the smoothness of object motion, by analyzing the constancy or change of objects' perceptible properties such as shape, color, and texture, or by recognizing objects as instances of familiar kinds.

These findings suggest that there exists a basic process for perceiving object identity that is independent of knowledge of the kind of object under consideration. Contrary to Gestalt theory, this process does not appear to depend on a general propensity to organize events into units that are maximally simple and regular. Rather, infants appear to perceive object identity in accord with the principle that objects exist continuously and move on paths that are connected over space and time (see Figure 8.8c). This principle of *continuity* appears to operate prior to the emergence of knowledge of many object kinds. It may form part of the initial capacities that make possible the development of the latter knowledge.

Putting together the findings from studies of perception of object boundaries and studies of perception of object identity, young infants appear to organize visual arrays into bodies that move cohesively (preserving their internal connectedness and their external boundaries), that move together with other objects if and only if the objects come into contact, and that move on paths that are connected over space and time.

Cohesion, contact, and continuity are highly reliable properties of inanimate, material objects: objects are more likely to move on paths that are connected than they are to move at constant speeds, for example; and they are more likely to maintain their connectedness over motion than they are to maintain a rigid shape. Infants' perception appears to accord with the most reliable constraints on objects.

8.4   Developmental Changes in Object Perception

The principles of cohesion, contact, and continuity, nevertheless, leave many object boundaries unspecified. They fail to specify the border between two stationary objects that touch, the connectedness of a stationary object whose center is occluded, or the distinctness of moving objects that emerge from the same place at different times. Given that adults usually perceive object boundaries under these conditions, object perception must undergo developmental change, but the nature of that change is unclear. It is possible that adults perceive objects in accord with further general principles, such as the Gestalt principles of good continuation, similarity, and good form. Alternatively, adults may perceive objects by recognizing objects of known kinds. For example, adults may perceive the bowl in Figure 8.1 as complete and bounded, not by organizing the scene into the simplest shapes but by recognizing the bowl as a familiar object with a known shape.

Studies of the development of object perception might serve to distinguish between these possibilities. If adults perceive objects in accord with a single, general tendency to confer the simplest organization on perceptual experience, then there may be a single time in development when this tendency emerges. In contrast, if the organizational phenomena that Gestalt psychologists described depend on processes of object recognition, then these phenomena should appear earlier in development when children are presented with objects that are more common and familiar.

Studies of developmental changes in perception of stationary, adjacent objects provide preliminary support for the latter view. In three experiments, infants of different ages were presented with adjacent objects that differed in color, texture, and form (Figure 8.10). The objects in one experiment included a solid block; infants evidently perceived it as distinct from the object adjacent to it by eight months of age (Needham and Baillargeon 1994). Infants may have perceived two objects in this study, because blocks come to be recognized early in development. The objects in a second study were toy animals and vehicles; infants began to perceive them as distinct objects between ten and twelve months (Xu and Carey 1994). Infants may have perceived two objects in this study when cate-

(a)

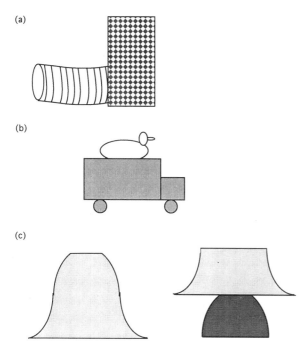

(b)

(c)

Figure 8.10
Displays for studies of infants' perception of adjacent objects. (Redrawn by permission from
A. Needham and R. Baillargeon, Object segregation in eight-month-old infants, 1994,
manuscript submitted for publication; F. Xu and S. Carey, Infants' ability to individuate and
trace the identity of objects, 1994, paper presented at the International Conference on
Infant Studies, Paris, June; and E. S. Spelke.)

gories such as *truck* and *duck* (or, perhaps *toy vehicle* and *toy animal*)
emerged (see Mandler and McDonough 1993). The objects in a third
study had simple, smooth conical shapes but were not clearly recognizable;
the ability to perceive their distinctness began to emerge between three
and five months but was not complete at nine months: nine-month-old
infants' perception of the boundaries of the two-object display appeared
to be indeterminate (Spelke et al. 1993). Because the shapes bore a weak
resemblance to a number of objects that may be familiar to infants but no
strong resemblance to any one object kind, the developmental change in
perception in this study may have occurred over a protracted period, with
considerable variability across infants. At any given age, therefore, some
infants may perceive the conical objects as familiar whereas others do not.

Our interpretations of the above studies are tentative, for no experiment has tested directly the effect of familiarization with an object category on perception of object boundaries. These studies nevertheless give

some plausibility to the view that perception of objects in accord with Gestalt relations depends, at all ages, on object recognition. Adults may appear to apply Gestalt principles to unfamiliar arrays of objects because our vocabulary of object models is sufficiently extensive to encompass displays designed to be unfamiliar as well as those composed of familiar objects. Processes of categorization may underlie the organizational phenomena that the Gestalt psychologists described.

Little is known about infants' abilities to recognize objects. Although a wealth of research provides evidence that they are capable of a variety of perceptual categorizations (e.g., Quinn, Eimas, and Rosenkrantz 1993), this research does not reveal whether infants view the members of a single perceptual category as different individuals of the same kind (for discussion, see Mandler and McDonough 1993). Despite a paucity of evidence, however, it is highly likely that infancy and early childhood are the times when we acquire the bulk of our knowledge of object kinds and the corresponding visual representations by which we recognize each kind of object. Studying the processes by which infants come to recognize objects may shed light on the nature of object recognition processes: processes that have received extensive study in other areas of cognitive science (as this volume attests) but rather less attention from students of perceptual development.

8.5   From Infants to Adults

Studies of infancy suggest a picture of the origins and development of object perception. In infancy, object perception is a late perceptual process taking as input a representation of three-dimensional surface arrangements and motions. Surfaces are grouped into objects, over both space and time, in accord with principles capturing three fundamental properties of material bodies: Such bodies are cohesive, they interact only on contact, and they move on connected paths. By applying these principles to visible surface layouts, infants perceive objects as distinct from the surfaces behind them, as separated from the objects they touch, as continuing in places where they are hidden, and as existing and moving continuously between successive perceptual encounters.

In each case we have considered, we have found limits to infants' abilities to perceive objects, and these limits appeared to be overcome during development by developing abilities to recognize objects of known kinds. Processes of object recognition may allow older infants to perceive object boundaries where younger infants fail to find them, and they may allow infants to trace object identity in circumstances that, for young infants, are ambiguous. These findings suggest that general principles of object perception become enriched by kind-specific principles as infants grow. We close by asking whether the same general and kind-specific processes underlie object perception in adults.

8.5.1   Core and Peripheral Processes of Object Perception

Although existing studies of adults are open to multiple interpretations, a number of hints suggest that adults' perception depends on the same general and kind-based processes we find in infants. One hint comes from the studies of object perception we reviewed in the introduction: Neither general principles of perceptual organization nor kind-specific processes of object recognition appear to provide a complete account of adults' perception of objects. Further hints come from two sources we have not yet considered: chronometric studies of attention and visual representation and interview studies of intuitions about object persistence and change.

8.5.2   Visual Attention and Object Representations

Although this chapter cannot do justice to the extensive literature on the time course of the processes by which we construct and act on visual representations (see Chapter 2), one series of studies is especially relevant to the problem of object perception. When an event suddenly occurs in the visual field, human perceptual systems appear to make a very rapid decision: Has a new object appeared, or has a previously visible object changed state or position? Studies using a variety of methods focusing on the speed with which we can identify an object or indicate its location have explored the nature of this decision process. Because these studies typically use two-dimensional displays of alphabetic characters, their findings cannot be compared directly to the findings of studies of infants' perception of the identity of real objects moving through three-dimensional scenes. The convergence between the findings of these very different studies is nonetheless striking: adults appear to identify objects in rapidly changing displays by taking account of relations such as cohesion and continuity, and not by taking account of relations such as sameness in color and form.

In an experiment by Kahneman, Treisman and Gibbs (1992; Figure 8.11), for example, a letter appeared in one of eight stationary boxes, and subjects named it quickly. Immediately before this event, the letter to be named and a different letter appeared in two boxes. Subjects named the target letter more slowly if a different letter previously occupied its box, providing evidence that a representation of the earlier display influenced processing of the target display. This effect was then used to investigate subjects' representations of displays of moving boxes. When the boxes moved between the two letter presentations, subjects were slower to name

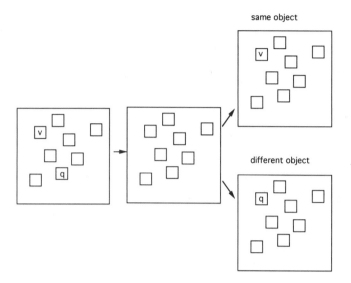

Figure 8.11
Displays and events in two conditions of a study of rapid identification of object identity
and change. (See Kahneman et al. 1992, for a description of the complete study).

the target letter if a different letter previously occupied the target's box
than if a different letter previously occupied a different box at the same
distance from the target. A representation of the persisting box evidently
influenced subjects' responding.

Kahneman et al. (1992) next used this procedure to assess the factors
that influence adults' representation of identity and change over space
and time. Like infants, adults' representation of identity was influenced by
information about the cohesiveness and continuity of object motion and
not by the constancy or change of an object's shape, color, or category
membership. When a set of visual elements appeared or moved continuously together, adults saw them as a single unit. Experiments using other
chronometric methods suggest that such object representations survive
periods of occlusion (Tipper, Brehaut, and Driver 1990), as do the object
representations studied in infants. These experiments suggest that a basic
process for representing the boundaries of objects and for tracing object
identity through time exists in adults, independent of processes for cate-
gorizing objects as instances of known kinds. This process bears some
resemblance to the process by which infants perceive objects.

8.5.3   Intuitions about Identity

The most extensive discussions of human intuitions about object boundaries and persistence have taken place within philosophy and linguistics.

These discussions appear to pose serious problems for the view that mature apprehension of objects depends on any general process.

Consider, for example, Cornell University. Much of the university is located in Ithaca, but its medical school is in New York City. Because Cornell is located both in Ithaca and in New York City but not along any route that connects these places, it is not a cohesive object. (Neither is the United States.) Consider, next, a church that occupies an old building at one time and then moves to a new building (Hume 1739). Because no continuous motion connects the old building to the new, the church is not a spatiotemporally continuous object. (Neither is Estonia.) Consider, finally, a duck. Like all animals, a duck can direct its motion from within, guided by its perception of objects at a distance. At the level of perceptible objects, therefore, the duck's motion is not subject to the contact principle.

These and other examples suggest that concepts such as *university*, *church*, and *duck* come to influence our intuitions about object persistence and change, overcoming in some cases the principles of cohesion, continuity, and contact. It is possible, however, that the latter principles continue to guide our intuitions about objects as well. In learning new categories of objects, we may be strongly predisposed to single out, as potential category members, things that are cohesive, continuous, and subject to the constraints of contact mechanics. Moreover, our intuitions about the boundaries and persistence of objects in any category may continue to be influenced by these principles: we may tend to conceive of physical bodies as persisting as long as they maintain their cohesiveness, continuity, and powers to act on contact with other objects, and we may overcome this predisposition only with difficulty and in special cases.

A recent study provides support for this view (Gutheil, Spelke and Hayes, 1995). College students read scenarios in which a table or a ship was introduced and then a transformation of the object was described. In different scenarios, the transformation changed the object's appearance (for example, the table's shape was changed from rectangular to round), its function (for example, the table was used as a piano bench) or its cohesiveness (for example, the table was cut into pieces and then reassembled). A number of questions probed subjects' intuitions about the effect of each transformation on the object's existence and category membership.

Although judgments about the persistence of an object sometimes coincided with judgments about its category membership, these judgments diverged in many cases. For example, 60 percent of the subjects judged that the table that emerged from the disassembly/reassembly process was not the same object as the original table, whereas less than 5 percent of the subjects judged that it was not the same kind of object. Conversely, 76 percent of the subjects judged that the table-turned-piano bench was no longer a table, whereas less than 5 percent judged that it was no longer the same individual. These findings suggest that mature, commonsense intuitions about object persistence are not inextricably tied to intuitions about object kind.

Further findings from this experiment suggest that the cohesion principle is central to intuitions about object persistence. A majority of subjects judged that the table or ship no longer existed after transformations that destroyed its cohesiveness. Over the set of transformations tested in these studies, loss of cohesion led to more frequent judgments that the object ceased to exist than did loss of original shape or function. These findings are consistent with the thesis that a general conception of objects as cohesive bodies underlies many of our intuitions about object persistence and change. Common principles, therefore, may guide representations in infants and adults, and common processes may underline both immediate perceptions and reflective judgments of object identity.

8.5.4   From Perceptual Development to Cognitive Science

Object perception is a wonderful subject for studies in cognitive science because of the long and intense tradition of work on this problem within philosophy, computer science, linguistics, and psychology. Studies of object perception in these disciplines have often proceeded independently of each other: For example, philosophers concerned with the metaphysical question of what a physical body is, or with the epistemological question of how we perceive and gain knowledge of objects, have rarely consulted research on mechanical systems for parsing visual scenes, or the reverse. The problem of how we conceive of object persistence and change has seemed quite different from the problem of how we perceive objects in immediately visible scenes.

Research on the early development of object perception is beginning to suggest, however, that philosophical investigations assessing common sense intuitions about object persistence and change, linguistic analyses of the primitive notions underlying the meanings of words and expressions, chronometric experiments probing the extremely rapid processes by which we detect changes in objects, and psychophysical and computational studies of image segmentation and object recognition may reveal common cognitive capacities. All of these abilities may be rooted in a single system of knowledge that emerges early in human infancy and underlies our first perceptions of objects. Studies within all these disciplines, in concert with studies of early perceptual development, may serve to probe the nature of this knowledge system.

Suggestions for Further Reading

An excellent introduction to the philosophical problems of object perception is provided by Hirsch 1982, who proposes a solution to the problem of physical identity that combines general rules and category-based perspectives. For a briefer introduction to this and other problems of metaphysics from a cognitive science perspective, see Goldman 1993.

On the problem of dividing the world into objects from the perspective of computational vision, see Marr 1982 (Chapter 3). Marr suggests that the parsing problem is solved by processes of model-based object recognition; general discussions of such procedures can be found in Lowe 1985 and Ullman 1989. For rule-based approaches to the parsing problem similar to the approach of the Gestalt psychologists, see Witkin and Tenenbaum 1983 and Palmer and Rock 1994.

A good original source on Gestalt psychology is Koffka 1935. Excellent modern discussions of the Gestalt approach to object perception are provided by Hochberg 1974 and Rock 1983. For an example of a contemporary approach to perception within the Gestalt tradition, see Kanizsa 1979. Hatfield and Epstein 1985 provide a general discussion of the Gestalt simplicity principle.

Atkinson and Braddick 1989 provide a good introduction to the development of basic visual functions in infancy, including sensitivity to contrast and color, binocularity, and eye movement control. A collection of recent papers on perceptual and cognitive development in infancy can be found in Granrud 1993. In particular, the chapter by Kellman discusses the early development of perception of objects and motion, and the chapter by Arterberry, Craton, and Yonas discusses intriguing recent work on perception of depth and occlusion. On infants' sensory and perceptual capacities, see Salapatek and Cohen 1987. On methods of studying perception in infancy, see Gottlieb and Krasnegor 1985.

Although our chapter focused exclusively or perception of inanimate objects, considerable research has investigated infants' perception of human faces. Johnson and Morton 1991 discuss this research from both biological and cognitive science perspectives.

It is likely that early-developing abilities to perceive objects underlie the child's earliest learning of words for objects. For research on early word learning that addresses this possibility, see Markman 1989. For analyses of the conceptual primitives underlying mature language, see Jackendoff 1983 and Volume 1 of the present series.

Problems

8.1   Some investigators have proposed that young infants learn to see partly occluded objects as continuing behind their occluders by repeatedly experiencing visual arrays in which objects move in and out of view (e.g., Edelman 1987; Smith and Katz, in press). What preexisting capacities must an infant have, in order for such learning to be possible? How could one study whether these experiences are necessary?

8.2   Just as a major task of the young child is to acquire the lexicon of his or her language, a major task of early childhood is to acquire a vocabulary of object categories and models for object recognition. How could one investigate when children begin to develop models of objects, and (more important) the nature of the models they develop? Note that demonstrations that children respond to a given array as familiar, or that they respond to a group of arrays as similar to one another, are not sufficient to establish that children carve the arrays into objects and categorize each object as a member of a certain kind.

Questions for Further Thought

8.1   Goodale (in Chapter 5) discusses two distinct visual pathways involved in the representation of objects. Which of these pathways do you think might underlie infants' perception of objects? How could the functional separation of the pathways underlying object recognition and object-directed action be investigated further with infants? How might the general process for perceiving objects, independently of processes of object recognition, be studied with neurologically impaired patients?

8.2   Studies of normal and neurologically impaired adults suggest that face recognition depends on distinct mechanisms from other forms of object recognition (Chapter 3). Would you expect infants to show the same perceptual abilities (and limitations) perceiving faces as they show perceiving inanimate objects?

8.3   Pashler (in Chapter 2) discusses evidence that in certain situations people divide attention more easily among two features of one object than among two features belonging to two different objects. What do you think defines a single *object* in these experiments? For example, suppose subjects were presented with a map of the world and asked to detect events that could occur in either of two places. Would attention divide as easily between events in Hawaii and California as between events occurring in two equally distant parts of the continental United States? Which pair of events would be detected faster: events in Washington and Alaska, or events in Washington and British Columbia?

References

Arterberry, M. E., L. G. Craton, and A. Yonas (1993). Infants' sensitivity to motion-carried information for depth and object properties. In C. E. Granrud, ed., *Visual perception and cognition in infancy,* vol. 23, *Carnegie-Mellon Symposia on Cognition.* Hillsdale, NJ: L. Erlbaum.

Atkinson, J., and O. Braddick (1989). Development of basic visual functions. In A. Slater and G. Bremner, eds., *Infant development.* Hillsdale, NJ: L. Erlbaum.

Banks, M. S. and E. Shannon, (1993). Spatial and chromatic visual efficiency in human neonates. In C. E. Granrud, ed., *Visual perception and cognition in infancy.* Hillsdale, NJ: Erlbaum.

Clifton, R. K., P. Rochat, R. Litovsky, and E. Perris (1991). Object representation guides infants' reaching in the dark. *Journal of Experimental Psychology: Human Perception and Performance* 17, 323–329.

Edelman, G. (1987). *Neural Darwinism.* New York: Basic Books.

Goldman, A. I. (1993). *Philosophical applications of cognitive science.* Boulder, CO: Westview Press.

Gottlieb, G., and N. Krasnegor, eds. (1985). *Measurement of audition and vision in the first year of postnatal life.* Norwood, NJ: Ablex.

Granrud, C. E. (1993). *Visual perception and cognition in infancy,* vol. 23, *Carnegie-Mellon Symposia on Cognition.* Hillsdale, NJ: L. Erlbaum.

Granrud, C. E., R. J. Haake, and A. Yonas (1985). Infants' sensitivity to familiar size: The effect of memory on spatial perception. *Perception and Psychophysics* 37, 459–466.

Gunderson, V. M., A. Yonas, P. L. Sargent, and K. S. Grant-Webster (1993). Infant macaque monkeys respond to pictorial depth. *Psychological Science* 4, 93–98.

Gutheil, D. G., E. S. Spelke, and A. Hayes (1995). *Revisiting the ship of Theseus: Commonsense understanding of object identity.* Manuscript in preparation.

Hatfield, G., and W. Epstein (1985). The status of the minimum principle in the theoretical analysis of visual perception. *Psychological Bulletin* 97, 155–186.

Hirsch, E. (1982). *The concept of identity.* Cambridge, MA: Harvard University Press.

Hochberg, J. (1974). Organization and the Gestalt tradition. In E. C. Carterette and M. P. Friedman, eds., *Handbook of perception,* vol. 1, *Historical and philosophical roots of perception,* 180–211. New York: Academic Press.

Hofsten, C. von, and E. S. Spelke (1985). Object perception and object-directed reaching in infancy. *Journal of Experimental Psychology: General* 114, 198–212.

Hume, D. (1739). *A Treatise of human nature.* Oxford: Clarendon.

Jackendoff (1983). *Semantics and cognition.* Cambridge, MA: Bradford/MIT Press.

Johnson, M. H. (1990). Cortical maturation and the development of visual attention in early infancy. *Journal of Cognitive Neuroscience* 2, 81–95.

Johnson, M. H., and J. Morton (1991). *Biology and cognitive development: The case of face recognition.* Oxford: Blackwell.

Johnson, S. P. and R. Aslin (in press). Perception of object unity in two-month-old infants. *Developmental Psychology.*

Johnson, S. P., and J. E. Nanez (1995). Young infants' perception of object unity in two-dimensional displays. *Infant Behavior and Development,* 18, 133–143.

Kahneman D., and A. Treisman (1984). Changing views of attention and automaticity. In R. Parasuraman and D. A. Davies, eds., *Varieties of attention.* New York: Academic Press.

Kahneman, D., A. Treisman, and B. J. Gibbs (1992). The reviewing of object files: Object-specific integration of information. *Cognitive Psychology* 24, 175–219.

Kanizsa, G. (1979). *Organization in vision.* New York: Praeger.

Kellman, P. J., H. Gleitman, and E. S. Spelke (1987). Object and observer motion in the perception of objects by infants. *Journal of Experimental Psychology: Human Perception and Performance* 13, 586–593.

Kellman, P. J., and E. S. Spelke (1983). Perception of partly occluded objects in infancy. *Cognitive Psychology* 15, 483–524.

Kestenbaum, R., N. Termine, and E. S. Spelke (1987). Perception of objects and object boundaries by three-month-old infants. *British Journal of Developmental Psychology* 5, 367–383.

Koffka, K. (1935). *Principles of Gestalt psychology.* New York: Harcourt, Brace & World.

Lowe, D. G. (1985). *Perceptual organization and visual recognition.* Boston: Kluwer.

Mandler, J. M., and L. McDonough (1993). Concept formation in infancy. *Cognitive Development* 8, 291–318.

Markman, E. M. (1989). *Categorization and naming in children.* Cambridge, MA: Bradford/MIT Press.

Marr, D. (1982). *Vision.* San Francisco, CA: Freeman.

Michotte, A. (1963). *The perception of causality.* New York: Basic Books.

Needham, A., and R. Baillargeon (1994). Object segregation in eight-month-old infants. Manuscript submitted for publication.

Palmer, S., and I. Rock (1994). Rethinking perceptual organization: The role of uniform connectedness. *Psychonomic Bulletin and Review* 1, 29–55.

Peterson, M. A. (1994). Object recognition processes can and do operate before figure-ground organization. *Current Directions in Psychological Science.* 3, 105–111.

Quinn, P. C., S. Burke, and A. Rush (1993). Part-whole perception in early infancy: Evidence for perceptual grouping produced by lightness similarity. *Infant Behavior and Development,* 16, 19–42.

Quinn, P. C., P. D. Eimas, and S. L. Rosenkrantz (1993). Evidence for representations of perceptually similar natural categories by 3-month-old and 4-month-old infants. *Perception* 22, 463–476.

Rock, I. (1983). *The logic of perception.* Cambridge, MA: MIT Press.

Salapatek, P., and L. Cohen (1987). *Handbook of infant perception,* vol. 1, *From sensation to perception.* Orlando, FL: Academic Press.

Slater, A., V. Morison, M. Somers, A. Mattock, E. Brown, and D. Taylor (1990). Newborn and older infants' perception of partly occluded objects. *Infant Behavior and Development* 13, 33–49.

Smith, L. B., and D. B. Katz (in press). Activity-dependent processes in perceptual and cognitive development. To appear in R. Gelman and T. Au, eds., *Handbook of Perception and Cognition: Cognitive Development.*

Spelke, E. S. (1988). The origins of physical knowledge. In L. Weiskrantz, ed., *Thought without language.* New York: Oxford University Press.

Spelke, E. S., K. Breinlinger, K. Jacobson, and A. Phillips (1993). Gestalt relations and object perception: A developmental study. *Perception* 22, 1483–1501.

Spelke, E. S., R. Kestenbaum, D. Simons, and D. Wein (1995). Spatio-temporal continuity, smoothness of motion, and object identity in infancy. *The British Journal of Developmental Psychology.* 13, 113–142.

Spelke, E. S., and G. Van de Walle (1993). Perceiving and reasoning about objects: Insights from infants. In N. Eilan, R. McCarthy, and W. Brewer, eds., *Spatial representation.* Oxford: Basil Blackwell.

Teller, D. Y., and M. H. Bornstein (1987). Infant color vision and color perception. In P. Salapatek and L. Cohen, eds., *Handbook: of infant perception,* vol. 1. *From sensation to perception.* Orlando, FL: Academic Press.

Tipper, S. P., J. C. Brehaut, and J. Driver (1990). Selection of moving and static objects for the control of spatially directed action. *Journal of Experimental Psychology: Human Perception and Performance,* 16, 492–504.

Ullman, S. (1989). Aligning pictorial descriptions: An approach to object recognition. *Cognition 32,* 193–254.

Van de Walle, G. A., and E. J. Spelke (1993). L'integration spatio-temporelle dans la perception des objets chez le bebe. *Psychologie Francaise,* 38, 75–83.

Waxman, S. R., and D. G. Hall (1993). The development of a linkage between count nouns and object categories: Evidence from fifteen- to twenty-one-month-old infants. *Child Development* 64, 1224–1241.

Wertheimer, M. (1923). "Principles of perceptual organization." *Psychologische Forschungen* 4, 301–350. (A translation appears in D. C. Beardslee and M. Wertheimer, eds., *Readings in Perception.* New York: Van Nostrand.)

Wiggins, D. (1980). *Sameness and substance.* Oxford: Basil Blackwell.

Witkin, A. P., and J. M. Tenenbaum (1983). On the role of structure in vision. In J. Beck, B. Hope, and A. Rosenfeld, eds., *Human and machine vision,* 481–543. New York: Academic Press.

Xu, F., and S. Carey (in press). Infants' metaphysics: The case of numerical identity. *Cognitive Psychology.*

Xu, F., and S. Carey (1994). Infants' ability to individuate and trace the identity of objects. Paper presented at the International Conference on Infant Studies, Paris, June.

Yonas, A., and C. E. Granrud (1985). Reaching as a measure of infants' spatial perception. In G. Gottlieb, and N. Krasnegor, eds., *Measurement of audition and vision in the first year of postnatal life.* Norwood, NJ: Ablex.

Chapter 9
Meaningful Perception
Fred Dretske

One can perceive an object without knowing or understanding what it is. One sees, one might even taste, a poisonous mushroom without realizing that it is poisonous—perhaps even without knowing it is a mushroom. Nonetheless, despite such ignorance, despite not perceiving *what* it is, one still perceives (sees and tastes) it. The poisonous mushroom is the object one perceives.

Meaningful perception refers, not to the objects one perceives, but to *how* one perceives them. It is perception that embodies a judgment or belief, some degree of recognition or identification of what one is perceiving. Meaningful perception requires more than good eyesight. It requires the kind of conceptual skills needed to classify and sort perceptual objects into distinct categories. At the most basic level, it involves perceiving that one thing is, in certain respects, the same as another. Even the ignorant can see poisonous mushrooms. What they cannot do is what experienced observers can do: see that they are poisonous or (at a more basic level) that they are mushrooms or (a a yet more basic level) that they are the same kind of thing as those other objects.

We begin by distinguishing perception of objects (mushrooms) from perception of facts about these objects (that they are mushrooms). The perception of objects—what I will call *sense perception*—is that early phase of the perceptual process that culminates in sense experience (visual, auditory, tactile, etc.) of the object. Perception of facts about these objects, on the other hand—that which constitutes *meaningful perception*—is a more inclusive process. Besides sense perception, meaningful perception includes a knowledge (at least a judgment or belief) about the object being experienced.

9.1 Perceiving Objects and Perceiving Facts

When they speak of visual perception, it seems reasonable to suppose that cognitive scientists are referring to something we normally use the verb *to see* to describe. Seeing the cat on the sofa is to visually perceive the cat on the sofa.

To avoid misunderstanding, however, one should ask, at the very beginning, whether the term *visual perception* (or *seeing*) is to be reserved for objects, facts, or something else. After all, we normally speak of seeing objects (like cats and sofas), the properties of objects (e.g., the color of a cat, the size of the sofa), events (the cat's jumping onto the sofa), states of affairs (the cat's being on the sofa), and facts (that the cat is on the sofa). If these are all to be counted as instances of visual perception, as they appear to be in ordinary language, then care must be taken in a scientific study of visual perception to specify what is being perceived, an object, a property, an event, a state of affairs, or a fact. For it is not at all clear that the same processes, mechanisms, and results are, or need be, involved in the perception of these different things. Quite the contrary.

Consider, for example, a small child glancing at the sofa and mistaking a sleeping cat for an old sweater. Does she see an object? Yes, of course. Besides the sofa there is an object, the black cat on the sofa, that the child mistakenly believes to be a black sweater. Although she does not recognize the cat (*as* a cat), she must, in some sense, see the cat in order to mistake it for a sweater. Nevertheless, though she sees a black cat on the sofa, sees an object fitting the description of a cat, she does not realize that this is a correct description of what she sees. She thereby fails to see the corresponding fact: that there is a black cat on the sofa. Shall we say, then, that the child perceives the black cat on the sofa? The answer to this question will obviously depend on whether one is thinking of objects (black cats) or facts (that they are black cats).

One can, of course, merely stipulate that visual perception is to be reserved for the way of seeing objects that involves, in some essential way, knowledge of the object seen. So when a child—or, indeed, any other kind of animal (an unsuspecting mouse, a bird, or even an inattentive adult)—sees a cat on the sofa without realizing what it is, without learning or coming to know that it is a cat, then this way of seeing the cat will not count as perceiving the cat. To *perceive* a cat is, according to this way of using words, to come to know, by visual (auditory, tactile, etc.) means, by the use of one's senses, that it is a cat. Perception is perceiving facts of the form: that object (on the sofa) is a cat.

We are free to use words as we please. There is nothing to prevent our restricting visual perception to *visual cognition*, a coming-to-know-by-visual-means. It would seem that this particular restriction is, in fact, rather widespread in cognitive psychology. Interested, as they are, in what subjects learn in their perceptual encounters with objects, cognitive psychologists tend to focus on a subject's recognition or identification of objects, ways of seeing (hearing, smelling) things that require some knowledge of what is seen (heard, smelled). So, for instance, recognizing a geometric figure as a triangle requires the subject to realize, to come to know, upon seeing it, that it is a triangle. If he or she, upon seeing it, doesn't know what kind of figure it is, doesn't at least distinguish it from other sorts of geometric figures, then he or she does not really perceive (i.e., recognize) the triangle. *Recognitional perception*, in turn, is a way of seeing facts of the sort: *that* is a triangle.

We are, as I say, free to use words as we please. But this proposed restriction of *visual perception* to the perception of facts, to recognition, to a way of seeing things that requires a knowledge of the thing seen, has unfortunate consequences. For we now have no convenient way of describing the child who mistakes the cat for a sweater. Since the child doesn't know it is a cat, the child does not, in this way of using words, perceive the cat. What, then, is the relation between the child and the cat? The child isn't blind. Light rays, reflected from the cat, are entering the child's eyes and, in some perfectly normal way, causing within the child a visual experience that would be quite different if the cat weren't there. This being so, it seems most natural to say, from a commonsense standpoint, that the child sees the cat but doesn't realize that this is what she sees. If, because of the way we have decided to use the word perception, this doesn't count as perceiving the cat, it must surely count as seeing the cat. Using the word *perception* in this restricted way, then, would have the unfortunate result of not letting us classify, as visual perception, a person seeing cats in normal circumstances.

It seems preferable, therefore, to distinguish between seeing objects and seeing facts, not (as above) by artificially reserving the word *perception* for one way of seeing, the way of seeing that requires knowledge of the thing seen (i.e., seeing facts), but rather by distinguishing two forms of perception, two ways of seeing. We are then free to speak of seeing objects (like black cats) without necessarily realizing (knowing or believing) what they are, and, on the other hand, a recognitional way of seeing objects, seeing facts about them, that requires some level of identification. I will call the first *sense perception* (of objects) and the latter *meaningful* or (sometimes) *cognitive perception* (to emphasize the involvement of such factors as memory and conception in the process). This brings our use of the term *perception* (including as it now does both meaningful and sense perception) into closer harmony with such ordinary verbs as *see, taste,* and *hear.* It allows us to preserve the important distinction between seeing black cats and tasting poisonous mushrooms on the one hand (sense perception) and seeing and tasting what they are—that they are black cats and poisonous mushrooms —on the other (meaningful perception).

Given this way of using words, we can now describe the work of cognitive scientists as studying the processes underlying these forms of perception. It seems reasonable to assume, for instance, that the early- and intermediate-level processes described in Chapters 1 and 2 are essential to our perception of objects (and, therefore, whatever facts we end up perceiving about these objects). Later vision, the sort of processes described in Chapters 4, 5, and 8, however, applies only to meaningful perception. If this is so, then one must be careful not to conclude that there is no perception of an object simply because there is no meaningful perception of that object. Perception of facts about objects (meaningful perception of objects) may require object perception, but the converse is not true. *Object perception*, the early phase of the perceptual process, may well occur without a normal cognitive (meaningful) completion. Indeed, there are certain visual deficits (agnosias) in which there is perception of objects without any meaningful recognition or identification (see Chapter 5; also Farah 1990).

A number of different issues require a clear separation of sense perception from cognitive or meaningful perception. As we shall see, debates about whether perceptual processes are top-down or bottom-up, about whether they are inferential or constructive in character, about whether they are massively parallel or sequential, and about their comparative modularity, are topics that can be given sharper focus by distinguishing the kind of perception—sensory or meaningful—the debate is a debate about. Discussions of perceptual learning and development (see Chapter 8), including questions about whether perception is relative to linguistic and conceptual factors (topics to which I will return), can also benefit by a close observance of the difference between perceiving objects and perceiving facts about them.

It is for this reason that throughout this chapter I consistently distinguish sensory from meaningful perception of objects. The first, once again, is a way of perceiving cats, mushrooms, and triangles that does not require (though it may in fact be accompanied by) knowledge, judgment, or belief that it is a cat, a mushroom, or a triangle that is seen. This is what I have been calling object perception. Cognitive or meaningful perception of cats, mushrooms, and triangles will be reserved for that way of perceiving these objects that necessarily involves a coming to know, a cognition of some fact about the object being perceived. This cognition may occur at various levels of specificity. At the most basic level it is simple discrimination or generalization—that this differs from, or is the same as, that. One may see a cat and recognize it as Felix, the neighbor's cat. Or simply as a cat. Or, less specifically, as an animal of some sort. Or, simply, as the kind of thing one has seen before (familiarity). Meaningful perception is a cognitive skill that, like all skills, is a matter of degree. Two people can perceive the same object but perceive quite different facts about it. One person may see more facts, or perhaps more significant facts about it than another person. The number and importance of the facts perceived is a measure of the perception's meaningfulness.

Perceiving an object *as* a so-and-so (a cat as a rumpled sweater, for instance) is a hybrid form of perception. It falls somewhere between sense perception and meaningful perception. Like meaningful perception it requires a fairly specific cognitive or judgmental attitude or tendency on the part of the perceiver: the perceived object is classified or identified in some way. Unlike meaningful perception, however, this judgment or belief need not qualify as knowledge or recognition. The judgmental outcome of the perception need not be veridical. One sees a stick (object perception) as a snake—thus, coming to believe (mistakenly) that it is a snake. The stick, obviously, doesn't have to be a snake for one to see it as a snake. Hence, this is not recognition (not, at least, if we think of recognition as a form of knowledge). It is, however, a species of identification—*mis*identification. One perceives a stick and misidentifies it as, mis-takes it to be, a snake. The knowledge required for meaningful perception (perceiving that the X one sees is an X) is here replaced by some variant of belief: one believes, or is inclined to believe, or would believe if one didn't know better, of the object (it may or may not be an X) that it is an X. For present purposes, although the outcome need not be veridical, we will take perceiving-as to be a form of meaningful perception because it requires some level or conceptualization of the object being perceived. The behavioral consequences of perceiving something as a poisonous mushroom (whether or not it is a poisonous mushroom) are the same as meaningful perception of the fact that something is a poisonous mushroom. One doesn't eat it.

9.2   What We Perceive: The Objects of Perception

Meaningful perception exhibits a hierarchical structure. Many, perhaps most, of the facts we come to know by perceptual means are mediated in some way. Our perception of fact F_1 depends on, and derives from, our perception of fact F_2. We see that we need gas (F_1) by seeing that our fuel gauge registers empty (F_2). We see that there is a finch at the bird feeder (F_1) by observing the distinctive color (F_2), markings (F_3), and profile (F_4) of the bird. In such cases, we see one fact (that we need gas, that the bird is a finch) by seeing other facts (that our gauge registers empty, that the bird has a reddish coloration, etc.). This is a pervasive and common phenomenon. We see *by the newspapers* that there has been a tragic plane crash, *by the tracks* that the animal went this way, *by her frown* that she is displeased, and *by the thermometer* that the patient has a fever.

Given this dependence of some visually known facts on other visually known facts, the question naturally arises whether some facts are basic, in the sense of being known directly and without this kind of dependence on other visually known facts. If my perceptual knowledge that the bird is a finch is derived (or somehow depends on) my perceptual knowledge of its distinctive features, if my knowledge of what other people are thinking and feeling is somehow derived or inferred from what I see of their observable behavior and expression, are the latter pieces of knowledge themselves derived from some yet more fundamental, even more basic, kind of knowledge? Is it possible, for instance, that one's perceptual knowledge of objective conditions is derived from, or dependent on, a more fundamental form of perceptual knowledge about how such things affect the senses or the brain—on how things subjectively seem to be? If, when I recognize a finch, my perceptual knowledge that it is a finch is ultimately based on what features I recognize the bird to have, isn't my perceptual knowledge of these features ultimately based on how the light reflected from the bird affects my eyes and, eventually, my brain? If this is so, then all meaningful perception of the objective world would seem to depend on a prior knowledge of subjective events—how things seem. Meaningful perception of the world would thus begin with meaningful perception (or awareness) of the self (the brain).

These speculations are about meaningful perception, about the structure of fact perception. Are there some facts we know that are fundamental—*foundational* as philosophers sometimes like to put it—in the sense that all other things we know are derived from them? Is our knowledge of reality —that is a finch—derived from, dependent on our knowledge of appearances—that it looks like a finch?

The answer to this question depends on the answer to a somewhat different question, a question about sense perception. What objects do we see? Do we see cats, sofas, birds, newspapers, and people? If not, then it would seem that our knowledge of these things, the fact, for instance, that the newspapers said there was a plane crash and the fact that Susan is frowning (the facts on the basis of which I know there was a plane crash and that Susan is upset) must derive from our factual knowledge about yet other things—objects that we do see. From the point of view of common sense, my knowledge of the plane crash derives from my knowledge of the newspapers. I did not see the plane crash; I saw only the newspaper. Hence, whatever facts I learn about the plane crash, including the fact that there was a plane crash, must derive from facts I learn about the newspaper —that it said there was a plane crash. But if it should turn out, for whatever reason, that, contrary to common sense, I do not directly see the newspaper itself but only (let us suppose) a subjective image of the newspaper, then whatever facts I come to know about the newspaper must derive from facts I come to know, through perception, about this image.

What facts we see, and which of these facts are fundamental, depend, therefore, on what objects we see. If you don't see the gas tank, then your visual knowledge of the tank, that it still contains gas, must derive from your visual knowledge about whatever objects you do see—in this case, typically, your fuel gauge. You see that you have some gas left by seeing what the gauge registers, and this dependence among meaningful perceptions (your knowledge of the gauge being primary) derives from facts about sense perception, from the fact, namely, that it is the gauge, not the tank, that you see.

Even when we speak of perceiving one object *by*, *through* or *in* perceiving another—in the way we speak, for instance, of seeing the game on television or seeing someone in a movie (or photograph)—knowledge of the game or person is secondary relative to our knowledge of the electronic or photographic image. Insofar as we regard the image appearing on our television or movie screen as the *primary*, or real, object of perception, we regard facts about these images as cognitively primary. Facts about the people and events being represented are *secondary*. So, for instance, we learn (see) that a player scored a goal by observing the behavior of the electronically produced images of the player, the ball, and the goal posts appearing on our television screen. We could be seeing that he scored a goal hours or even days after he scored it (tape delay).

Hence, a question about the structure of cognitive perception—whether, in fact there is a fundamental level of visual knowledge, and, if so, whether this is knowledge of objective or subjective facts—awaits the answer to a prior question: what is the structure of sense perception? What objects do we see? The answer to this question will constrain, if not fully determine, the answer to the questions about cognitive perception. If we do not see physical objects, if we are (in sense perception) always aware of mental images (representations) of external objects (as some philosophers and psychologists believe), then our knowledge of objective reality (if, indeed, we have such knowledge) will necessarily be derivative from and secondary to our knowledge of our own mental states.

Discussions of these issues are often clouded by failure to appreciate the difference between meaningful perception and sense perception. It is sometimes argued, for instance, that we do not perceive ordinary physical objects because, for whatever reason (the reason is usually philosophically skeptical in character) we do not know, or cannot know for certain, that there are physical objects. For all we know, all experience, even the experience we take to be of a real external world, may be illusory. It could all be a dream. The facts about what is out there (in the objective world) are beyond our perceptual grasp. All one can be directly aware of is the experience itself, the sort of images and representations that occur in both dreams and (if there are any) veridical perceptions.

This argument, although it has a distinguished history, confuses meaningful perception of objects (seeing that there is a cat on the sofa) with sense perception of objects (seeing a cat on the sofa). It mistakenly assumes that perception of objective reality requires meaningful perception of that reality, that perceiving the cat on the sofa (object perception) cannot take place unless there is perception of the fact that there is a cat on the sofa (meaningful perception of the cat). This doesn't follow. One doesn't have to know, let alone know for certain (whatever that might mean), that there are physical objects in order to see (sense perception) physical objects. Such knowledge is only required for the perception of the fact that there are such objects. Just as the child, described above, saw a cat on the sofa without knowing what it was, it may turn out that we see ordinary physical objects (including cats and sofas) every moment of our waking life without ever being able to know (if the skeptic is right) that this is what we are seeing. Questions about what objects we perceive are quite distinct from questions about what facts we perceive.

Failure to maintain the distinction between the sensory and the meaningful perception of objects also tempts students into mistakenly supposing that if our knowledge of physical objects somehow depends on the way they appear to us, on the way they look, then, what we really perceive, what we are directly aware of, when (as we commonly say) we see a cat, is an internal mental image of the cat. We see (as it were) the look or appearance of the cat. Such an inference is fallacious because even if cognitive perceptions rest on subjective foundations (on the way things look to us), our sense perceptions need not rest on similar foundations. Even if we know that there are physical objects by the way they appear to us (so that meaningful perception has, in this sense, a subjective basis), our sense perception of objects might itself be direct and unmediated. Even if we come to know (see) it is a cat (a fact) by the way it looks, the only object we actually see may be the cat itself (not its appearance).

Aside from the these possible confusions, there are various positions that have been, and continue to be taken on the directness of perception. Although these theories, in both their classical and modern form, are often hard to classify because of their failure to be clear about whether it is fact or object perception they intend to be talking about, they can be roughly characterized as follows.

Direct realism (sometimes said to be the view of the person-on-the-street) holds that (1) there is a real (hence, *realism*) physical world whose existence is independent of our perception of it; and (2) under normal conditions observers are, in a direct and unmediated way, perceptually aware of the objects and facts that constitute this world (hence, *direct* realism). The direct realist holds that the objects we are directly aware of in sense perception (unlike headaches or afterimages) continue to exist when we are no longer aware of them.

Representative realism (also called the causal theory of perception) shares with direct realism (and common sense) the first of these two doctrines; that is, that there is an independently existing reality. It disagrees about the second. According to representative realism, our perception of physical objects is indirect, mediated by a more direct apprehension of internal representations (hence the name *representative* realism) of external physical objects. These mental representations have been given various names: sensations, ideas, impressions, percepts, sense data, and experiences. But the idea is almost always the same. Just as we see what is happening on a distant playing field by seeing what is happening on our television screen (so that our knowledge of the game, when seen on television, is indirect), so knowledge of even the most obvious physical fact, the fact, say, that there is a table (or, indeed, a television set) in front of us, is itself indirect. We see that there is a table in front of us by seeing or (if *seeing* is the wrong word to use with respect to mental events) being aware of the table's internal representation. When watching a game on television, then, our knowledge of the actual game is doubly indirect: we know about a game occurring one thousand miles away by knowing what is happening on a television screen a few feet away; and we come to know what is happening a few feet away by becoming aware of what is happening (presumably no distance away) in our own minds. In the last analysis, then, all our knowledge of objective (physical) fact, all meaningful perception, rests on a knowledge of subjective (mental) fact because the only objects directly perceived are mental: the appearances (percepts, sense data).

Going beyond these forms of realism are various forms of *idealism* (sometimes called *phenomenalism*), theories that deny an objective physical reality altogether. Everything that exists depends for its existence (like a headache or an afterimage) on someone's awareness of it; hence, everything is in the nature of a mental entity like an idea (hence, *idea*lism). Since these extreme views have few, if any, serious advocates within the philosophical (not to mention cognitive science) community today, I leave them without further comment.

As indicated earlier, one might be a direct realist on sense perception but an indirect (representational) realist on meaningful perception. The objects we see are physical objects, but we know about them through their effect on us (the way they appear to us) in sense perception. The problem with this mixed position is that of how one might come to know how objects look—which, according to some theorists, is a knowledge of how, in sense perception, we internally represent them—without thereby becoming aware of (thus perceiving) the internal representations themselves (thereby becoming an indirect realist on sense perception also). To put it crudely, how can one know how things look without perceiving, or somehow being aware of their look?

The debate between direct and indirect realists becomes very technical at this point. Indirect realists maintain that we are directly aware of mental objects—images—in hallucinations and dreams. Aside from the cause of the experience, however, there is no reason to distinguish between these illusory experiences and our ordinary veridical perceptions of (physical) objects. In both cases we are directly aware of the internal mental representation. When we speak, as we commonly do, of seeing an ordinary object (like a cat) we are, if we speak truly, being caused to experience some catlike image by a real cat (a real cat that we do not directly perceive). When we hallucinate or dream of a cat, there is no such external cause; hence, we speak of these experiences as illusory. In all cases, though, it is the image that we directly experience; only the cause of the experience is different. Direct realists try to counter this, and related, arguments by insisting that although sensory perception of real objects requires the having (and thereby the existence) of internal mental representations, and though such representations in fact determine the way these objects look or appear to us, there is no reason to suppose that we perceive these representations themselves. We perceive a cat by (internally) representing a cat, yes, but not (the direct realist argues) by perceiving (being aware of) an internal cat representation.

9.3   How We Perceive: The Perceptual Process

The debate about the objects of perception is related to a debate (not always clearly distinguished from it) about the kind of processes underlying perception. Do perceptual processes exhibit the properties of intelligence? Do they have an inferential or computational character, moving from premises about the proximal stimulus to perceptual conclusion (deductive reasoning), or from data about what is happening at or in the sensory transducers to explanatory hypothesis (inductive reasoning), in something like the way human agents consciously reason and solve problems? When I see a cat on the sofa, or that there is a cat on the sofa, does my visual system do something similar to what clever detectives do when they infer, on the basis of certain clues and signs, that a certain state of affairs not directly apprehended must be the case?

We can, of course, metaphorically describe the operations of anything, even the simplest machine, in thoughtlike, semicognitive terms. We are especially fond of doing this with computers. We say they know, that they remember, recognize, infer, and conclude. If one counts arithmetical operations as forms of computation, even dime-store calculators perform (or are described as performing) impressive feats of reasoning—multiplying, taking square roots, and calculating percentages in fractions of a second.

We even speak of such comparatively humble devices as thermostats and electric eyes in such quasi-perceptual terms as, for example, *sensing* a drop in room temperature or the approach of a person and responding by turning the furnace on or opening a door. The question, then, is not whether we can speak this way, not even whether it is sometimes useful to talk this way (to adopt what Dennett 1987 calls the intentional stance), but whether this is anything more than a metaphorical device, a figure of speech that conceals our ignorance about underlying causal processes and mechanisms. Do visual systems ever literally solve problems, infer that something is so, formulate (on the basis of sensory input) hypotheses about the distant source of stimulation in the way that rational agents do this at the conscious level?

Helmholtz, the great nineteenth-century physiologist, thought so, and many investigators today (see, for example, Gregory 1974a, 1978; Rock 1977, 1983; Ullman 1980) are inclined to agree. At least they view the processing of visual information as a form of problem solving and, hence, as a form of reasoning that, though unconscious, exhibits enough of the essential properties of fully rational thought and judgment to make it, in a fairly literal sense, an instance of problem solving itself. The light reaching the receptors (sometimes called the *proximal stimulus*) carries information —fragmentary and impoverished (and thereby ambiguous) information to be sure, but information nonetheless—about distant situations (the *distal stimulus*). The visual system's function is to take this data and to construct, as best it can, a reasonable conjecture (hypothesis, judgment) about the distal source of this stimulation. It begins with premises describing receptor activity—data concerning the distribution and intensity of energy reaching the receptor surface—and is charged with the task of arriving at useful conclusions about the distal source of this stimulation. The conclusions it reaches (e.g., it must be a cat out there causing this pattern of retinal activity) constitutes the subject's perception of a cat. If the visual system reaches a different conclusion—that, for instance, it is probably an old black sweater—the subject sees an old black sweater instead of a fluffy black cat. If the perceptual system can't make up its mind, or keeps changing its mind—it's a cat; no, on second thought, it's probably a sweater; no, that can't be right, it's probably a cat—the subject sees first a cat, then a sweater, then a cat again. Such flip-flopping seldom occurs when we are looking at real cats (because, in normal circumstances, light from real cats is generally richer in information—hence, less ambiguous— about the kind of object that has structured the light). It does sometimes happen, however, with specially constructed figures viewed under restricted (e.g., monocular) conditions—Necker cubes, for instance. Since so much emphasis is placed on the visual system's efforts at constructing a reasonable interpretation or hypothesis (about the distal stimulus) from information reaching the receptor surfaces, this approach to perceptual processing is often described as a *constructivist* or *computational* approach to visual perception.

Since constructivists regard sensory stimulation, even in the best of viewing conditions, as inherently ambiguous (there is always a variety of distal arrangements that could have produced whatever pattern of proximal stimulation arrives at the receptor surfaces), they view perceptual processing as primarily a matter of adding information to the stimulus (or supplementing the information available in the stimulus) to reach a perceptual outcome: seeing a cat. Since the proximal stimulation does not unequivocally specify the distant object as a cat, and since we nonetheless (under optimal viewing conditions) see a cat (the visual system reaches this conclusion), the perceptual system must exploit some other source of information to reach this judgment, adding or supplementing (through some inductive inference) the information contained in the stimulus.

There has been a vigorous challenge to this (more-or-less) orthodox position in the last forty years. Gibson, in a series of influential books (1950, 1966, 1979) and articles (1960, 1972) has argued that the stimulus, properly understood, contains all the information needed to specify the distal state of affairs. If the proximal stimulus is understood, not as a static distribution of energy occurring on the receptor surfaces at a time, but as the total dynamic pattern of stimulation reaching a mobile observer over time, there is no need for inference, reasoning, and problem solving. There is sufficient information in the stimulus (thus broadly conceived) to specify (i.e., unambiguously determine) the character of the distal object. Why reason about what is out there when the stimulus tells you what is out there? Why suppose, as constructivists do, that perceptual systems are smart detectives when all they really have to be (given reliable informants, i.e., information-rich stimuli) is good listeners, good extractors of the information in the signals reaching the receptors? Since this approach tends to eliminate all intervening cognitive (indeed, all intervening psychological) mechanisms from the processes resulting in our perception of objects, it is often referred to as a direct theory of perceptual processing.

Relevant to the question of whether perceptual systems are more like intelligent detectives doing their best with ambiguous data (*constructivism*) or more like good listeners faithfully registering stimulus information (*direct theory*) is what Fodor (1983) describes as the modularity of information-processing systems. A system is (comparatively) modular when it is (comparatively) insulated from information available to other parts of the total system. If I am told (and thereby know) it is an animal on the sofa, for instance, does this, can this, affect my visual perception of the cat? If not, my visual system exhibits modularity with respect to this kind of information (information available to the central processor from auditory sources).

If this collateral information is capable of affecting the way what I see looks to me, then the visual system (understood as that subsystem responsible for my seeing what I see) is not modular in relation to this kind of information.

If the visual system is modular, its operation (and therefore presumably what the subject perceives) is unaffected by what other information may be available to other parts of the system (or what the subject may know as a result of information received from these other parts). Modular systems are, therefore, described as stimulus driven (the processing is bottom-up rather than top-down): it is the stimulus itself (information at the bottom, as it were), not the system's (possibly variable) hypotheses about that stimulus (information available at the top) that guides the processing of incoming signals and thereby determines what the subject perceives. Modular systems are therefore most naturally thought of in the second of the two ways described above—as good extractors of preexistent information, information that is already in the stimuli, not as good detectives or problem solvers about the best interpretation of informationally ambiguous stimuli. There is no point in supposing a process of reasoning is occurring in modular systems when the process, being modular, is not allowed to use information (other than what is in the stimulus itself) to generate perceptual conclusions. Modular systems are not intelligent. They don't have to be. They have no problems to solve. They just do what the stimulus tells them to do.

It is by no means obvious that these two approaches to the analysis of perceptual processes are incompatible. It may turn out, for example, that although the stimulus, properly understood, is rich in information about distal objects, rich enough (let us suppose) to unambiguously determine what distal objects produced it, it still requires inferential (reasonlike) processes to decode the signal, to extract this information from the stimulus. Fingerprints, being unique to their bearers, may unambiguously determine or specify (in an information-theoretic sense) who held the gun. It nevertheless takes a good deal of problem solving after one has discovered the incriminating prints to figure out who held the gun. One has to know which people go with which prints, and this may take memory, inference, and prior learning (the sort of cognitive work that organizations like the FBI invest in the creation of a fingerprint file). As Ullman (1980) puts it, the role of processing may not be to create information but to extract it, integrate it, make it explicit and usable.

There are, then, a variety of ways of expressing questions about the nature of those processes underlying our perception of the world. But these questions should not be confused, as they often are, with questions about the objects of perception (the questions discussed above in section 9.2). Gibson's views have been described (by both Gibson and others) as a theory of direct perception. This can be misleading. It certainly is confusing. The sense (if any) in which this theory is direct is much different from the sense in which direct realism (as described above) is direct. Direct realism is a theory about the objects of perception, about what we see. The kind of direct realism we are now talking about, the kind associated with Gibson's work, is a theory about the processes underlying perception, about how we see what we see. There is no reason why one cannot be a direct realist about the objects of perception, holding that we directly perceive physical objects (not sensations or mental intermediaries), and remain a constructivist about the processes underlying our (direct) perception of these objects. One can suppose that intelligence, some kind of thoughtlike process, is involved in the construction of internal representations without supposing that one thereby sees (or in any way perceives or becomes aware of) the constructed representations. One can, in other words, be a direct realist about the objects of perception and an indirect realist, a constructivist, about the processes underlying this direct relationship.

Once again, however, controversy about the intelligence (or lack of it) of perceptual processes is often muddled by failure to be clear about exactly which processes are in question. It should be obvious that cognitive perception, our perception of facts, our seeing that (and hence coming to know that) there is a cat on the sofa, is the result of a process that is strongly influenced by higher-level cognitive factors. Meaningful perception is clearly not modular. A person who does not already know what a cat is or what a cat looks like—a small child or an inexperienced animal, say—will be unable to see (recognize) what is on the sofa, unable to see that there is a cat there (to be carefully distinguished from an ability to see the cat there). For meaningful perception of the cat on the sofa, in contrast to object perception of the cat, requires not only the appropriate concepts (for cat and sofa), but some intelligence in the application of these concepts to the objects being perceived (the cat and the sofa). The upshot of meaningful perception is some known (or, at least, believed) fact (e.g., that there is a cat on the sofa); and such facts are not learned without the cooperation of the entire cognitive system. By changing a subject's cognitive set—changing what the subject knows or believes about the way things look, for instance—one easily changes what the subject learns, comes to know, hence perceives in a cognitive way, about the objects it sees (in a sensory way). Some form of constructivism or computationalism is therefore inevitable for meaningful perception, for seeing facts.

The real question is, or should be, whether that part of the visual system given over to sense perception, to seeing objects (like cats and sofas), is also intelligent. Does it exhibit some (any? all?) of the marks of rational inference and judgment? Are these early processes, those culminating in an experience of an object, modular?

The answer to this question will clearly depend on just which processes in the total perceptual process are responsible for object perception. It seems fairly safe to say that the events comprising very early vision exhibit fewer marks of intelligence than those farther "up" the visual pathway. The events occurring on the retina, for example, the activation (via a system of intermediate neurons) of ganglion cells (the axons of which form the optic nerve to the brain) by the photosensitive rods and cones is a fairly rigid, reasonably modular, phase of the total perceptual process. But how far up does one have to go to reach the percept that— conscious experience whose occurrence constitutes the (sense) perception of the object reflecting light into the eyes? If the upshot or outcome of meaningful perception is some known (or believed) fact—that there is a cat on the sofa—what or where is the terminus or culmination of sense perception? When, at exactly what stage in the processing of incoming information do we see the cat on the sofa? If recognizing the cat as a cat isn't necessary to the perception of the cat (as it is to the meaningful perception of the cat), what are the necessary elements of cat perception? Since we can see a cat at a distance, in bad lighting, or in unusual conditions (circumstances in which it does not even look like a cat), we cannot suppose, following Gibson, that to see a cat is simply to have information in the stimulus that specifies the cat as a cat. For in such cases there may be little or no information in the stimulus about what it is we see. That doesn't prevent our seeing it.

It is true, but not terribly illuminating, to be told that the perception of objects occurs at that point (if there is a point) when the visual system constructs a sensory representation of an object. For this merely raises questions about the nature of sensory representations and the difference between them and cognitive representations of the same objects. If meaningful perception of X does not occur until there is a cognitive representation of X (some kind of judgment or belief about X), exactly what is a sensory representation of X, the kind of representation whose occurrence will constitute the (sense) perception of X? Are sensory representations what philosophers and psychologists used to call *sensations* (raw, uninterpreted, sensory givens)? Or are they more like what they (or some of them) now call *percepts*, cognitively enriched (more fully interpreted) experiences of the object? Or, to use even more fashionable jargon, are sensory representations more like what David Marr (1982) and his associates call a two-and-a-half dimensional sketch of the object?

Until these questions are answered, one can expect little progress on issues concerning the nature and determinants of perception itself. How can one tell if perception of objects is best thought of in terms of a clever detective or a good listener if one cannot say what the final product, what kind of representation, this kind of perception is supposed to produce?

9.4   Perceptual Change: The Relativity of Meaningful Perception

Do we learn to perceive objects? Does prolonged experience of the world change what we perceive or, perhaps, only how we perceive it? Do people with radically different languages, and, thus, with different ways of describing the world, perceive their surroundings differently? Do completely different worldviews—what Kuhn (1962), for instance, calls incommensurable scientific theories—generate differences in what people can observe and, therefore, in the data on which their theoretical differences rest?

Such questions have fascinated philosophers and psychologists, linguists and anthropologists, for centuries. The answers to these questions are rot easy. Nevertheless, some things seem reasonably clear—if not the final answers themselves, then at least the sorts of considerations that must inform the search for final answers.

The first point, a point that has been made repeatedly in this chapter, is that before rushing in with answers to any of these questions, one should first be very clear about the question. What kind of perception is the question a question about?

As a case in point, the question about whether we learn to see things has a reasonably straightforward answer if it is a question about meaningful perception, about the facts we come to know by visual means. The first time I saw an armadillo (it was on a Texas road at dusk) I did not know what it was. Having no experience or knowledge of armadillos, being ignorant of what they were, I did not recognize what I saw as an armadillo. There was no meaningful perception of the armadillo, at least not as an armadillo. (I did not even recognize it as an animal of some kind; I thought it might be tumbleweed.) Now, however, I am quite expert in this kind of identification. When I see them, I recognize them as armadillos. Since that first encounter I have learned something about armadillos. This has changed what I believe, what I come to know, when I see them. My perception of armadillos is now much more meaningful than it was on that first occasion. Learning of this kind is a pervasive and familiar phenomena.

But, if the question about perceptual learning and development is a question about sensory perception, about the objects we see, about whether we learn to see armadillos (and not just the fact that they are armadillos), the answer appears to be quite different. I did not learn to see armadillos. I did that—saw the armadillo (though not, of course, as an armadillo)—the very first time. That, in fact, is why I swerved to avoid hitting it as it crossed the road. What I learn to do is to identify the things I see, things I (therefore) must be able to see before I identify them. Sensory perception of objects normally comes before meaningful perception of these same objects. If it didn't, there would be no way to learn what objects look like. How can you learn what objects look like if you can't see them? Human subjects do not see things at the moment of birth, of course. Certain physiological changes must occur before we can focus on objects and coherently process information contained in light. But these maturational processes are not to be classified as learning in any ordinary sense. We no more learn to see solid objects than we learn to digest solid food.

This is not to deny that there may be changes in object perception after, and as a result of, experience. Nor is it to deny that developmental processes improve the sensitivity of vision, thus enabling an adult to see better (and thus *more*) than a newborn infant. Significant changes occur in a human being's perceptual apparatus as he or she ages. The changes are sometimes peripheral. The cones in an infant's eyes, for example, are very immature. They improve several orders of magnitude by the time he or she is an adult. Adult cones simply capture more photons than immature cones (van Sluyters et al. 1990). Perhaps, too, objects start looking different after they become familiar, or after we know certain facts about them. Does a familiar face—that of a loved one, say—look different from the way it looked the first time you saw it? Do coins look larger to poor children than they do to rich children? After one has adapted to tinted or distorting glasses, does the world still look different (you just don't notice it anymore)? Or does it, once again, come to look normal (the way it looked before you put the glasses on)? Do the lines that look different lengths in the Mueller-Lyer illusion start looking the same after you learn (by measuring them) that they are the same length?

These questions are questions about the way things look, about our visual experience of them, that we earlier dubbed (without really knowing or explaining what it was) a sensory representation of objects. They are not questions about the facts we perceive, about what we believe or know, or about meaningful perception. For changes and differences in cognitive representations, in what we come to believe and know as a result of perception, are, as I said, an obvious and familiar fact of life. That such changes exist is not worth arguing about (although the changes themselves are certainly worth studying). Changes, if any, in our sensory representations are not so obvious. Quite the contrary. To document such changes one has to be very clear about what sensory representations are and what constitutes a change in them. To answer questions about whether we learn to see in this sense—about whether object perception (as opposed to fact perception) is sensitive to (and, thus, relative to) learning and development—requires a clear, at least a much better, understanding of the nature of sensory representation, of the internal response to an external object that constitutes our seeing the object. One has to pry apart (if this is possible) the way things look or seem (the sensory aspect of perception) from the way they are judged to be (or judged to look and seem)—the cognitive aspect of perception. It may well turn out that learning changes the cognitive dimension of perception (that component having to do with meaningful perception), while leaving the sensory aspect (that part having to do with object perception) unchanged. If this is so, then learning what armadillos are did not change the way armadillos looked to me or the kind of experience seeing them caused me to have. It only changed the kinds of judgments I am prepared to make about what I saw.

Similar remarks can be made about various forms of perceptual relativity. Is perception relative? Well, meaningful perception is certainly relative. It is relative to all those things capable of influencing what one comes to believe. If not having a word for X or a theory about X means I cannot come to have certain beliefs about X—the belief, for instance, that it is X—then not having a word (or a theory) for X will prevent me from seeing (the fact) that something is X. It will impair meaningful perception of X. Without a way of talking about oxygen, without an appropriate language or theory for saying or judging that something is oxygen, young children, for example, can hardly be expected to see when oxidation is occurring (see that it is occurring), even when the flames are literally in front of their nose. They just won't recognize what they see as oxidation. So the perception of facts about oxidation—and, therefore, meaningful perception of oxidation—is relative to those factors (including knowledge of chemical theory) that are essential to a knowledge.

A person having badly mistaken astronomical views will not be able to perceive what others perceive when he views the moon moving into the earth's shadow. He will not see, will not be able to see, that a lunar eclipse is in progress; holding mistaken views about what is happening (he thinks, perhaps, that the gods are showing displeasure by extinguishing the moon), he will not recognize, will not judge, what others better informed about the arrangement of sun, earth, and moon recognize and judge when they witness this event: that the earth is casting a shadow on the moon (i.e., that an eclipse is occurring).

Yet even though meaningful perception, our perception of facts, is obviously relative in this way, there is no reason to think—in fact, a lot of reason not to think—that sense perception, our perception of objects, is similarly relative. An ignorant person may not see (the fact) that an eclipse is occurring, but he certainly sees an event we correctly describe as an eclipse. That, in fact, is what frightens him. Thus, like the child who saw a cat but did not realize that this was a correct description of what she saw (the child took it to be a rumpled sweater), the ignorant man sees an eclipse without realizing that this, in fact, is a correct description of what he sees (taking it as a divine omen). For exactly the same reason, although the chemically uninformed can hardly be expected to see that oxidation is occurring, they can, given normal eyesight, witness the oxidation, the blazing fire, as well as more knowledgeable people. What is relative in these cases is the meaningful perception of objects and events, not the perception of the objects and events themselves.

To say that perception is relative to a certain factor is to say that our perception of things depends on that factor. Change that factor (enough) and you change what is perceived or, possibly, whether anything at all is perceived. To suggest, then, that object perception is not relative to a host of factors affecting perception of facts is a way of saying that sense perception is comparatively modular. It is not sensitive to those cognitive influences (e.g., one's language, conceptual scheme, scientific worldview) that affect one's perception of facts.

The topic of perceptual relativity, and the topics of perceptual change and learning, then, also require a clear separation of the kind of processes, and their respective products, that we mean to be describing when we describe the perceptual achievements of organisms. Meaningful perception, the perception of facts about the objects we perceive, is the important kind of perception, the kind of perception required for intelligent planning and timely action. Without it, the eyes and ears are useless. It not useful to see an attacking predator or a nourishing meal unless these are seen as, are judged to be, a predator (or at least dangerous) and nourishing (at least edible). As a result of this comparative importance, meaningful perception is the kind of perception to which all animals, including human beings, aspire. Furthermore, the perception of facts has come to dominate our thinking, both philosophical and psychological, about perception. In the minds of some investigators, perception has come to mean meaningful perception, the sensory acquisition of knowledge.

It has been one of the purposes of this chapter to correct this imbalance, to emphasize the role that object perception, perceiving the cat on the sofa, plays in one's perceptual relation to the world. Fact perception, seeing that it is a cat, is important, undeniably so; it constitutes our meaningful perception of the world around us. Nevertheless, the way meaningful perception is achieved, the way we have of perceiving facts, is by perceiving the objects embedded in these facts. One invites only confusion by ignoring the distinction.

Suggestions for Further Reading

For more detailed treatments of the distinction between the perception of objects and the perception of facts and a defense of the idea that sense perception does not require fact perception—that seeing is not (not necessarily anyway) believing—see Dretske 1969, 1978, 1981, Sibley 1971 and Warnock 1955. For opposing viewpoints defending the idea that *all* perception involves, if not knowledge, than some kind of judgment or belief, read Armstrong 1973, Hamlyn 1957, Heil 1983, and Pitcher 1971.

For treatments of direct (naive) and representative realism (section 9.2), whether we directly perceive physical objects or, instead, some mental surrogate, see Dretske 1969 and 1981, Goldman 1977, Sanford 1976, and Chisholm 1957 for direct theories; and Jackson 1977, Ayer 1956 and 1962 and Price 1932 for indirect theories. Perhaps the best recent book defending a representative theory is Perkins 1983; it contains informed discussion of each of the senses (not only, as is usual in philosophical treatments, vision). A (by-now) classical article on causal theories of perception is Grice 1961.

As to the question of whether perceptual processes are constructive or not (section 9.3), an exchange that brings out most of the issues can be found in Ullman 1980, a constructivist, and his commentators, many of whom defend a direct theory. For vigorous exposition and defense of constructivism see Gregory 1974a, 1974b, and 1978, Rock 1977 and 1983, and Fodor and Pylyshyn 1981. Authors generally supportive of a direct theory of perceptual processing (and therefore sympathetic to many of Gibson's ideas) include Turvey 1977, Mace 1977, Michaels and Carello 1981, and many of the contributors to Shaw and Bransford 1977 and Macleod and Pick 1974. For further discussion, including evaluations of the empirical status of these two approaches, see Hayes-Roth 1977, Johansson, Von Hofstein, and Jansson 1980 and Epstein 1973.

For perceptual learning, change and development (some of the topics discussed in section 9.4), consult the references in Chapter 8. For perceptual relativity, see Churchland 1979 and 1988 and Dretske 1969, Chapter 5. For discussion of the interaction between theory and observation in science, see Hanson 1958, Kuhn 1962, Brown 1987, and Shapere 1982.

Questions for Further Thought

9.1   Is it possible to perceive an object in a completely meaningless way? That is, is it possible to perceive an object while perceiving *no* facts about it? Can one see a cat, for instance, without thereby coming to know (or believe) something (not necessarily that it is a cat) about it? If not, does this mean that some kind of conceptual ability (whatever cognitive capacities are needed to know or believe something) is necessary for vision—the ability to see things?

9.2   Do animals see the same things we do? Do they have beliefs? Do they make judgments? If so, do they have the same kind of beliefs, make the same kind of judgments we do? Does every animal with eyes (and therefore, presumably, vision—the ability to see things) have the cognitive capacities to perceive objects meaningfully? If not, is this a case (see question 1) of completely meaningless perception?

9.3   Is it possible to perceive facts (meaningful perception) while perceiving no objects— to have cognitive perception without sense perception? What is the best way to describe what happens when one detects a change in overall illumination (that the lights went out, say) with one's eyes closed? Is this a case cf seeing a fact (that the lights went out) without seeing any objects?

9.4   Blindsight is a phenomenon of persons with brain damage who have no experience of objects in a certain part of the visual field; that is, they say they see nothing in that part of their visual field. Yet, (despite not "seeing" anything) it turns out that in forced-choice situations in which they are asked to guess about objects in the "blind" part of their visual field, these subjects perform above chance level and, sometimes, describe certain features of the objects there quite well. Is this to be classified as perception? If so, what kind? Do they see objects? Facts? Or nothing? If nothing, how does one explain their correct answers? (They get too many answers correct to be *just* guessing all the time.)

9.5   Is seeing an event (the cat's jumping on the sofa) and a state of affairs (the cat's being on the sofa) more like seeing objects (the cat on the sofa) or more like seeing facts (that the cat is on the sofa)? An example was used in section 9.4 about perceiving an eclipse. Is an eclipse an object? Or is it more like an event (the cat's jumping on the sofa)?

9.6   What is required to see the properties of objects—for example, the size or color of a cat? Can one see the properties of objects without seeing the objects themselves? Does one see the color of a black cat when one sees another (different) object of the same color (e.g., a black ball)?

9.7   Are objects and facts seen in dreams and hallucinations? Or does one merely dream (or hallucinate) that one is seeing an object or a fact? Is this difference, the difference between seeing an object in one's dream and dreaming one sees an object, a real difference? Are the things (if they are things) of which one is aware in dreams and hallucinations in the mind? Are there round red things in the brain if and when one dreams of something red and round?

9.8   If a star explodes and disappears when the light from it is still on its way to earth, does one nonetheless still see the star when the light reaches the earth (and enters one's eyes) many years later? If so, does this mean that one can see things that no longer exist? If not, what (if anything) does a person see when the light from the star enters the eyes and gives rise to an "experience" of a twinkling spot of light?

9.9   Are experts in a given field—auto mechanics (on cars), cooks (on food), and seamstresses (on fabrics), for instance—able to see things the nonexpert, the layperson, cannot see? How is one to interpret a mechanic's claim that he can hear things you cannot hear—that, for example, your car's valves need adjusting or that your car needs a tune-up. What kind of perception is this?

References

Armstrong, D. M. (1961). *Perception and the physical world*. London: Routledge and Kegan Paul.

Ayer, A. J. (1956). *The problems of knowledge*. London: Penguin Books.

Ayer, A. J. (1962). *The foundations of empirical knowledge*. London: Macmillan.

Brown, H. (1987). *Observation and objectivity*. Oxford: Oxford University Press.

Chisholm, R. (1957). *Perceiving: A philosophical study*. Ithaca, N.Y.: Cornell University Press.

Churchland, P. M. (1979). *Scientific realism and the plasticity of mind*. Cambridge: Cambridge University Press.

Churchland, P. M. (1988). Perceptual plasticity and theoretical neutrality: A reply to Jerry Fodor. *Philosophy of Science* 55, 167–187.

Dennett, D. (1987). *The intentional stance*. Cambridge: MIT Press.

Dretske, F. (1969). *Seeing and knowing*. Chicago: University of Chicago Press.

Dretske, F. (1978). The role or the percept in visual cognition. In Wade Savage, ed., *Minnesota Studies in the Philosophy of Science: Perception and Cognition*, vol. 9. Minneapolis: University of Minnesota Press.

Dretske, F. (1981). *Knowledge and the flow of information*. Cambridge: MIT Press/Bradford.

Epstein, W. (1973). The process of "taking-into-account" in visual perception. *Perception* 2, 267–285.

Farah, M. M. (1990). *Visual agnosia*. Cambridge: MIT Press.

Fodor, J. (1983). *The modularity of mind*. Cambridge: MIT Press.

Fodor, J., and Z. Pylyshyn (1981). How direct is visual perception?: Some reflections on Gibson's "ecological approach." *Cognition* 9, 139–196.

Gibson, J. J. (1950). *The perception of the visual world*. Boston: Houghton Mifflin.

Gibson, J. J. (1960). The concept of the stimulus in psychology. *American Psychologist* 15, 694–703.

Gibson, J. J. (1966). *The senses considered as perceptual systems.* Boston: Houghton Mifflin.

Gibson, J. J. (1972). A theory of direct visual perception. In J. R. Royce and W. W. Rozeboom, eds., *The psychology of knowing.* New York: Gordon and Breach.

Gibson, J. J. (1979). *The ecological approach to visual perception.* Boston: Houghton Mifflin.

Goldman, A. (1977). Perceptual objects. *Synthese* 35, 257–284.

Gregory, R. (1974a). Choosing a paradigm for perception. In E. C. Carterette and M. P. Friedman, eds., *Handbook of perception,* vol. 1, *Historical and philosophical roots of perception.* New York: Academic Press.

Gregory, R. (1974b). Perceptions as hypotheses. In S. C. Brown, ed., *Philosophy of psychology.* London: Macmillan.

Gregory, R. (1978). *Eye and brain: The psychology of seeing,* 3rd. ed. New York: McGraw-Hill.

Grice, P. (1961). The causal theory of perception. *Proceedings of the Aristotelian Society,* suppl. vol. 35.

Hamlyn, D. W. (1957). *The psychology of perception.* London: Routledge and Kegan Paul.

Hanson, N. R. (1958). *Patterns of discovery.* London: Cambridge University Press.

Hayes-Roth, F. (1977). Critique of Turvey's "Contrasting orientations to the theory of visual information processing." *Psychological Review* 84, 531–535.

Heil, J. (1983). *Perception and cognition.* Berkeley: University of California Press.

Jackson, F. (1977). *Perception.* Cambridge: Cambridge University Press.

Johansson, G., C. von Hofstein, and G. Jansson (1980). Event perception. *Annual Review of Psychology* 31, 27–63.

Kuhn, T. S. (1962). *The structure of scientific revolutions.* Chicago: University of Chicago Press.

Mace, W. M. (1977). James Gibson's strategy for perceiving: Ask not what's in your head, but what your head is inside of. In R. Shaw and J. Bransford, eds., *Perceiving, acting and knowing.* Hillsdale, NJ: L. Erlbaum.

Macleod, R. B., and H. L. Pick, Jr., eds. (1974). *Perception: Essays in honor of James J. Gibson.* Ithaca, N.Y.: Cornell University Press.

Marr, D. (1982). *Vision: A computational investigation into the human representation and processing of visual information.* San Francisco: W. H. Freeman.

Michaels, C. F., and C. Carello (1981). *Direct perception.* Englewood Cliffs, NJ: Prentice Hall.

Perkins, M. (1983). *Sensing the world.* Indianapolis: Hackett.

Pitcher, G. (1971). *A theory of perception.* Princeton: Princeton University Press.

Price, H. H. (1932). *Perception.* London: Methuen.

Rock, I. (1977). In defense of unconscious inference. In W. Epstein, ed., *Stability and constancy in visual perception.* New York: Wiley.

Rock, I. (1983). *The logic of perception.* Cambridge: MIT Press.

Sanford, D. (1976). The primary objects of perception. *Mind* 85, 189–208.

Shapere, D. (1982). The concept of observation in science and philosophy. *Philosophy of Science* 49, 485–525.

Shaw, R., and J. Bransford, eds. (1977). *Perceiving, acting and knowing.* Hillsdale, NJ: L. Erlbaum.

Sibley, F. N. (1971). Analyzing seeing. In F. N. Sibley, ed., *Perception.* London: Methuen.

Turvey, M. T. (1977). Contrasting orientations to the theory of visual information processing. *Psychological Review* 84, 67–88.

Ullman, S. (1980). Against direct perception. *Behavioral and Brain Sciences* 3, 373–415.

van Sluyters, R. C., J. Atkinson, M. S. Banks, R. M. Held, K. P. Hoffman, and C. J. Shatz (1990). The development of vision and visual perception. In L. Spillmann and J. S. Werner, eds., *Visual perception: The neurophysiological foundation.* San Diego: Academic Press.

Warnock, J. (1955). Seeing. *Aristotelian Society Proceedings* 55.

Index